Mitochondrial Case Studies

Mitochondrial Case Studies

Underlying Mechanisms and Diagnosis

Edited by

Russell P. Saneto
Department of Neurology/Division of Pediatric Neurology, Seattle Children's Hospital/University of Washington, Seattle, WA, USA

Sumit Parikh
Cleveland Clinic Lerner College of Medicine & Case Western Reserve University, Cleveland, OH, USA;
Neurogenetics, Metabolism and Mitochondrial Disease Center, Cleveland Clinic, Cleveland, OH, USA

Bruce H. Cohen
Northeast Ohio Medical University, Rootstown, OH, USA;
The NeuroDevelopmental Science Center and Divison of Neurology, Department of Pediatrics, Children's Hospital and Medical Center of Akron, Akron, OH, USA

ELSEVIER

AMSTERDAM • BOSTON • HEIDELBERG • LONDON
NEW YORK • OXFORD • PARIS • SAN DIEGO
SAN FRANCISCO • SINGAPORE • SYDNEY • TOKYO

Academic Press is an imprint of Elsevier

Academic Press is an imprint of Elsevier
125 London Wall, London EC2Y 5AS, UK
525 B Street, Suite 1800, San Diego, CA 92101-4495, USA
225 Wyman Street, Waltham, MA 02451, USA
The Boulevard, Langford Lane, Kidlington, Oxford OX5 1GB, UK

ISBN: 978-0-12-800877-5

British Library Cataloguing-in-Publication Data
A catalog record for this book is available from the British Library

Library of Congress Cataloging-in-Publication Data
A catalog record for this book is available from the Library of Congress

For information on all Academic Press publications
visit our website at http://store.elsevier.com/

Publisher: Janice Audet
Acquisition Editor: Janice Audet
Editorial Project Manager: Pat Gonzalez
Production Project Manager: Julia Haynes
Designer: Mark Rogers

Typeset by TNQ Books and Journals
www.tnq.co.in

Printed in United States of America

Contents

Contributors

Rashid Alshahoumi Department of Pediatrics, McMaster University, Hamilton, Ontario, Canada

Emanuele Barca Columbia University Medical Center, Department of Neurology, New York, NY, USA

Christopher Beatty Department of Neurology/Division of Pediatric Neurology, Seattle Children's Hospital/University of Washington, Seattle, WA, USA

Sirisak Chanprasert Department of Molecular and Human Genetics, Baylor College of Medicine, Houston, TX, USA; Texas Children's Hospital, Houston, TX, USA

Patrick F. Chinnery Department of Clinical Neuroscience, University of Cambridge, Cambridge, UK; MRC Mitochondrial Biology Unit, Cambridge Biomedical Campus, Cambridge, UK

John Christodoulou Genetic Metabolic Disorders Research Unit, Children's Hospital at Westmead, Westmead, NSW, Australia; Western Sydney Genetics Program, Children's Hospital at Westmead, Westmead, NSW, Australia; Disciplines of Paediatrics and Child Health and Genetic Medicine, Sydney Medical School, University of Sydney, Sydney, NSW, Australia

Bruce H. Cohen Northeast Ohio Medical University, Rootstown, OH, USA; The NeuroDevelopmental Science Center and Divison of Neurology, Department of Pediatrics, Children's Hospital and Medical Center of Akron, Akron, OH, USA

James E. Davison Metabolic Medicine, Great Ormond Street Hospital for Children NHS Foundation Trust, London, UK

Suzanne D. DeBrosse Departments of Genetics and Genome Sciences, Pediatrics, and Neurology, Case Western Reserve University School of Medicine, University Hospitals Case Medical Center, Cleveland, OH, USA

Adela Della Marina Pediatric Neurology, University of Essen, Essen, Germany

Beatriz García Díaz Columbia University Medical Center, Department of Neurology, New York, NY, USA

Salvatore DiMauro Columbia University Medical Center, New York, NY, USA

Simon Edvardson Pediatric Neurology Unit, Hadassah-Hebrew University Medical Center, Jerusalem, Israel

Marni J. Falk Department of Pediatrics, University of Pennsylvania Perelman School of Medicine, Philadelphia, PA, USA; Division of Human Genetics, The Children's Hospital of Philadelphia, Philadelphia, PA, USA

Xiaowu Gai Center for Biomedical Informatics, The Children's Hospital of Philadelphia, Philadelphia, PA, USA; Ocular Genomics Institute, Ophthalmology, Massachusetts Eye and Ear Infirmary, Harvard Medical School, Boston, MA, USA

Amy Goldstein Neurogenetics & Metabolism, Division of Child Neurology, Children's Hospital of Pittsburgh, Pittsburgh, PA, USA

Leon Grant University of Hawaii, Kapi'olani Medical Center for Women and Children, Honolulu, HI, USA

R.H. Haas Department of Neurosciences, University of California San Diego, La Jolla, CA, USA; Rady Children's Hospital San Diego, San Diego, CA, USA; Department of Pediatrics, University of California San Diego, La Jolla, CA, USA; Metabolic & Mitochondrial Disease Center, University of California San Diego, San Diego, CA, USA

Michio Hirano Columbia University Medical Center, Department of Neurology, New York, NY, USA

Rita Horvath The John Walton Muscular Dystrophy Research Centre, MRC Centre for Neuromuscular Diseases, Institute of Genetic Medicine, Newcastle University, Newcastle upon Tyne, UK

Pirjo Isohanni Department of Child Neurology, Children's Hospital, Helsinki University and Helsinki University Hospital, Helsinki, Finland; Research Programs Unit, Molecular Neurology, Biomedicum Helsinki, University of Helsinki, Helsinki, Finland

Douglas S. Kerr Departments of Pediatrics, Biochemistry, Nutrition and Pathology, Case Western Reserve University School of Medicine, University Hospitals Case Medical Center, Cleveland, OH, USA

Mary Kay Koenig University of Texas Medical School at Houston, Department of Pediatrics, Division of Child and Adolescent Neurology, Endowed Chair of Mitochondrial Medicine, Houston, TX, USA

Tuula Lönnqvist Department of Child Neurology, Children's Hospital, Helsinki University and Helsinki University Hospital, Helsinki, Finland

Mariana Loos Columbia University Medical Center, Department of Neurology, New York, NY, USA

S.E. Marin Department of Neurosciences, University of California San Diego, La Jolla, CA, USA

Shana E. McCormack Division of Endocrinology and Diabetes, Department of Pediatrics, The Children's Hospital of Philadelphia, Philadelphia, PA, USA; Department of Pediatrics, University of Pennsylvania Perelman School of Medicine, Philadelphia, PA, USA

Elizabeth M. McCormick Divisions of Human Genetics and Child Development and Metabolic Disease, Department of Pediatrics, The Children's Hospital of Philadelphia, Philadelphia, PA, USA

Robert K. Naviaux The Mitochondrial and Metabolic Disease Center, Departments of Medicine, Pediatrics, and Pathology, University of California, San Diego School of Medicine, San Diego, CA, USA; Veterans Affairs Center for Excellence in Stress and Mental Health (CESAMH), La Jolla, CA, USA

Anders Paetau Department of Pathology, HUSLAB, Helsinki University Hospital and University of Helsinki, Helsinki, Finland

Sumit Parikh Cleveland Clinic Lerner College of Medicine & Case Western Reserve University, Cleveland, OH, USA; Neurogenetics, Metabolism and Mitochondrial Disease Center, Cleveland Clinic, Cleveland, OH, USA

Emily Place Ocular Genomics Institute, Ophthalmology, Massachusetts Eye and Ear Infirmary, Harvard Medical School, Boston, MA, USA; Division of Human Genetics, The Children's Hospital of Philadelphia, Philadelphia, PA, USA

Catarina M. Quinzii Columbia University Medical Center, Department of Neurology, New York, NY, USA

Shamima Rahman Metabolic Medicine, Great Ormond Street Hospital for Children NHS Foundation Trust, London, UK; Mitochondrial Research Group, Genetics and Genomic Medicine, UCL Institute of Child Health, London, UK

Lisa Riley Genetic Metabolic Disorders Research Unit, Children's Hospital at Westmead, Westmead, NSW, Australia

Ann Saada (Reisch) Department of Genetic and Metabolic Diseases & Monique and Jacques Roboh Department of Genetic Research, Hadassah-Hebrew University Medical Center, Jerusalem, Israel

Russell P. Saneto Department of Neurology/Division of Pediatric Neurology, Seattle Children's Hospital/University of Washington, Seattle, WA, USA

Fernando Scaglia Department of Molecular and Human Genetics, Baylor College of Medicine, Houston, TX, USA; Texas Children's Hospital, Houston, TX, USA

Ulrike Schara Pediatric Neurology, University of Essen, Essen, Germany

Kurenai Tanji Columbia University Medical Center, New York, NY, USA

Mark Tarnopolsky Department of Pediatrics, McMaster University, Hamilton, Ontario, Canada

Patrick Yu-Wai-Man Wellcome Trust Centre for Mitochondrial Research, Institute of Genetic Medicine, Newcastle University, UK; Newcastle Eye Center, Royal Victoria Infirmary, Newcastle upon Tyne, UK

Preface

Mitochondrial diseases are a heterogeneous group of disorders often affecting several organ systems in a variety of permutations. The primary mitochondrial disorders are caused by inherited or de novo mutations that affect one of over one thousand genes in either the nuclear and mitochondrial genome. Mitochondrial physiology, the interaction between two genomes and the influence of environmental factors add to the complexity of these diseases. Disease onset can be from the newborn period through late adulthood. The phenotypic expression of disease can range from a single disease expression with a narrow phenotype and an isolated biochemical abnormality to a complex set of symptoms that may present differently, even in related individuals. Diagnosing patients is often challenging as there is no single diagnostic finding completely specific for the presence of mitochondrial disease. The diagnosis has depended on a constellation of clinical, biochemical, and neuroimaging findings that meet a set of validated criteria for mitochondrial disease. Although the first mitochondrial syndromes have been associated with specific mutations in mitochondrial DNA, the vast majority of mitochondrial diseases have been diagnosed on clinical, biochemical and neuroimaging criteria. The advancement of genetic techniques for gene discovery and elucidation of genetic abnormalities has brought some clarity to diagnosis of specific mitochondria DNA-encoded and nuclear DNA-encoded diseases. However, we do not yet know all of the mitochondrial disease causing genes, which has hampered diagnosis in some patients. In addition, the clinical acumen of possible disease has varied from clinician-to-clinician due to the heterogeneity of the disorders, at times leading to mistaken diagnosis. Yet the projected prevalence has been estimated to be 1 in 4,000, suggesting the population of patients with mitochondrial disease is large.

Currently, much of mitochondrial medicine remains an "art" and clinical acumen remains paramount to diagnosis. In the days of the apprenticeship-mentor in medicine, the apprentice learned the "art" of medicine (e.g. correct history taking, methods of investigation, and possible differential diagnosis) from their mentor. The mastery of medicine was mostly dependent on the quality of the mentor and the ability of the student. The development of medical schools, residency and fellowship programs has replaced this system of medical learning and broadened the quality of the system to supplant a single mentor. But, as we in the "trenches" of clinical practice know, the true qualities of the mentor were not replaced by the "system." We have each found special mentors who have passed on the qualities of insightful approach to the diagnosis of disease. These mentors have also fostered an excitement to the possible unique expression of disease. They have promoted our self-motivated spirit of literature review and enhanced our physical diagnostic techniques. Most importantly, they have expanded our thought process to critically assess the differential diagnosis. These are attributes not found in textbooks or lecture halls. They are absorbed by the student. Clinically, this happens during rounds, night call, and asking those teachers and colleagues who we sought out to learn from. These special few became our mentors. In this learning paradigm, the patient is foremost the driver of our clinical acumen. We never forget that singular patient, unique disease, and the associated wonderment of diagnosis. We each have our mentors that we emulate because of their clinical skills and inquisitive mindset for the difficult and challenging diagnosis.

Our book is written to illustrate those special patients and demonstrate how the masters of mitochondrial medicine approach the evaluation and diagnosis. Some of the best clinicians in the field of mitochondrial medicine have described their approach to making the diagnosis for each case. We hope we created the "medium for the message" of how to approach the possible mitochondrial patient. The book is not meant to be read once, but repeatedly over time, as maybe the tools may change, but the mindset for putting the pieces together for the diagnosis remains in the thought process used by each expert to solve the case; being the "doctor" remains central to the clinician.

Russell P. Saneto, DO, PhD, Sumit Parikh, MD, and Bruce H. Cohen, MD

Chapter 1

Introduction: Mitochondrial Medicine

Bruce H. Cohen
Northeast Ohio Medical University, Rootstown, OH, USA; The NeuroDevelopmental Science Center and Divison of Neurology, Department of Pediatrics, Children's Hospital and Medical Center of Akron, Akron, OH, USA

INTRODUCTION

Although case descriptions of cases with mitochondrial dysfunction had been described in the literature for decades before Luft's report in 1962, the underlying mitochondrial pathophysiology was not documented until the description of a 30-year-old woman with an illness present since childhood. Not only did this case represent the first illness linked to mitochondrial dysfunction, it led to the now-established concept of an illness caused by dysfunction of an organelle. In the ensuing half of a century, a remarkable story unfolded, which has included the elucidation of a new tiny but essential fragment of human DNA within the mitochondria, the interactions of this small piece of DNA with the DNA contained in the nucleus, the advances in laboratory medicine allowing measurement of the enzymatic and polarographic function of the respiratory chain (RC), the descriptions of dozens of phenotypic disorders linked to the biochemical derangements within the mitochondria, and the discovery of the myriad of genetic mutations linked to human disease.

Many of the clinical diseases were initially associated with a descriptor that was defined by histologic features, typically seen in muscle. The initial microscopic findings included the description of a nonspecific light microscopy histochemical finding of the ragged red fiber, described in 1963 by Engel and coworkers, followed 2 years later by Shy and Gonatas' report of ultrastructural (electron microscopic) mitochondrial findings. For example, case's illnesses were defined as *ragged red fiber myopathy*. Over time, these visual features were supported by complementary enzymatic dysfunction such as the techniques described to measure pyruvate dehydrogenase complex (1970, Blass et al.), complex III (1970, Spiro et al.), carnitine palmitotransferase (1973, DiMauro et al.), and complex IV (1977, Willems et al.). Again, a case's illness was often defined by the associated biochemical defect such as cytochrome

Mitochondrial Case Studies. http://dx.doi.org/10.1016/B978-0-12-800877-5.00001-2

c oxidase deficiency. The term *mitochondrial encephalomyopathy* was coined by Yehuda Shapira in 1977. Although both microscopy and enzymology could be performed in many tissues, for well over three decades, the *muscle biopsy* became the primary method of detection and confirmation of mitochondrial disease. The advent of molecular genetic diagnostics allowed for some cases to achieve a diagnosis without a muscle biopsy, but not until the full mitochondrial DNA (mtDNA) genome could be sequenced quickly and the nuclear DNA (nDNA) genes were being discovered (both occurring in the 2007–2008 era) did the molecular genetic technology become part of the day-to-day diagnostic evaluation. Finally, the application of massive parallel sequencing technology (so-called Next-Gen sequencing) allowed for detection of mutations in both known and putative gene candidates.

OVERVIEW OF MITOCHONDRIAL STRUCTURE AND FUNCTION

The mitochondria are a complex organelle responsible for the vast majority of cellular energy production. The final process in energy metabolism involves the phosphorylation of adenosine diphosphate (ADP) to adenosine triphosphate (ATP). ATP can be thought of the energy currency for almost all cellular functions, where hydrolysis of ATP releases energy along with ADP and P_i, which are returned to the mitochondria to again be re-phosphorylated into ATP. About 1500 enzymatic and structural proteins are necessary for normal mitochondrial function, and most of the genes that encode for these proteins reside in the nDNA. Pathogenic mutations in these genes may be passed along from parent to child with the same Mendelian and non-Mendelian rules that apply for other nuclear genes. As with other single nuclear gene disorders, de novo mutations may arise and cause disease. Mitochondria are unique organelles as they contain a small piece of circular DNA, which is referred to as mtDNA, discovered by Nass and Nass in 1963. Human mtDNA contains 37 genes encoding the translational machinery (2 rRNAs and 22 tRNAs) and 13 enzymatic proteins that make up only a portion of the electron transport chain (ETC; also known as the respiratory chain). These 13 proteins include subunits of complex I (7 proteins), complex III (1 protein), complex IV (3 proteins), and complex V (2 proteins). Unlike the nDNA, mutations in mtDNA are inherited by maternal transmission, meaning that mutations are passed from the mitochondria contained in the oocyte to all offspring. There is no paternal contribution of mtDNA to the offspring. The mitochondria contained along the midshaft of the spermatozoa either do not enter the oocyte or are immediately destroyed by proteases on the inner wall of the oocyte. Most pathogenic mutations present in the germline of the mother will be inherited or passed on to all children. Another feature of mtDNA is the concept of heteroplasmy. With nDNA, aside from notable exceptions that are beyond the scope of this discussion, one allele is inherited from each parent, so there are three possible states with respect to pathogenic inheritance: mutant homozygote, wild homozygote, or heterozygote (and hemizygote for X-linked disorders). Mutations in mtDNA tend not to be *all or none*, meaning that mutations occur in a variable percentage, the concept known as heteroplasmy. A mutation may occur in only a few percentage of mtDNA copies (low-level mutant heteroplasmy) or a large percentage of mtDNA (high-level mutant heteroplasmy). Aside from mutations causing Leber hereditary optic neuropathy (LHON), in which the mutations occur with 100% homoplasmy, most of the mtDNA mutations occur with variable levels of heteroplasmy. Higher percentages of mutant heteroplasmy are generally associated with earlier disease onset and more severe presentations. Because there is variability in mutant heteroplasmy in the mature oocyte, each offspring may have, on average, a different percent of mutant heteroplasmy than the mother. Furthermore, the mutations may segregate differently during embryogenesis, thus resulting in different ultimate organ dysfunction later in life.

The physical structure of the mitochondria is constantly changing in vivo. The organelle is composed of an outer mitochondrial membrane (OMM) and a heavily infolded inner mitochondrial membrane (IMM). One major property of the IMM that is not found in any other cellular membrane is the presence of the phospholipid cardiolipin. Unlike the other phospholipids that are composed of two hydrocarbon tails, cardiolipin has four, and this structure likely is responsible for the complex tight curves in the membrane that allow for the infolding. Proper cardiolipin content is also necessary to allow the proper stoichiometry and supercomplex structure of the ETC elements. There are contact points riveting the IMM and OMM containing the complex group of proteins that form the voltage-dependent anion channel and mitochondrial permeability transition pore. This structure is responsible for several critical functions that include the adenosine nucleotide transporter, which exchanges ATP exchange for ADP, as well as components for calcium exchange and apoptosis.

Embedded within the IMM are the five multicomplex proteins referred to as the RC or ETC. It is the reducing equivalents derived from the tricarboxylic acid cycle and fatty acid beta-oxidation (nicotinamide adenine dinucleotide and flavin adenine dinucleotide) that donate electrons into complexes I and II, and they supply the electrical force required to pump protons from the matrix into the intermembrane space. The intermembrane space has a small volume and functions to store the electrochemical capacitance. The protons can flow back into the matrix through a pore in complex V, which results in the physical movement of a molecular rotor, condensing a phosphate moiety onto a molecule of ADP, to form ATP. This electrochemical charge not only results in ATP formation by complex V, but also is an affector and effector of calcium regulation, and, ultimately, the most critical factor initiating intrinsic apoptosis if this potential is lost. Coenzyme Q10 is a mobile electron carrier, which shuttles electrons in a redox cycle between complexes I and III and complexes II and III. Cytochrome *c* shuttles electrons in a redox reaction between complex III and complex IV. In addition to its function as a proton pump, complex IV is the site in the ETC where molecular oxygen is reduced tetravalently to water. The process of proton translocation across the IMM, the chemical reduction of molecular oxygen to water, and condensation of phosphate onto ADP to form ATP is termed *oxidative phosphorylation.*

The mitochondrial matrix is the volume within the IMM and contains hundreds of enzymes necessary to the tricarboxylic acid cycle, the urea cycle, the functional cascade of proteins necessary for apoptosis, the synthesis of a number of amino acids, the breakdown of long-chain fatty acids, and the oxidation of eight amino acids. Also found within the matrix of the individual mitochondria are 2–10 copies of mtDNA.

The structure of the mitochondria depends on the specific cell type and if the cell is living or preserved. Within fixed cells, the individual mitochondria assume a cylindrical shape about 1 μm in length. It is estimated that most fixed cells contain 1000 or so mitochondria. In the living cell, the mitochondrial structure differs significantly between individual tissues, but it generally forms a syncytial network that is constantly changing shape. As part of this dynamic structure, both budding formations (created by mitochondrial fission) and reorganization of separate mitochondria (mitochondrial fusion) occur constantly. Failure of mitochondrial fission and fusion is known to cause human disease.

The mtDNA is contained within the matrix. It is a circular molecule with 16,569 nucleotide pairs, which contain 37 genes. These genes encode for 13 structural proteins that are contained in complexes I, III, and IV as well as the 22 transfer RNAs and two ribosomal RNAs required for mtDNA translation. The translational components (mt-rRNA and mt-tRNA) differ structurally from those in the nucleus of the cell and are structurally similar to that of the bacteria, the ancestral forerunner to the mitochondria. Some of the protein coding language of the mtDNA differs from its nuclear counterparts as well. The remainder of the mitochondrial structure and enzymes are encoded by nDNA. This includes all the other subunits of the ETC; the IMM and OMM components; the proteins required for unfolding, chaperoning, and refolding the nuclear gene products into the mitochondria; the assembly proteins that will assemble the ETC; the metal cores (iron-sulfur and copper) of the ETC structures; the matrix enzymes; and the system responsible for mtDNA replication. It is believed that over the eons most of the genes encoding the ETC proteins moved from the mtDNA into the nDNA, with those currently remaining in the mtDNA encoding for the highly hydrophobic proteins that would be difficult to transport through the mitochondrial membranes for assembly. The catalog of these 1013 human (and 1098 mouse) genes has been labeled the MitoCarta. Although the function of most of the gene products of the MitoCarta are known, this is not true for all known mitochondrial proteins.

A BRIEF HISTORY OF CLINICAL MITOCHONDRIAL MEDICINE AND CLINICAL FEATURES

In 1962, Luft and coworkers described the case of a 30-year-old woman with an illness beginning in childhood. She was euthyroid but had had an excessive calorie intake, extreme perspiration, heat intolerance, polydipsia with polyuria, a clinical myopathy, and a basal metabolic rate 180% of normal. Light microscopy found excessive numbers of large mitochondria, and electron microscopy showed electron-dense inclusions, which were later coined as paracrystalline inclusions. Polarographic studies showed that oxidation and phosphorylation were uncoupled, meaning that oxygen was being consumed without the normal, linked production of ATP. Although only one other case with this presentation has been reported, the field of mitochondrial medicine was born. At the time this first case was reported, biochemical characterization of disease was in its infancy, and molecular genetic elucidation was not possible and, at the time, probably incomprehensible. In the subsequent half of century, the field has evolved; with first the development of biochemical assays of ETC and soluble enzyme function, the seemingly exponential growth of the clinical descriptions followed along.

Decreased energy production is believed to result in the symptoms of mitochondrial disease. Along with generating energy, mitochondria have at least three other critical functions: they are the major source of free radical generation, are essential in calcium homeostasis, and also play a critical role in apoptosis as well as being involved in innate immunity and inflammation. Although the role of energy production has been the primary explanation for causing illness, all properties of normal mitochondrial function are necessary for health.

The clinical features of mitochondrial disorders spanned the organ systems that tended to be postmitotic at birth. These included brain, the special senses (retina, optic nerve, and inner ear), muscle (heart, skeletal, and smooth), nerve, pancreas, liver, and renal tubular cells (Table 1). Until the advent of genetic testing, most cases were diagnosed after careful consideration of their clinical findings, with supporting evidence from a combination of radiographic findings, abnormalities in blood, CSF and urine analytes, pathologic findings of usually muscle tissue, and abnormal ETC function as tested using enzymology or polarographic techniques. Most of these technologies were developed and put into clinical practice in the 1970s and 1980s. It was during this time that a few previously characterized clinical disorders including Leigh syndrome and Kearns-Sayre syndrome (described in 1951 and 1958, respectively) were accepted as mitochondrial diseases. In the mid-1980s, the syndromes of MERRF (myoclonus epilepsy and ragged red fiber disease) and MELAS (mitochondrial myopathy, encephalopathy, lactic acidosis, and stroke-like episodes) were defined.

TABLE 1 Clinical Features by Tissue in Mitochondrial Diseases

Brain	Seizures, stroke and stroke-like episodes, weakness due to pyramidal dysfunction, cortical blindness, ataxia and other cerebellar dysfunction, dystonia and other extrapyramidal disorders, developmental regression and dementia, intellectual disability, neuropsychiatric symptoms and mood disturbance, atypical migraine
Muscle	Myopathy resulting in weakness and exercise intolerance and fatigability, ptosis, ophthalmoplegia, pain, cramping, hypotonia
Nerve	Neuropathy (weakness, pain), dysautonomia (gastroesophageal reflux, bowel pseudo-obstruction), orthostatic hypotension, abnormal sweating, aberrant temperature regulation
Kidney	Proximal renal tubular dysfunction (Fanconi syndrome)
Cardiac	Cardiac conduction defect and cardiomyopathy
Hepatic	Hepatocellular dysfunction, nonalcoholic steatohepatitis
Special senses	Optic atrophy, retinitis pigmentosa, sensory-neural hearing loss, aminoglycoside-induced hearing loss
Pancreas	Diabetes and exocrine pancreatic failure
Systemic and other	Failure to gain weight, linear growth failure, other endocrinopathy, sideroblastic anemia

About a quarter of a century after Luft's report, the genetic understanding of mitochondrial disease was first identified. In 1988, a large-scale deletion in the mtDNA was associated with Kearns-Sayre syndrome (KSS). In the approximate quarter of a century since this report, this specific deletion has remained the most common cause of KSS and clinically, the most important deletion causing mitochondrial disease. Later that same year, the first point mutation causing a mitochondrial disease was reported: a mtDNA-encoded complex I gene associated with LHON [1]. In the next year, PEO (progressive external ophthalmoplegia), PEO+, and KSS were all linked to mtDNA deletions. In 1990, MERRF was linked to a mutation in $tRNA^{Lys}$ and MELAS to a mutation in $tRNA^{Leu(UUR)}$. By 2001, over 115 point mutations in the mtDNA were linked to disease, and this number increased to over 200 by 2006.

The nDNA element to the mitochondrial story lagged a bit behind mtDNA, but by 1994, the nDNA story of mitochondrial illness began to unfold: the biochemical features along with the genetic findings of multiple mtDNA deletions were described in cases with mitochondrial neurogastrointestinal encephalomyopathy (MNGIE). Yet the mystery remained as to the cause of the multiple mtDNA deletions. In 1996, *POLG* gene encoding polymerase gamma was discovered, although the full clinical implications for human disease were not uncovered for 5 years. In 1999, the group that demonstrated the multiple mtDNA deletions in MNGIE discovered the root cause of the disease—mutations in *EGCF1,* the gene encoding thymidine phosphorylase, resulted in elevated thymidine levels. This resultant change in the nucleotide pool resulted in enough infidelity in mtDNA replication to result in the multiple mtDNA deletions, as reported 5 years earlier. Two years later, the first mutations in *POLG* causing PEO were described, and in 2004, a different set of mutations in the same gene were associated with Alpers-Huntenlocher syndrome. Hundreds of mutations in dozens of nuclear gene discoveries have occurred over the following decade.

Developing a nomenclature for mitochondrial disease had been troublesome, in part because the diagnosis depended on being able to define the pathology, and the technology to do so did not exist in most medical centers. DiMauro's 1985 general classification aided in compartmentalizing the pathophysiology, breaking down the disorders into ones of substrate transport, substrate utilization, tricarboxylic acid cycle defects, defects in the ECT, and defects in coupling of oxidation and phosphorylation. A number of research centers developed algorithms to estimate the certainty of a mitochondrial disease, acknowledging the difficulty in translating the work performed outside of these centers to those practicing outside the major diagnostic laboratories. The algorithms were based mainly on the clinical presentation of a classic available technology of the day, but at the time these were published, clinical molecular genetic technology was in its infancy. These algorithms provided a rigid framework for estimating the certainty of diagnosis, and they were very useful to clinicians practicing outside the research centers but were limited because not all cases could be studied with all the testing needed to apply the algorithm. And, if all the fine print was not appropriately applied to each case, the algorithm could not be expected to be accurate.

By the late 1980s and early 1990s, a number of clinical laboratories offering enzymatic testing expanded, and large numbers of cases were able to undergo this testing, although applying this technology was not without problems [2]. The first commercially available genetic tests were made available in the early 1990s, and they included a handful of point mutations for the known mtDNA disorders and Southern blot of the mtDNA, evaluating large-scale deletions and duplications of the mtDNA. It was not until the late 1990s that the first commercial tests were available to sequence the entire mtDNA. The first nuclear disorders were described in the late 1990s and early 2000s, with the commercial testing for these single gene disorders being offered in the latter half of the 2000s. Although panels of the nuclear genes were available in 2010, the testing was performed using Sanger sequencing of single genes. At this point in time, most cases diagnosed with mitochondrial disorders came to the diagnosis using clinical, biochemical, and pathological findings. It was not until 2011 and 2012 that the Next-Gen sequencing technology, sequencing multiple genes simultaneously, first offered testing for very large panels of genes, which lead to most clinicians in the field performing fewer and fewer muscle biopsies, and becoming less reliant on biochemical data. As could be predicted, having enormous molecular sequencing capabilities has led to unintended consequences that include an inability to determine whether a genetic variant is pathogenic. Proving pathogenicity can be a complex task, requiring both older technology (e.g., enzymatic function) and newer technology (e.g., proteinomic investigation).

Despite all the advances, several problems vex the clinician. All of the major phenotypic disorders (e.g., MELAS, LHON, MERRF, POLG-spectrum disorders, and so forth) with clear genetic and biochemical explanation for mitochondrial dysfunction, despite all having phenotypic expression of mitochondrial dysfunction, all have vastly different clinical presentations. The disorders vary in their age of onset, rate of progression, and even in families with the identical mutations, the illness can have different manifestations. It is not clear why some cases with specific mutations in mtDNA present with a classic phenotype for that specific mutation, but other cases have similar although distinct phenotypes. In the case of illness caused by mtDNA mutations, the heteroplasmic variability may play a role in some of the observed differences, but the future holds many mysteries in terms of modifying mutations and the role of epigenetics. Just as the first 50 years of mitochondrial medicine have revealed many exciting discoveries, the next 50 years will do so as well.

REFERENCES

[1] Wallace DC, Singh G, Lott MT, Hodge JA, Schurr TG, Lezza AM, et al. Mitochondrial DNA mutation associated with Leber's hereditary optic neuropathy. Science 1988;242:1427–30.

[2] Thorburn DR, Smeitink J. Diagnosis of mitochondrial disorders: clinical and biochemical approach. J Inherit Metab Dis 2001;24:312–6.

FURTHER READING

[1] Bernier FP, Boneh A, Dennett X, Chow CW, Cleary MA, Thorburn DR. Diagnostic criteria for respiratory chain disorders in adults and children. Neurology November 12, 2002;59(9): 1406–11.

[2] Cohen BH. Neuromuscular and systemic presentations in adults: diagnoses beyond MERRF and MELAS. Neurotherapeutics April 2013;10:227–42.

[3] DiMauro S, Bonilla E, Zeviani M, Nakagawa M, DeVivo DC. Mitochondrial myopathies. Ann Neurol 1985;17:521–38.

[4] DiMauro S, Schon EA. Mitochondrial respiratory-chain diseases. N Engl J Med June 26, 2003;348:2656–68.

[5] DiMauro S, Garone C. Historical perspective on mitochondrial medicine. Dev Disabil Res Rev 2010;16:106–13.

[6] Goto Y, Nonaka I, Horai S. A mutation in the tRNA(Leu)(UUR) gene associated with the MELAS subgroup of mitochondrial encephalomyopathies. Nature 1990;348:651–3.

[7] Hakonen AH, Davidzon G, Salemi R, Bindoff LA, Van Goethem G, Dimauro S, et al. Abundance of the POLG disease mutations in Europe, Australia, New Zealand, and the United States explained by single ancient European founders. Eur J Hum Genet 2007;15:779–83.

[8] Haas RH, Parikh S, Falk MH, et al. The in-depth evaluation of suspected mitochondrial disease. Mol Genet Metab 2008;94:16–37.

[9] Hirano M, Silvestri G, Blake DM, Lombes A, Minetti C, Bonilla E, et al. Mitochondrial neurogastrointestinal encephalomyopathy (MNGIE): clinical, biochemical, and genetic features of an autosomal recessive mitochondrial disorder. Neurology 1994;44:721–7.

[10] Holt IJ, Harding AE, Morgan-Hughes JA. Deletions of muscle mitochondrial DNA in patients with mitochondrial myopathies. Nature 1988;331:717–9.

[11] Koopman WJ, Willems PH, Smeitink JA. Monogenic mitochondrial disorders. N Engl J Med 2012;366:1132–41.

[12] Luft R, Ikkos D, Palmieri G, et al. A case of severe hypermetabolism of nonthyroid origin with a defect in the maintenance of mitochondrial respiratory control: a correlated clinical, biochemical, and morphological study. J Clin Invest 1962;41:1776–804.

[13] Morava E, van den Heuvel L, Hol F, de Vries MC, Hogeveen M, Rodenburg RJ, et al. Mitochondrial disease criteria: diagnostic applications in children. Neurology 2006;67:1823–6.

[14] Nass MM, Nass S. Intramitochondrial fibers with DNA Characteristics I. Fixation and electron staining reactions. J Cell Biol 1963;19:593–611.

[15] Naviaux RK, Nguyen KV. POLG mutations associated with Alpers' syndrome and mitochondrial DNA depletion. Ann Neurol 2004;55:706–12.

[16] Nishino I, Spinazzola A, Hirano M. Thymidine phosphorylase gene mutations in MNGIE, a human mitochondrial disorder. Science 1999;283:689–92.

[17] Ropp PA, Copeland WC. Cloning and characterization of the human mitochondrial DNA polymerase, DNA polymerase gamma. Genomics 1996;36(3):449–58.

[18] Shapira Y, Harel S, Russell A. Mitochondrial encephalomyopathies: a group of neuromuscular disorders with defects in oxidative metabolism. Isr J Med Sci 1977;13:161–4.

[19] Shoffner JM, Lott MT, Lezza AM, Seibel P, Ballinger SW, Wallace DC. Myoclonic epilepsy and ragged-red fiber disease (MERRF) is associated with a mitochondrial DNA tRNA(Lys) mutation. Cell 1990;61:931–7.

[20] Wolf NI, Smeitink JA. Mitochondrial disorders: a proposal for consensus diagnostic criteria in infants and children. Neurology 2002;59:1402–5.

[21] Van Goethem G, Dermaut B, Löfgren A, Martin JJ, Van Broeckhoven C. Mutation of POLG is associated with progressive external ophthalmoplegia characterized by mtDNA deletions. Nat Genet 2001;28:211–2.

[22] Walker UA, Collins S, Byrne E. Respiratory chain encephalomyopathies: a diagnostic classification. Eur Neurol 1996;36:260–7.

Part I

Mitochondrial DNA Encoded Diseases

Chapter 2

Mitochondrial Myopathy, Encephalopathy, Lactic Acidosis, and Stroke-Like Episodes (MELAS)

S.E. Marin[1], R.H. Haas[1,2,3,4]

[1]Department of Neurosciences, University of California San Diego, La Jolla, CA, USA; [2]Rady Children's Hospital San Diego, San Diego, CA, USA; [3]Department of Pediatrics, University of California San Diego, La Jolla, CA, USA; [4]Metabolic & Mitochondrial Disease Center, University of California San Diego, San Diego, CA, USA

CASE PRESENTATION

This boy was born at 40 weeks gestational age by repeat cesarean section, due to prior cesarean delivery. The pregnancy and delivery were unremarkable, and he was discharged home at 48 h without complications. His family history included a healthy mother, five maternal siblings (three sisters and two brothers) without health concerns, a maternal grandmother with a history of migraines and diabetes, and a healthy 17-year-old brother.

He first presented at 4 years with blurred vision and focal clonic seizures. His physical examination revealed ptosis, hearing loss, fatigue, and short stature. Plasma lactate was elevated, with levels of 2.6–10 mmol/L (normal range 0.7–2.1 mmol/L). Magnetic resonance imaging (MRI) of the brain revealed hyperintense signal in bilateral lentiform nuclei and within both occipital lobes (not confined to the arterial territories) on T2-weighted imaging without restricted diffusion on diffusion-weighted imaging (DWI). The clinical and radiological features were felt to be suggestive of mitochondrial myopathy, encephalopathy, lactic acidosis, and stroke-like episodes (MELAS). He was placed on thiamine, coenzyme Q10, and L-arginine.

He continued to have intractable seizures. At the age of 8 years, he was admitted to the hospital with four left-sided clonic seizures, headache, vomiting, left-sided weakness, and visual changes with no apparent trigger. A repeat MRI revealed new regions of T2 signal abnormality in the putamena and caudate nuclei, periventricular regions, and temporo-occipital lobes of

Mitochondrial Case Studies. http://dx.doi.org/10.1016/B978-0-12-800877-5.00002-4

both hemispheres. There was associated cerebral volume loss and restricted diffusion in bilateral basal ganglia. Intravenous L-arginine (300 mg/kg/day) was given for 5 days. An echocardiogram revealed mild left ventricular hypertrophy.

One year later, he developed acute word finding difficulty, altered level of consciousness, and acute right hemiplegia. A computerized topography (CT) scan revealed a large area of hypodensity in the left temporal lobe. An electroencephalogram (EEG) revealed frequent electrographic seizures originating in the left parieto-occipital lobe without clear clinical manifestations. Subsequently, he developed clonic jerking in the fingers of his right hand consistent with epilepsia partialis continua. He was treated with a combination of fosphenytoin, phenobarbital, and lacosamide leading to seizure resolution. His right-sided hemiplegia improved slowly over the course of a few months.

At 10 years of age, he presented with acute left-sided hemiplegia and gait unsteadiness after a 2-day episode of low-grade fever and decreased oral intake. He was admitted to the hospital, and a repeat MRI revealed new foci of restricted diffusion within the right external capsule and throughout the cortex in bilateral occipital, parietal, and temporal lobes. An EEG revealed multifocal epileptiform discharges and diffuse background slowing. He received IV L-arginine for 5 days followed by oral L-arginine 0.1 g/kg/day, and his symptoms slowly improved over the period of 1 month.

A repeat MRI done 1 year later revealed significant interval increase in cortical atrophy and cerebellar atrophy, new foci of T2 hyperintensity within the left frontal lobe, left parietal lobe, and right temporal lobe without restricted diffusion. T1 hyperintensity was noted throughout the cortical ribbon, thought to represent cortical necrosis or dystrophic calcification. New areas of restricted diffusion were noted in bilateral putamina. His magnetic resonance spectroscopy (MRS) showed lactate peaks in multiple voxels. It was identified that he had the m.3243A>G mutation at a heteroplasmy level of 40–50% in his blood. His asymptomatic mother was found to carry the m.3243A>G mutation with 15% heteroplasmy in blood.

At 13 years of age, he is in grade seven in a special education class. His most recent hospital admission was just prior to his 13th birthday for status epilepticus, during which an MRI revealed new areas of restricted diffusion in the left caudate nuclei and putamen, posterior parasagittal frontal lobes, posterior bank of the left precentral gyrus, posterior right insula, and lateral right temporal lobe. MRS confirmed elevated lactate and reduced *N*-acetyl aspartate, indicative of neuronal loss. His most recent physical examination reveals short stature, marked hirsutism, bilateral ptosis, bilateral sensorineural hearing loss, axial hypotonia, appendicular spasticity, generalized muscle weakness, muscle stretch hyperreflexia, and exercise tolerance. He requires a gastrostomy tube for feeding. His current medication regimen includes baclofen for spasticity, lamotrigine, lacosamide, and levetiracetam for seizures, and a mitochondrial cocktail composed of riboflavin, levocarnitine, citrulline 0.2 g/kg/day, creatine monohydrate, and vitamin B complex. He is receiving experimental treatment with EPI-743, provided for compassionate use. Ubiquinol was discontinued when EPI-743 treatment was started as per protocol of the study.

INTRODUCTION

The first description of a case with reversible alexia, mitochondrial myopathy, and lactic acidemia was in 1979 [1]. The multisystem phenotype was characterized, and the acronym MELAS was coined in 1984 by Pavlakis and colleagues [2]. The clinical spectrum of MELAS (OMIM 540000) has broadened since its original description; however, the core features that define the syndrome remain unchanged. The main features required for the diagnosis of MELAS include stroke-like episodes prior to 40 years of age, encephalopathy characterized by seizures, dementia, or both, and biochemical and/or structural abnormalities (lactic acidosis and/or ragged red fibers on muscle biopsy) with two of (1) normal early psychomotor development, (2) recurrent headache, and (3) recurrent vomiting [3].

The most common initial symptoms include seizures, recurrent headaches, anorexia, exercise intolerance, and recurrent vomiting. Stroke-like episodes, seen in nearly 100% of cases, typically present acutely after a stressor, such as a febrile illness, with vomiting, headache, and new or worsening focal neurological signs [4]. The most common symptoms occurring with the stroke-like episodes include cortical blindness, visual field abnormalities, and hemiparesis or hemiplegia. Although there may be full return to baseline after individual episodes, ultimately cumulative residual effects result, which gradually impair motor abilities, vision, and cognition by adolescence or young adulthood. Seizures, also present in a majority of cases with MELAS, commonly occur concurrently with stroke-like episodes [4,5]. Various types of seizures have been documented in cases with MELAS including focal seizures (recurrent or as part of epilepsia partialis continua), bilateral generalized tonic-clonic seizures, and myoclonic seizures [6]. Status epilepticus may occur in up to 19% of cases [4–6]. Other symptoms seen in cases with MELAS include the following: short stature (82%), diabetes mellitus (12–26%), sensorineural hearing loss, headaches (typically migrainous, 50%), myopathic weakness (37%), exercise intolerance (100%), ataxia (33%), psychiatric manifestations, cognitive impairment (90%), ophthalmologic manifestations (including optic atrophy, pigmentary retinopathy, and ophthalmoplegia), gastrointestinal dysmotility, hirsutism, and cardiomyopathy or arrhythmias [7–14].

The age of onset of symptoms in MELAS is variable, but it is usually in childhood, commonly between 2 and 10 years old [7,8,15]. It is becoming evident that both juvenile- and adult-onset forms exist, with the former more commonly having a short stature and a more severe and rapidly deteriorating course to early mortality and the latter presenting with hearing loss, diabetes, and a more insidious course [4,5]. It is currently estimated that 40% of MELAS cases present in adulthood [5]. The difference in phenotype is believed to result from a greater initial burden of dysfunctional mitochondria in those with earlier onset.

The natural history of MELAS is widely variable, even in individuals who harbor the same mutation. In a majority, there is progressive, and often stepwise, cognitive and neurological impairment in conjunction (often associated with stroke-like events) with worsening biochemical profiles (specifically, increasing CSF lactate) and neuroimaging abnormalities [16]. The annual mortality has been estimated at 5–8% with an estimated overall median survival time of 16.9 years from onset of focal neurologic disease [4,16–18]. The average age of death is approximately 34.5 ± 19 years (range 10.2–81.8 years), with approximately 22% occurring in childhood (<18 years) [16].

MELAS is a prototypical mitochondrial disorder in that it exhibits a maternal pattern of inheritance due to mutations in mitochondrial DNA (mtDNA), has a variable proportion of mutated mitochondria in different tissues (heteroplasmy), and requires a certain level of affected mitochondria in different tissues before clinical features manifest (the threshold effect). These factors contribute to the clinical heterogeneity of the disorder. The earliest recognized genetic alteration causing MELAS syndrome is the point mutation m.3243A>G in the tRNA leucine (UUR) gene (*MT-TL1*) of the mitochondrial genome (OMIM 590050) [19,20]. The m.3243A>G mutation is also the most common genetic alteration resulting in a MELAS phenotype, with approximately 80% of cases harboring this point mutation [19,21–23]. The prevalence of the m.3243A>G mutation in the adult Finnish population was reported to be 10.2/100,000, but based on the assumption that all first-degree maternal relatives of a verified mutation carrier also harbor the mutation, true prevalence is approximately 16/100,000 [24]. Recent population studies suggest a carrier rate of the m.3243A>G mutation of 1/400 [25,26]. In cases that do not harbor the m.3243A>G mutation, more than 40 less common mitochondrial DNA mutations have been associated with a MELAS phenotype [21,27]. The more common mutations are as follows [8]: m.13513G>A in *MT-ND5* (15%), m.3271T>C in *MT-TL1* (7.5%), and m.3252A>G in *MT-TL1* (5%).

PATHOPHYSIOLOGY

Although 30 years have passed since the first clinical and pathological descriptions of MELAS, the pathogenesis of many of the clinical features remains unclear. The most common mutation (m.3243A>G) causes a decreased lifespan of $tRNA^{Leu(UUR)}$. Once a threshold level of heteroplasmy is reached, defective protein synthesis reduces oxidative phosphorylation and ATP synthesis [28,29]. As expected, complexes I and IV are particularly affected as the number of mtDNA encoded subunits are highest in these complexes (7 and 3, respectively) [20,30].

The hallmark clinical feature of MELAS, stroke-like episodes, has been the most investigated clinical feature from a pathophysiologic standpoint. The stroke-like episodes have been hypothesized to represent a metabolic stroke due to a deficiency of adenosine triphosphate, limited oxidative glucose metabolism, increased lactic acid production, a deficiency of L-arginine, and/or mitochondrial angiopathy [31]. Through various pathological studies and therapeutic findings, it has become evident that endothelial dysfunction plays a significant role in stroke-like episodes in MELAS. Mitochondrial angiopathy with segmental dilatation and degenerative changes in small arteries and arterioles in the brain has been reported in autopsy cases of MELAS cases [32,33]. Mitochondrial proliferation and accumulation has been documented in the endothelium and smooth muscle cells of cerebral arterioles and capillaries in a manner similar to ragged red fibers (RRFs) [30]. The affected blood vessels have been designated strongly succinate dehydrogenase-reactive vessels (SSVs) [34,35]. Unlike RRFs and SSVs seen in myoclonus epilepsy with ragged red fibers (MERRF) and Kearns-Sayre syndrome (KSS), which are typically cytochrome *c* oxidase (COX) negative, those seen in MELAS are typically positive, known as the MELAS paradox [36].

More recently, the role of L-arginine in stroke-like episodes has been a focus of research. L-arginine, a nonessential amino acid, is a precursor for the synthesis of ornithine and urea (via arginase), nitric oxide (NO) (via nitric oxide synthase [NOS]), creatine (via arginine:glycine aminotransferase [AGAT]), and agmatine (via arginine decarboxylase). Cases with MELAS have significantly decreased levels of ictal and interictal L-arginine [29]. It has been hypothesized that the decreased levels of L-arginine may partially explain the segmental impairment in vasodilatation in intracerebral small arteries identified in this population [29,37–39], considering its vital role in endothelial-dependent vasodilation through the generation of NO [39,40].

There is a growing body of evidence supporting the hypothesis that there is a nitric oxide (NO) deficiency underlying MELAS pathophysiology. Lower concentrations of NO metabolites have been documented during stroke-like episodes in cases with MELAS [39]. Additionally, NO synthesis rates have been documented to be lower in individuals with MELAS independently of stroke-like episodes [41]. Assessment of COX-negative fibers in cases with MELAS has revealed a reduced level of sarcoplasmic NOS [42]. In addition to a secondary deficiency in nitric oxide due to deficient L-arginine, NO may be further decreased through the inhibition of NOS through the accumulation of free radicals and high NADH/NAD ratio that results from defective respiratory chain enzyme activities [29]. Additionally, hyperactive COX (as described in the MELAS paradox) may further decrease regional NO concentration as NO can bind to the active site of COX and displace heme-bound oxygen [36].

Pathophysiology of epilepsy in cases with MELAS is also poorly understood. It has been hypothesized that inhibition of mitochondrial oxidative phosphorylation leading to ATP deficiency may result in increased neuronal excitability and epileptogenesis through multiple mechanisms [43,44].

Further studies are necessary to elucidate the mechanisms underlying the other clinical features of MELAS, but energy failure is likely etiologically important.

DIAGNOSTIC APPROACH

MELAS should be considered in any case with any of the classical symptoms, as outlined by Hirano et al. [3]. However, the clinical spectrum has been recognized to be much more diverse and heterogeneous compared to original reports [15,22,26,45]. In the most recent cohort study, only 8.6% of cases with a documented MELAS-associated mutation fulfilled invariant criteria for MELAS [15]. In the largest cohort study of cases with the most common m.3243A>G mutation, 10% of cases exhibited the classical MELAS phenotype, 49% had a combination of MELAS/maternally inherited diabetes and deafness and other overlap phenotypes, while 28% demonstrated clinical features that were not at all consistent with the classical syndrome [26]. Based on these findings, screening for MELAS should be considered in cases with three or more of the following clinical features: cardiomyopathy, deafness, developmental delay or cognitive decline, diabetes mellitus, epilepsy, gastrointestinal disturbance, migraine, progressive external ophthalmoplegia, and retinopathy.

Investigation of cases is first aimed at excluding other causes of seizures and stroke in young people. This typically requires serology, cerebrospinal fluid (CSF) analysis, and neuroimaging. Genetic testing may be completed noninvasively on saliva, urine, and blood. However, muscle biopsy may be required in cases without recognized mutations to demonstrate the characteristic pathological changes of a mitochondrial cytopathy.

Metabolites

There are no pathognomonic patterns of metabolic derangement noted on routine blood investigations; however, elevation of lactate (>2.1 mM) is common and indicative of an abnormal redox ratio. Lactate may be normal outside of the setting of an acute stressor, therefore a normal lactate does not eliminate from consideration the diagnosis of a mitochondrial disease. Additionally, it should be cautioned that spurious elevation of lactate is common and may be secondary to a struggling case, use of a tourniquet, or inappropriate possessing of the blood sample. Elevation of plasma alanine (>450 uM) may be found resulting from transamination of accumulated pyruvate [46]. Plasma arginine is often decreased. Urine organic analysis, when completed outside of acute symptoms, has a low sensitivity for detecting mitochondrial disease [46]. Additionally, lactate elevation on urine organic acid analysis is insensitive for mitochondrial cytopathies [47]. Increased excretion of tricarboxylic acid cycle intermediates malate and fumarate, and organic acids ethylmalonic acid, and 3-methyl glutaconic acid, commonly occur in mitochondrial disease, but they are rarely diagnostic of a specific mitochondrial disorder [46,48].

Cerebrospinal Fluid

Lumbar puncture is commonly performed due to the acute nature of the presentation in cases with MELAS. Routine studies, including biochemistry, virology, and microbial studies, are normal. CSF lactate is commonly elevated in cases with MELAS at rates up to 97–100% [20]. CSF protein, which is commonly elevated in other mitochondrial disorders such as KSS, may be elevated in MELAS cases (usually < 100 mg/dL).

Neuroimaging

MRI has superseded CT as the neuroimaging modality of choice in most cases; however, CT is widely available and indicated in the acute presentations of seizures and stroke-like neurological deficits. While there is no pathognomonic finding on CT, calcification of the basal ganglia is frequently observed, even before the occurrence of stroke-like episodes [7,49]. As such, basal ganglia calcifications should lead clinicians to consider an underlying mitochondrial disorder in the presence of other appropriate clinical features [4]. During stroke-like episodes, focal areas of apparent infarction may be seen. CT angiogram is commonly performed to eliminate from consideration large or medium vessel vasculitis, and it is normal in the setting of MELAS [49].

Although MRI is considered the gold-standard neuroimaging tool in cases with MELAS, there are no pathognomonic features seen, although occipital stroke-like lesions are particularly common (Figure 1(a)). Conventional MRI completed in the acute phase of a stroke-like episode reveals hyperintensity on T2-weighted or fluid attenuation inversion recovery sequences in non-arterial territories. The affected areas involve cortical and subcortical white matter with sparing of the deep white matter [50]. The affected areas may fluctuate, migrate, or even disappear during the time when seen on serial MRI imaging [4,29]. More commonly, the lesions affect the occipital and parietal lobes [51].

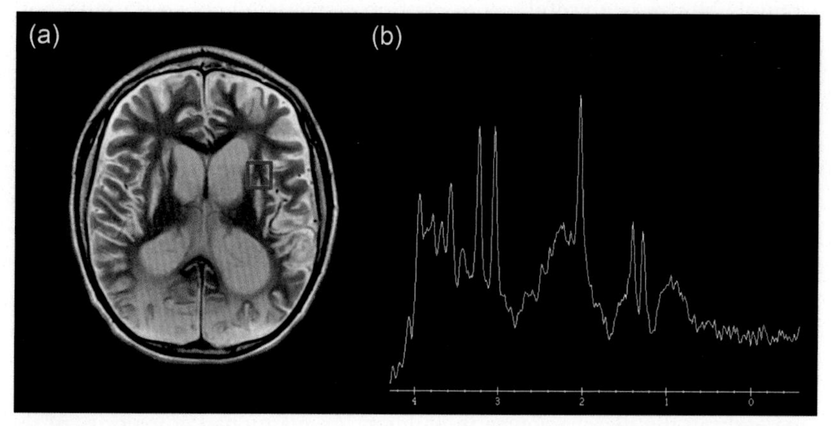

FIGURE 1 MRI and MRS of 12-year-old boy with MELAS due to m.3243A>G mutation at 90% in blood. (a) T2 MRI shows atrophy with ventriculomegaly, caudate nucleus and putaminal bright signal and bilateral occipital subacute metabolic strokes in a nonvascular distribution, and parietal and left frontal stroke lesions. (b) MRS using a short TE (voxel in over left caudate and adjacent white matter) shows high lactate doublet (1.3 ppm) and a decreased *N*-acetylaspartate (NAA) (2.0 ppm)/creatine (3.2 ppm) ratio.

Deep gray matter, particularly the thalamus, is often involved, reflecting the high metabolic demand of these areas. If serially evaluated over weeks, slow spreading of the stroke-like lesion has been documented [52]. On diffusion-weighted imaging, in the acute period, DWI typically reveals a high signal in the areas of change seen on conventional imaging and normal or increased signal on the apparent diffusion coefficient, suggesting vasogenic rather than cytotoxic edema [53–55]. However, many case reports exist in which diffusion restriction is seen and has been persistent over time, suggesting mitochondrial energetic impairment leading to irreversible cellular failure with cytotoxic edema [56,57]. Recent studies suggest that there is heterogeneity in MRI findings in the acute setting with concurrent cytotoxic and vasogenic edema [58]. Like CT angiogram, MR angiogram does not reveal large-vessel pathology in a majority of cases. MRS has emerged as a useful test in the evaluation of MELAS to assess in vivo brain metabolites. The most important finding is an elevated lactate in areas outside what is acutely ischemic, confirming metabolic derangement and poor functioning of the respiratory chain (Figure 1(b)). Phosphorus MRS studies reveal decreased levels of high-energy phosphate compounds in the brains of MELAS cases [59].

Although infrequently used as a clinical tool, single emission computed tomography reveals abnormal tracer uptake both ictally and interictally, implying inappropriate cerebral circulation [37,60]. Positron emission tomography scanning reveals decreased oxygen consumption relative to glucose utilization, implying impairment of oxidative phosphorylation [61–63].

Muscle Biopsy

A muscle biopsy is a useful test to confirm a suspected case with mitochondrial disease. Newer techniques using a suction-modified Bergström needle are enabling the completion of less invasive muscle biopsies compared to the traditional open biopsy technique while maintaining the integrity of the muscle sample [64]. Non-cell-dividing muscle tissue tends to retain higher levels of mutant load compared to continually dividing leukocytes in blood samples; therefore, it should be considered as part of the diagnostic evaluation, particularly in the setting of normal genetic testing on blood, urine, or saliva. The hallmarks of mitochondrial diseases on muscle biopsy include RRFs on modified gomori trichrome stain or ragged blue fibers on succinate dehydrogenase stain, increased neutral lipid staining within muscle fibers, and railway track or parking lot paracrystalline inclusions on electron microscopy [4]. However, these are not always present in cases with MELAS, particularly when done early in the disease course in children. As mentioned, a majority of the RRFs stain positively for COX activity, unlike in other mitochondrial disorders [65]. An additional morphologic feature that is characteristic (though not pathognomonic) of MELAS is the overabundance of mitochondria in smooth muscle and endothelial cells of intramuscular blood vessels, best revealed with the SDH stain (strongly succinate dehydrogenase-reactive blood vessels, or SSVs), as previously discussed [34,35]. Biochemical analysis of respiratory chain enzymes in muscle extracts usually shows multiple partial defects, especially involving complex I and/or complex IV. However, biochemical results can also be normal.

Genetic Testing

Genetic testing should be completed looking for the most common MELAS-associated mutations. Initial screening should investigate for the m.3243A>G mutation, as this is present in 80–90% of cases with MELAS [19,21–23]. As such, targeted mutation analysis of *MT-TL1* is usually the first step in diagnosis. Mutations are usually present in all tissues and can be detected in mtDNA from blood leukocytes; however, the occurrence of heteroplasmy may result in varying tissue distribution of mutated DNA. Targeted PCR analysis of blood can miss mutation percentages less than 15%. Therefore, screening more than one tissue type may be needed. Additionally, the heteroplasmic level in blood samples has been shown to decrease with age [66–68], and fibroblasts, muscle, urine sediment, and hair follicle samples typically have a higher level of heteroplasmy compared to blood [15].

Depending on the laboratory, other mutations in *MT-TL1* are commonly included in the mutation analysis, including m.3243A>G, m.3217T>C, m.3256C>T, and m.3252A>G. If screening for the common *MT-TL1*-associated mutations is negative, consider *MT-TL1* sequence analysis, followed by investigating for other mutations known to be involved in a MELAS phenotype (including the *MT-ND5* point mutation m.13513G>A). Mutations known to cause MELAS have been identified in other mitochondrial tRNA genes including *MT-TC, MT-TK, MT-TV, MT-TF, MT-TQ, MT-TS1, MT-TS2*, and *MT-TW*, and in protein-encoding genes *MT-CO1, MT-CO2, MT-CO3, MT-CYB, MT-ND1, MT-ND3,* and *MT-ND6.* Due to the occurrence of overlap syndromes, such as MELAS/MERRF, MELAS/CPEO, and MELAS/LHON, screening should be undertaken for other known point mutations if initial genetic screening is noncontributory (including m.8344A>G, m.8356T>C, m.8993T>G, m.8993T>C, m.3460G>A, m.11778G>A, and m.14484T>C).

Recently, the Mitochondrial Medicine Society released guidelines to assist clinicians in determining the best route for diagnosis if an mtDNA mutation is suspected [69]. It has been suggested that massively parallel sequencing or next-generation sequencing (NGS) of the mtDNA genome in blood is preferred over testing of specific point mutations. Applying NGS technologies for mitochondrial disorders has become an area of increasing interest in recent years, and there will be further development of diagnostic testing for mitochondrial disorders on the whole genome level using this technology. Studies investigating NGS in mitochondrial diseases have shown high concordance (98%) with Sanger sequencing [70]. The emergence of new technologies will help provide molecular diagnosis for cases of mitochondrial disorders that have yet to be resolved. If testing of blood is negative, another tissue should be assessed (particularly urine considering the high content of mtDNA in renal epithelial cells or muscle tissue, if available) [69]. Heteroplasmy should be pursued in cases with mutations in mtDNA. Heteroplasmy analysis in urine may be more informative and accurate than testing in blood alone, particularly in cases with MELAS [69].

Despite the availability of genetic testing, the diagnosis of MELAS can be difficult. This is particularly the case given the phenomenon of heteroplasmy, an apparent threshold effect, the number of pathological mutations, and the variation in findings on muscle biopsy. A negative genetic test or a normal muscle biopsy does not definitively eliminate the condition from consideration.

Other Investigations

Electrocardiogram completed in cases with MELAS may reveal evidence of cardiomyopathy, preexcitation, or incomplete heart block [8]. Electromyography and nerve conduction studies are often consistent with a myopathic process, but a concurrent neuropathy, typically axonal and sensory, may exist in up to 22% [18,71].

DIFFERENTIAL DIAGNOSIS

Acute stroke-like episodes are the clinical feature that typically prompts evaluation for a possible mitochondrial disorder. Importantly, these metabolic strokes must be differentiated from causes of ischemic strokes secondary to thrombosis or embolism in a young person including heart disease, vasculopathies, sickle cell disease, carotid or vertebral dissection, coagulopathies, and moyamoya disease. Stroke-like episodes in MELAS differ from ischemic strokes in many important ways. These episodes are atypical in that they affect a younger case population than most ischemic strokes and are often triggered by a stressor (e.g., a febrile illness, migraine-like headache, seizure, physiological stress, or dehydration). Additionally, there is often a history of other features consistent with a mitochondrial disorder, including sensorineural hearing loss, short stature (greater than two standard deviations below the mean for age), epilepsy, and/or migraine-like headaches either in the proband or maternal relatives. Moreover, stroke-like episodes are often recurrent in cases with MELAS at a frequency higher than the recurrence rate for ischemic stroke (outside of untreated syndromes, such as sickle cell disease, that place individuals at a very high risk of recurrence). Considering that neuroimaging is imperative to the diagnosis of ischemic stroke, it is important to note that stroke-like episodes in MELAS can be differentiated from ischemic strokes based on neuroimaging. Migrating or expanding lesions on recurrent imaging, which are not restricted to distinct arterial territories, strongly suggest a metabolic or mitochondrial disorder [4,31,72].

In cases that present with seizures, a history of features suggestive of a mitochondrial disease should be sought. The consideration of MELAS (or other mitochondrial disorders) in the differential diagnosis of seizures is imperative when considering therapy. One of the most widely used anticonvulsants used in the pediatric population is valproic acid (VPA). VPA should be avoided in mitochondrial disorders, including MELAS. The pathogenesis of VPA toxicity in mitochondrial disorders includes (but is not limited to) interference of VPA (or its metabolites) with mitochondrial β-oxidation, acetyl coA production and carnitine sequestration, oxidative stress due to compromised free-radical scavenging activity and/or enhanced production of reactive oxygen species, impaired proton-pumping ability of complex IV, and inhibition of dihydrolipoyl dehydrogenase activity leading to impaired 2-oxoglutarate-driven oxidative phosphorylation [73–75]. The use of VPA may result in worsening of the symptomatology, including seizures, when used in cases with mitochondrial disease [74,76].

If atypical clinical symptoms combined with biochemical and/or pathological features of mitochondrial disease are present, other mitochondrial syndromes should be considered in the differential diagnosis. All of these disorders may have findings of an elevated serum and/or CSF lactate and pathological findings on muscle biopsy including RRFs on Gomori trichrome stain and paracrystalline inclusions on electron microscopy. Many of the features present in cases with MELAS, such as short stature, hearing loss, seizures, and dementia, are also present in other mitochondrial disorders, including KSS and MERRF. KSS typically has additional features, including pigmentary retinopathy, ophthalmoplegia, ptosis, and cardiac conduction defects, while MERRF presents with predominantly myoclonic seizures. However, the presence of stroke-like episodes is not typically seen in these disorders and should prompt the diagnostician to consider MELAS as more likely. Additionally, cytochrome *c* oxidase (COX) negative fibers are more typical in cases with KSS and MERRF than with cases with MELAS [36]. When gastrointestinal dysmotility, cachexia, and neuropathy are prominent with widespread MRI changes involving cerebral white matter but sparing of the cortex, mitochondrial neurogastrointestinal encephalomyopathy needs to be considered [77]. Additionally, if a case presents with a MELAS-like phenotype and an autosomal dominant or autosomal recessive inheritance is evident based on family history, nuclear encoded genes that affect mitochondrial replication, fission, or fusion should be sought, although rarely a cause of classic MELAS. One individual with MELAS was found to have compound heterozygous mutations in the nuclear DNA gene *POLG*, encoding the catalytic subunit of polymerase gamma [8].

Basal ganglia calcification noted on CT scanning, which is commonly present in cases with MELAS, has its own differential diagnosis. Basal ganglia calcifications are seen in approximately 1% of older individuals and are idiopathic, but they may also be present in younger individuals in the setting of Fahr disease, carbon monoxide poisoning, lead intoxication, prior radiation therapy, congenital infections (including CMV, toxoplasmosis, and HIV), neurocysticercosis, tuberculosis, disease of the parathyroid gland, and other inherited metabolic conditions, such as Cockayne syndrome. The clinical phenotype is important in differentiating other causes of basal ganglia calcifications from MELAS and other mitochondrial disorders.

TREATMENT

At present, there is no proven therapy to prevent, attenuate, or treat established MELAS-related clinical episodes. For the most part, treatment is symptomatic. Cochlear implants should be considered for cases with hearing loss, seizures should be treated with anticonvulsant therapy (with the exception of valproic acid), standard analgesics should be used for migraine, and diabetes can be managed with dietary modification, oral hypoglycemic agents, or insulin.

Despite the Cochrane review conclusion that there was "no clear evidence supporting the use of any intervention in mitochondrial disorders," mitochondrial cocktails are frequently used in cases with MELAS [78]. All of the following have been tried in MELAS with variable success: coenzyme Q10 (ubiquinone), idebenone, dichloroacetate, creatine monohydrate, alpha-lipoic acid, L-carnitine, riboflavin, nicotinamide, sodium succinate, menadione (vitamin K3), phylloquinone (vitamin K1), ascorbate, exercise training, and ketogenic diets [18,78–83].

One of the most promising therapies for cases with MELAS is L-arginine. Due to the finding of low levels of arginine both ictally and interictally in MELAS cases and its known role in endothelial-dependent vascular relaxation, L-arginine has been used as a therapy in cases with MELAS and stroke-like episodes. L-arginine (0.5 g/kg/dose) infused during acute stroke-like episodes was found to rapidly decrease severity of stroke-like symptoms, reduce lactate accumulation, improve microcirculation, and reduce tissue injury from ischemia [39,84]. Oral L-arginine supplementation (0.15–0.3 g/kg/day) results in a lower frequency and severity of stroke-like episodes without serious adverse effects [29]. However, it should be cautioned that excessive arginine can lead to vasodilatory hypotension, severe hyponatremia (possibly due to natriuresis from a surge in NO production), and subsequent central pontine and/or extrapontine myelinolysis if rapid overcorrection of hyponatremia is attempted [29].

More recently, citrulline is being substituted for L-arginine for cases with MELAS [85]. Both citrulline and arginine result in increased NO production and hence potentially have therapeutic effects on NO-deficiency-related manifestations. However, citrulline is a more efficient NO donor than arginine, which results in a greater increase in de novo arginine synthesis and enhanced NO production [41].

Recently, a new potential therapeutic agent is currently undergoing studies in cases with mitochondrial disease. EPI-743 is a *para*-benzoquinone analog with potent antioxidant effects and a favorable safety profile [86]. It is believed to affect the cellular glutathione redox state. Results of controlled trials in

Leigh syndrome and Friedreich's ataxia are awaited; however, based on preliminary studies suggesting a favorable outcome, it has been offered to cases with MELAS on a compassionate-use basis.

CLINICAL PEARLS

- MELAS is a multisystem disease with classical features including stroke-like episodes prior to 40 years of age, encephalopathy (dementia and/or seizures), normal early psychomotor development, recurrent headaches, and recurrent vomiting.
- MELAS is secondary to mitochondrial DNA mutations, with 80% of cases being positive for the common m.3243A>G mutation in $tRNA^{Leu(UUR)}$. Diagnosis may be possible without muscle biopsy.
- RRFs on muscle biopsy are typically found with cytochrome *c* oxidase positivity, unlike many other mitochondrial disorders where COX staining is reduced.
- The cornerstone of treatment for cases with MELAS is arginine or citrulline, both daily (oral) and high-dose (intravenous) with acute stroke-like episodes. Coenzyme Q10 or ubiquinol and high-dose B vitamins are usually recommended.

REFERENCES

[1] Skoglund RR. Reversible alexia, mitochondrial myopathy, and lactic acidemia. Neurology 1979;29:717–20.

[2] Pavlakis SG, Phillips PC, DiMauro S, De Vivo DC, Rowland LP. Mitochondrial myopathy, encephalopathy, lactic acidosis, and strokelike episodes: a distinctive clinical syndrome. Ann Neurol 1984;16:481–8.

[3] Hirano M, Ricci E, Koenigsberger MR, et al. Melas: an original case and clinical criteria for diagnosis. Neuromuscular Disord 1992;2:125–35.

[4] Goodfellow JA, Dani K, Stewart W, et al. Mitochondrial myopathy, encephalopathy, lactic acidosis and stroke-like episodes: an important cause of stroke in young people. Postgrad Med J 2012;88:326–34.

[5] Yatsuga S, Povalko N, Nishioka J, et al. MELAS: a nationwide prospective cohort study of 96 patients in Japan. Biochim Biophys Acta 2012;1820:619–24.

[6] Hirano M, DiMauro S. Primary mitochondrial diseases. In: Engel J, Pedley TA, editors. Epilepsy: a comprehensive textbook. Philadelphia, PA: Lippincott-Raven Publisher; 1997.

[7] Hirano M, Pavlakis SG. Mitochondrial myopathy, encephalopathy, lactic acidosis, and stroke-like episodes (MELAS): current concepts. J Child Neurol 1994;9:4–13.

[8] DiMauro S, Hirano M. Melas. In: Pagon RA, Adam MP, Ardinger HH, et al. editors. GeneReviews(R). 1993. Seattle, WA.

[9] Anglin RE, Garside SL, Tarnopolsky MA, Mazurek MF, Rosebush PI. The psychiatric manifestations of mitochondrial disorders: a case and review of the literature. J Clin Psychiatry 2012;73:506–12.

[10] Petruzzella V, Zoccolella S, Amati A, et al. Cerebellar ataxia as atypical manifestation of the 3243A>G MELAS mutation. Clin Genet 2004;65:64–5.

[11] Menotti F, Brega A, Diegoli M, Grasso M, Modena MG, Arbustini E. A novel mtDNA point mutation in tRNA(Val) is associated with hypertrophic cardiomyopathy and MELAS. Ital Heart J 2004;5:460–5.
[12] Wortmann SB, Rodenburg RJ, Backx AP, Schmitt E, Smeitink JA, Morava E. Early cardiac involvement in children carrying the A3243G mtDNA mutation. Acta Paediatr 2007;96:450–1.
[13] Garcia-Velasco A, Gomez-Escalonilla C, Guerra-Vales JM, Cabello A, Campos Y, Arenas J. Intestinal pseudo-obstruction and urinary retention: cardinal features of a mitochondrial DNA-related disease. J Intern Med 2003;253:381–5.
[14] Chang TM, Chi CS, Tsai CR, Lee HF, Li MC. Paralytic ileus in MELAS with phenotypic features of MNGIE. Pediatr Neurol 2004;31:374–7.
[15] Chin J, Marotta R, Chiotis M, Allan EH, Collins SJ. Detection rates and phenotypic spectrum of m.3243A>G in the MT-TL1 gene: a molecular diagnostic laboratory perspective. Mitochondrion 2014;17:34–41.
[16] Kaufmann P, Engelstad K, Wei Y, et al. Natural history of MELAS associated with mitochondrial DNA m.3243A>G genotype. Neurology 2011;77:1965–71.
[17] Majamaa-Voltti KA, Winqvist S, Remes AM, et al. A 3-year clinical follow-up of adult patients with 3243A>G in mitochondrial DNA. Neurology 2006;66:1470–5.
[18] Kaufmann P, Engelstad K, Wei Y, et al. Dichloroacetate causes toxic neuropathy in MELAS: a randomized, controlled clinical trial. Neurology 2006;66:324–30.
[19] Goto Y, Nonaka I, Horai S. A mutation in the tRNA(Leu)(UUR) gene associated with the MELAS subgroup of mitochondrial encephalomyopathies. Nature 1990;348:651–3.
[20] Goto Y, Horai S, Matsuoka T, et al. Mitochondrial myopathy, encephalopathy, lactic acidosis, and stroke-like episodes (MELAS): a correlative study of the clinical features and mitochondrial DNA mutation. Neurology 1992;42:545–50.
[21] Goto Y, Nonaka I, Horai S. A new mtDNA mutation associated with mitochondrial myopathy, encephalopathy, lactic acidosis and stroke-like episodes (MELAS). Biochim Biophys Acta 1991;1097:238–40.
[22] Ciafaloni E, Ricci E, Shanske S, et al. MELAS: clinical features, biochemistry, and molecular genetics. Ann Neurol 1992;31:391–8.
[23] Vanniarajan A, Nayak D, Reddy AG, Singh L, Thangaraj K. Clinical and genetic uniqueness in an individual with MELAS. Am J Med Genet B Neuropsychiatr Genet 2006;141B:440–4.
[24] Majamaa K, Moilanen JS, Uimonen S, et al. Epidemiology of A3243G, the mutation for mitochondrial encephalomyopathy, lactic acidosis, and strokelike episodes: prevalence of the mutation in an adult population. Am J Hum Genet 1998;63:447–54.
[25] Manwaring N, Jones MM, Wang JJ, et al. Population prevalence of the MELAS A3243G mutation. Mitochondrion 2007;7:230–3.
[26] Nesbitt V, Pitceathly RD, Turnbull DM, et al. The UK MRC Mitochondrial Disease Patient Cohort Study: clinical phenotypes associated with the m.3243A>G mutation–implications for diagnosis and management. J Neurol Neurosurg Psychiatry 2013;84:936–8.
[27] Shanske S, Coku J, Lu J, et al. The G13513A mutation in the ND5 gene of mitochondrial DNA as a common cause of MELAS or Leigh syndrome: evidence from 12 cases. Arch Neurol 2008;65:368–72.
[28] Chomyn A, Martinuzzi A, Yoneda M, et al. MELAS mutation in mtDNA binding site for transcription termination factor causes defects in protein synthesis and in respiration but no change in levels of upstream and downstream mature transcripts. Proc Natl Acad Sci USA 1992;89:4221–5.
[29] Koga Y, Povalko N, Nishioka J, Katayama K, Kakimoto N, Matsuishi T. MELAS and L-arginine therapy: pathophysiology of stroke-like episodes. Ann N Y Acad Sci 2010;1201:104–10.

[30] Koga Y, Nonaka I, Kobayashi M, Tojyo M, Nihei K. Findings in muscle in complex I (NADH coenzyme Q reductase) deficiency. Ann Neurol 1988;24:749–56.
[31] Aurangzeb S, Vale T, Tofaris G, Poulton J, Turner MR. Mitochondrial encephalomyopathy with lactic acidosis and stroke-like episodes (MELAS) in the older adult. Pract Neurol 2014;14:432–6.
[32] Ohama E, Ohara S, Ikuta F, Tanaka K, Nishizawa M, Miyatake T. Mitochondrial angiopathy in cerebral blood vessels of mitochondrial encephalomyopathy. Acta Neuropathol 1987;74:226–33.
[33] Kishi M, Yamamura Y, Kurihara T, et al. An autopsy case of mitochondrial encephalomyopathy: biochemical and electron microscopic studies of the brain. J Neurol Sci 1988;86:31–40.
[34] Hasegawa H, Matsuoka T, Goto Y, Nonaka I. Strongly succinate dehydrogenase-reactive blood vessels in muscles from patients with mitochondrial myopathy, encephalopathy, lactic acidosis, and stroke-like episodes. Ann Neurol 1991;29:601–5.
[35] Sakuta R, Nonaka I. Vascular involvement in mitochondrial myopathy. Ann Neurol 1989; 25:594–601.
[36] Naini A, Kaufmann P, Shanske S, Engelstad K, De Vivo DC, Schon EA. Hypocitrullinemia in patients with MELAS: an insight into the "MELAS paradox". J Neurol Sci 2005;229–30:187–93.
[37] Koga Y, Akita Y, Junko N, et al. Endothelial dysfunction in MELAS improved by l-arginine supplementation. Neurology 2006;66:1766–9.
[38] Koga Y, Ishibashi M, Ueki I, et al. Effects of L-arginine on the acute phase of strokes in three patients with MELAS. Neurology 2002;58:827–8.
[39] Koga Y, Akita Y, Nishioka J, et al. L-arginine improves the symptoms of strokelike episodes in MELAS. Neurology 2005;64:710–2.
[40] Wang XL, Sim AS, Badenhop RF, McCredie RM, Wilcken DE. A smoking-dependent risk of coronary artery disease associated with a polymorphism of the endothelial nitric oxide synthase gene. Nat Med 1996;2:41–5.
[41] El-Hattab AW, Hsu JW, Emrick LT, et al. Restoration of impaired nitric oxide production in MELAS syndrome with citrulline and arginine supplementation. Mol Genet Metab 2012;105:607–14.
[42] Tengan CH, Kiyomoto BH, Godinho RO, et al. The role of nitric oxide in muscle fibers with oxidative phosphorylation defects. Biochem Biophys Res Commun 2007;359:771–7.
[43] Tzoulis C, Bindoff LA. Acute mitochondrial encephalopathy reflects neuronal energy failure irrespective of which genome the genetic defect affects. Brain 2012;135:3627–34.
[44] Folbergrova J, Kunz WS. Mitochondrial dysfunction in epilepsy. Mitochondrion 2012;12:35–40.
[45] Jean-Francois MJ, Lertrit P, Berkovic SF, et al. Heterogeneity in the phenotypic expression of the mutation in the mitochondrial $tRNA^{(Leu)(UUR)}$ gene generally associated with the MELAS subset of mitochondrial encephalomyopathies. Aust N Z J Med 1994;24:188–93.
[46] Mitochondrial Medicine Society's Committee on Diagnosis, Haas RH, Parikh S, et al. The in-depth evaluation of suspected mitochondrial disease. Mol Genet Metab 2008;94:16–37.
[47] Thorburn DR, Smeitink J. Diagnosis of mitochondrial disorders: clinical and biochemical approach. J Inherit Metab Dis 2001;24:312–6.
[48] Barshop BA. Metabolomic approaches to mitochondrial disease: correlation of urine organic acids. Mitochondrion 2004;4:521–7.
[49] Kim IO, Kim JH, Kim WS, Hwang YS, Yeon KM, Han MC. Mitochondrial myopathy-encephalopathy-lactic acidosis-and strokelike episodes (MELAS) syndrome: CT and MR findings in seven children. AJR Am J Roentgenol 1996;166:641–5.
[50] Sheerin F, Pretorius PM, Briley D, Meagher T. Differential diagnosis of restricted diffusion confined to the cerebral cortex. Clin Radiol 2008;63:1245–53.

[51] Sue CM, Crimmins DS, Soo YS, et al. Neuroradiological features of six kindreds with MELAS tRNA(Leu) A2343G point mutation: implications for pathogenesis. J Neurol Neurosurg Psychiatry 1998;65:233–40.
[52] Iizuka T, Sakai F, Kan S, Suzuki N. Slowly progressive spread of the stroke-like lesions in MELAS. Neurology 2003;61:1238–44.
[53] Ohshita T, Oka M, Imon Y, et al. Serial diffusion-weighted imaging in MELAS. Neuroradiology 2000;42:651–6.
[54] Ito H, Mori K, Harada M, et al. Serial brain imaging analysis of stroke-like episodes in MELAS. Brain Dev 2008;30:483–8.
[55] Yoneda M, Maeda M, Kimura H, Fujii A, Katayama K, Kuriyama M. Vasogenic edema on MELAS: a serial study with diffusion-weighted MR imaging. Neurology 1999;53:2182–4.
[56] Wang XY, Noguchi K, Takashima S, Hayashi N, Ogawa S, Seto H. Serial diffusion-weighted imaging in a patient with MELAS and presumed cytotoxic oedema. Neuroradiology 2003;45:640–3.
[57] Stoquart-Elsankari S, Lehmann P, Perin B, Gondry-Jouet C, Godefroy O. MRI and diffusion-weighted imaging followup of a stroke-like event in a patient with MELAS. J Neurol 2008;255:1593–5.
[58] Kim JH, Lim MK, Jeon TY, et al. Diffusion and perfusion characteristics of MELAS (mitochondrial myopathy, encephalopathy, lactic acidosis, and stroke-like episode) in thirteen patients. Korean J Radiol 2011;12:15–24.
[59] Kaufmann P, Shungu DC, Sano MC, et al. Cerebral lactic acidosis correlates with neurological impairment in MELAS. Neurology 2004;62:1297–302.
[60] Nishioka J, Akita Y, Yatsuga S, et al. Inappropriate intracranial hemodynamics in the natural course of MELAS. Brain Dev 2008;30:100–5.
[61] Ikawa M, Okazawa H, Arakawa K, et al. PET imaging of redox and energy states in stroke-like episodes of MELAS. Mitochondrion 2009;9:144–8.
[62] Nariai T, Ohno K, Ohta Y, Hirakawa K, Ishii K, Senda M. Discordance between cerebral oxygen and glucose metabolism, and hemodynamics in a mitochondrial encephalomyopathy, lactic acidosis, and strokelike episode patient. J Neuroimaging 2001;11:325–9.
[63] Molnar MJ, Valikovics A, Molnar S, et al. Cerebral blood flow and glucose metabolism in mitochondrial disorders. Neurology 2000;55:544–8.
[64] Tarnopolsky MA, Pearce E, Smith K, Lach B. Suction-modified Bergstrom muscle biopsy technique: experience with 13,500 procedures. Muscle Nerve 2011;43:717–25.
[65] DiMauro S, Bonilla E. Mitochondrial encephalomyopathies. In: Rosenberg RN, Prusiner SB, DiMauro S, Barchi RL, editors. The molecular and genetic basis of neurological disease. Boston, MA: Butterworth-Heinemann; 1997. p. 201–35.
[66] Frederiksen AL, Andersen PH, Kyvik KO, Jeppesen TD, Vissing J, Schwartz M. Tissue specific distribution of the 3243A->G mtDNA mutation. J Med Genet 2006;43:671–7.
[67] Hammans SR, Sweeney MG, Hanna MG, Brockington M, Morgan-Hughes JA, Harding AE. The mitochondrial DNA transfer RNALeu(UUR) A-->G(3243) mutation. A clinical and genetic study. Brain 1995;118(Pt 3):721–34.
[68] Rahman S, Poulton J, Marchington D, Suomalainen A. Decrease of 3243 A-->G mtDNA mutation from blood in MELAS syndrome: a longitudinal study. Am J Hum Genet 2001;68:238–40.
[69] Parikh S, Goldstein A, Koenig MK, et al. Diagnosis and management of mitochondrial disease: a consensus statement from the Mitochondrial Medicine Society. Genet Med 2015;17(9):689–701.
[70] Zaragoza MV, Fass J, Diegoli M, Lin D, Arbustini E. Mitochondrial DNA variant discovery and evaluation in human Cardiomyopathies through next-generation sequencing. PLoS One 2010;5:e12295.

[71] Karppa M, Syrjala P, Tolonen U, Majamaa K. Peripheral neuropathy in patients with the 3243A>G mutation in mitochondrial DNA. J Neurol 2003;250:216–21.
[72] Majamaa K, Turkka J, Karppa M, Winqvist S, Hassinen IE. The common MELAS mutation A3243G in mitochondrial DNA among young patients with an occipital brain infarct. Neurology 1997;49:1331–4.
[73] Luis PB, Ruiter JP, Aires CC, et al. Valproic acid metabolites inhibit dihydrolipoyl dehydrogenase activity leading to impaired 2-oxoglutarate-driven oxidative phosphorylation. Biochim Biophys Acta 2007;1767:1126–33.
[74] Lam CW, Lau CH, Williams JC, Chan YW, Wong LJ. Mitochondrial myopathy, encephalopathy, lactic acidosis and stroke-like episodes (MELAS) triggered by valproate therapy. Eur J Pediatr 1997;156:562–4.
[75] Silva MF, Aires CC, Luis PB, et al. Valproic acid metabolism and its effects on mitochondrial fatty acid oxidation: a review. J Inherit Metab Dis 2008;31:205–16.
[76] Lin CM, Thajeb P. Valproic acid aggravates epilepsy due to MELAS in a patient with an A3243G mutation of mitochondrial DNA. Metab Brain Dis 2007;22:105–9.
[77] Hirano M, Martí R, Spinazzola A, Nishino I, Nishigaki Y. Thymidine phosphorylase deficiency causes MNGIE: an autosomal recessive mitochondrial disorder. Nucleosides Nucleotides Nucleic Acids 2004;23(8-9):1217–25.
[78] Pfeffer G, Majamaa K, Turnbull DM, Thorburn D, Chinnery PF. Treatment for mitochondrial disorders. Cochrane Database Syst Rev 2012;4:CD004426.
[79] Rodriguez MC, MacDonald JR, Mahoney DJ, Parise G, Beal MF, Tarnopolsky MA. Beneficial effects of creatine, CoQ10, and lipoic acid in mitochondrial disorders. Muscle Nerve 2007;35:235–42.
[80] Napolitano A, Salvetti S, Vista M, Lombardi V, Siciliano G, Giraldi C. Long-term treatment with idebenone and riboflavin in a patient with MELAS. Neurol Sci 2000;21:S981–2.
[81] Oguro H, Iijima K, Takahashi K, et al. Successful treatment with succinate in a patient with MELAS. Intern Med 2004;43:427–31.
[82] Taivassalo T, Haller RG. Implications of exercise training in mtDNA defects–use it or lose it? Biochim Biophys Acta 2004;1659:221–31.
[83] Steriade C, Andrade DM, Faghfoury H, Tarnopolsky MA, Tai P. Mitochondrial encephalopathy with lactic acidosis and stroke-like episodes (MELAS) may respond to adjunctive ketogenic diet. Pediatr Neurol 2014;50:498–502.
[84] Kubota M, Sakakihara Y, Mori M, Yamagata T, Momoi-Yoshida M. Beneficial effect of L-arginine for stroke-like episode in MELAS. Brain Dev 2004;26:481–3. discussion 480.
[85] El-Hattab AW, Emrick LT, Craigen WJ, Scaglia F. Citrulline and arginine utility in treating nitric oxide deficiency in mitochondrial disorders. Mol Genet Metab 2012;107:247–52.
[86] Enns GM, Kinsman SL, Perlman SL, et al. Initial experience in the treatment of inherited mitochondrial disease with EPI-743. Mol Genet Metab 2012;105:91–102.

Chapter 3

MERRF: Myoclonus Epilepsy and Ragged Red Fibers

Bruce H. Cohen

Northeast Ohio Medical University, Rootstown, OH, USA; The NeuroDevelopmental Science Center and Divison of Neurology, Department of Pediatrics, Children's Hospital and Medical Center of Akron, Akron, OH, USA

CASE PRESENTATIONS

Case 1

The case was a 45-year-old woman presenting with a generalized tonic clonic seizure. Her past medical history was significant for severe bipolar disease, and she had worked in the past as a nurse prior to the seizure. A computed tomography demonstrated a 1-cm, enhancing right-sided high-parietal extra-axial mass attached to the meninges. She was started on valproic acid and underwent a craniotomy, at which time the mass (a benign meningioma) was completely resected. Her neurological examination after surgery was normal. She was followed with serial magnetic resonance images (MRIs), which never showed a regrowth of the neoplasm. She was maintained on divalproex sodium, as this seemed to help her bipolar symptoms. She was lost to follow-up after 10 years. She returned for evaluation 5 years later with the chief complaint of an unsteady gait that began about 2 years before, and she was concerned her brain tumor had returned. In the 5 years she had not been evaluated, she had developed diabetes mellitus type 2, gained considerable weight, and had a knee replacement. A repeat family history at that time determined that her mother had suffered from bipolar illness, but her adult daughter was healthy. She had remained on divalproex sodium. Her brain MRI was normal. Liver enzymes were normal and blood lactate was normal. The neurological examination was significant for a large multilobulated fatty mass over her posterior neck, truncal and limb ataxia, a mild proximal myopathy, and loss of muscle stretch reflexes. Testing for the two common MERRF mutations (m.8344 and m.8356) determined a heteroplasmic mutation in the m.8344A>G allele.

Mitochondrial Case Studies. http://dx.doi.org/10.1016/B978-0-12-800877-5.00003-6

Case 2

The case is a 30-year-old man with a long-standing history of myoclonus and seizures. He was well until about age 10 years, when he developed random jerking movements of his limbs. These were very mild at first, and the family chose not to seek medical attention. However, during the following few years, during adolescence, his linear growth and sexual maturity took place normally, but his body habitus became endomorphic, with loss of muscle mass and subcutaneous fat. His intellectual function remained normal. He finished high school and attended college, but he never earned his degree because of illness progression, with worsening myoclonus. At the age of 25, he had his first recognized seizure. His electroencephalograph (EEG) showed multifocal spikes. He was started on lamotrigine and levetiracetam, and clonazepam was added for symptomatic treatment of the myoclonus. His MRI demonstrated diffuse atrophy, and the suspicion of a mitochondrial disease lead to obtaining a blood lactic acid level, which was elevated at 5.4 mM (lab normal < 2.0 mM). A blood sample was sent for testing of the common MERRF mutations, showing the m.8344A>G mutation. His asymptomatic mother and sister also tested positive for this mutation, but heteroplasmy levels were not determined. In the last 5 years, the seizures have been fairly well controlled, with only a few grand mal seizures in that time. His myoclonus has worsened considerably to the point his family (or physicians) cannot decide whether the movements are myoclonus or seizures. His speech has become sparse, although he still participates in all family activities. His examination demonstrated the loss of social and language reciprocity, retinitis pigmentosa, ptosis without loss of extra-ocular motility, loss of facial musculature with facial weakness, loss of muscle bulk and subcutaneous fat with minimal weakness and myoclonus. There was no frank ataxia, but the myoclonus made it impossible to determine whether there was mild ataxia. The combination of worsening myoclonus with restricted verbal output led to the suspicion of subclinical seizures. An overnight video EEG showed multifocal spikes and spike-wave discharges, but no evidence of seizures during periods of excessive myoclonus or lack of responsiveness. Increasing the clonazepam has been helpful in treating his myoclonus.

DIFFERENTIAL DIAGNOSIS

These two cases were chosen to highlight the very different clinical presentations seen in cases with disease caused by MERRF mutations.

Case 1

Cervical lipomatosis (Ekbom syndrome) is a clinical feature of MERFF, and when occurring in conjunction with ataxia provides a narrow differential diagnosis. Subacute ataxia in an adult, when taken as a single clinical feature, has a large differential diagnosis. Given the subacute nature, intoxication is not likely.

Structural causes, such as a brain tumor, need to be investigated with neuroimaging, and in this case, did not reveal any clues. Ataxia, with or without neuropathy, can be a feature in the ataxia-neuropathy spectrum of *POLG* disease, but other supporting features, such as progressive external ophthalmoplegia or dysarthria, were not present. In this case, the loss of muscle stretch reflexes may have been due to diabetes mellitus type 2, or they could be part of the primary neurological illness. There are numerous neurogenetic disorders where ataxia is the prominent neurologic finding, such as in the spino-cerebellar ataxias. Finally, the seizures that lead to the initial discovery of the meningioma are most likely due to the meningioma, given the excellent response to the first anticonvulsant or resolution of seizures after the tumor surgery. However, it is not beyond possibility that the seizures are related to MERRF. As an interesting aside, whereas valproate and valproate derivatives are clearly contraindicated in the disorders of mitochondrial depletion, long-term therapy had no negative impact in this case.

Case 2

This case presented with progressive myoclonus beginning before adolescence with subsequent loss of subcutaneous fat and muscle atrophy. About a decade later, he began having seizures (which have been partially controlled by medication), worsening myoclonus, and features of dementia. This presentation is best classified as myoclonic epilepsy with dementia, which has a broad differential that includes mitochondrial disorders. As with most acquired neurological disorders, use of a search engine such as OMIM allows the clinician to focus on those disorders that fit the case's presentation. The *short list* includes Lafora body disease, the neuronal ceroid lipofuscinoses, Unverricht-Lundborg disease, sialidosis type 1, juvenile neuronopathic Gaucher disease, juvenile neuroaxonal dystrophy, and dentatorubral-pallidoluysian atrophy. Because the differential diagnosis includes mitochondrial disorders, the markedly elevated lactate strongly suggests a mitochondrial etiology. Although MERRF is the acronym that fits the clinical description, there are many mitochondrial disorders where myoclonus and myoclonic epilepsy may be the predominant features: mitochondrial encephalopathy, lactic acidosis, and stroke-like syndrome (MELAS); Kearn-Sayre syndrome (KSS); and POLG-spectrum disorders (including both childhood onset Alpers-Huttenlocher syndrome and the adult-onset myoclonic epilepsy myopathy sensory ataxia syndrome). Certainly other mitochondrial disorders can cause a similar phenotypic illness.

Diagnostic Approach

When MERRF is suspected, the clinician should probably begin the investigation by obtaining blood and urine analytic biomarkers suggestive of a mitochondrial illness, which would include plasma lactate, amino acids, and urine organic acids. Even if these are normal, if the clinical suspicion is high, genetic testing should be performed. Before the wide availability of genetic testing (the common MERRF mutations were identified by the early 1990s and commercial testing for these widely available in the USA since the mid-1990s), a muscle biopsy would be obtained to see if ragged red fibers were seen after immunohistochemical staining with the modified Gomori trichrome stain. However, a muscle biopsy is invasive, and the ragged red fibers are nonspecific and not completely sensitive if performed early in the course of the illness. Many mitochondrial disorders, including MELAS, KSS, and POLG disorders could result in the finding of ragged red fibers. Furthermore, ragged red fibers are seen in many muscle disorders including the muscular dystrophies, and they are, therefore, neither sensitive nor specific for MERRF or any of the mitochondrial muscle disorders. The decision of the clinician would be whether to order a small panel of common MERRF mutations—each diagnostic lab offers different panels for a phenotypic disease—or order a whole mitochondrial genome evaluation that generally would include long-range polymerase chain reaction (to look for deletions and duplications in the mtDNA molecule that are associated with KSS) and point mutations of the entire mtDNA genome, which would therefore include all known mtDNA mutations associated with MELAS and a host of other disorders. Genetic testing for mtDNA mutations can be performed in blood but may be more sensitive in DNA extracted from urinary sediment, buccal swabs, saliva, or muscle. If these tests are negative, the clinician would need to determine if further testing is warranted, and if so to consider a panel of nuclear genes associated with mitochondrial disease (current panels include 150 to over 1000 genes) or whole exome sequencing.

The cerebral spinal fluid (CSF) protein and lactate are often elevated in MERRF, which would aide in the diagnosis, but they do not influence therapy. If a lumbar puncture is performed, CSF folate should be measured. Although cardiac function is generally normal early in the illness, both cardiac conduction defects (WPW syndrome) and pump failure can occur; it is reasonable to screen cases with an EKG and echocardiogram. Diabetes and other endocrinopathies are common in mtDNA disorders, so screening with an HbA1c and a TSH is reasonable.

Cases with MERRF will need to be closely managed for their seizures. A routine EEG will be helpful in terms of defining the baseline EEG patterns, discerning if myoclonic movements are seizures, investigate if seizures occur at sleep onset and upon arousal, and if the case may be having subclinical seizures. Prolonged video EEG is often needed if the routine EEG is not able to answer these questions, if clinically relevant.

A brain MRI can help differentiate the MELAS syndrome, where there may be FLAIR and T2-weighted image changes in the parietal-occipital lobe and Leigh syndrome, where changes occur in the brainstem and deep gray masses. However, there are cases with the common MERRF mutations that have brain

MRIs that appear more like MELAS (or Leigh syndrome), underscoring the need to consider and manage these mixed genotype–phenotype cases.

TREATMENT STRATEGIES

There is no treatment for MERRF. Seizures and myoclonus should be treated with standard medications. Lamotrigine, zonisamide, levetiracetam, lacosamide, and the benzodiazepine class of medications are helpful for myoclonic seizures. Divalproex sodium is also useful for myoclonic seizures, and it was used in case 1 for decades without hepatic side effects, but it probably should be considered as a last-line therapy because of the concern of toxicity in the mitochondrial depletion disorders.

Many of the mitochondrial disorders that affect the brain can be associated with low CSF folate levels, so this should be checked as well. In most instances, clinicians would treat cases with low CSF folate with folinic acid (not folic acid). However, performing a lumbar puncture on cases is not universally performed, and there is no clear consensus if all cases without verification of low CSF folate should be treated with folinic acid.

The literature does not support the use of vitamins and cofactors in MERRF, but from a practical approach, it is difficult for clinicians and families not to use these supplements and medical foods. There are occasional cases that seem to benefit from these medical foods, and using the argument *lack of proof does not mean lack of efficacy,* most clinicians use these supplements in their practice. Because of the overlap in some cases with the MELAS syndrome, consideration of levoarginine and levocitrulline is reasonable, but again, without proof in MERRF.

LONG-TERM OUTCOME

The impact of MERRF is somewhat dependent on the exact nature of the phenotype and percent of mutant mtDNA (heteroplasmy). Even within families, the clinical severity and measured levels of mutant heteroplasmy vary widely.

PATHOPHYSIOLOGY

In general, most cases with MERRF are found to have mutations in $tRNA^{Lys}$. The most common mutation associated with MERRF is the m.8344A>G mutation, with a frequency of between 0 and 1.5:100,000 in populations from Western Europe. It is postulated that those mitochondrial proteins most rich in lysine are more prone to be affected in MERRF and are the basis of the illness, although the details of what proteins are affected and why these proteins cause the illness is not clear. Myoclonus, epilepsy, and normal early development are features of essentially all cases. Hearing loss occurs in 90%, dementia in 75%, neuropathy in 63%, short stature in about half, cardiomyopathy in one-third, Wolff-Parkinson-White in 22%, and retinopathy in 15%. Both ophthalmoparesis and lipomatosis are present in about 10%.

CLINICAL PEARLS

- MERRF should be considered in cases with progressive myoclonus with myoclonic epilepsy and medically refractory epilepsy.
- There may be considerable overlap of symptoms in cases with genetically confirmed MERRF often seen in other mitochondrial illnesses, especially MELAS, KSS, and Leigh syndrome.
- In general, MERFF is caused by mutations in the mitochondrial $tRNA^{Lys}$, and if testing in blood is negative, using another tissue such as urinary sediment, saliva, buccal swab, or muscle should be considered. Muscle is the best tissue in terms of sensitivity but requires an invasive procedure to obtain muscle.
- Epilepsy and myoclonus are the cornerstone of most case's illness, but dementia, visual loss, hearing loss, and cardiac dysfunction (both cardiac conduction defects and heart failure) may be affected and impact both quality of life and risk of early mortality.

FURTHER READING

[1] Abbott JA, Francklyn CS, Robey-Bond SM. Transfer RNA and human disease. Front Genet 2014;5:158. http://dx.doi.org/10.3389/fgene.2014.00158.

[2] Bindoff LA, Engelsen BA. Mitochondrial diseases and epilepsy. Epilepsia September 2012;53(Suppl. 4):92–7. http://dx.doi.org/10.1111/j.1528-1167.2012.03618.x. Review. PubMed PMID: 22946726.

[3] DiMauro S, Hirano M. MELAS. [Updated November 21, 2013]. In: Pagon RA, Adam MP, Ardinger HH, et al., editors. Gene reviews® [Internet]. Seattle (WA). Seattle: University of Washington; February 27, 2001. p. 1993–2015.

[4] Shoffner JM, Lott MT, Lezza AM, Seibel P, Ballinger SW, Wallace DC. Myoclonic epilepsy and ragged red fiber disease (MERRF) is associated with a mitochondrial DNA tRNA(Lys) mutation. Cell June 15, 1990;61(6):931–7. PubMed PMID: 2112427.

Chapter 4

Pearson Syndrome

Bruce H. Cohen
Northeast Ohio Medical University, Rootstown, OH, USA; The NeuroDevelopmental Science Center and Divison of Neurology, Department of Pediatrics, Children's Hospital and Medical Center of Akron, Akron, OH, USA

CASE PRESENTATION

The case was born healthy after a normal pregnancy. There were no issues initially, but at the age of 3 months, the child began having diarrhea. Poor weight gain was identified at 5 months of age. Laboratory evaluation revealed a hemoglobin of 5.5 g/dL (normal 11–16) and a bicarbonate value of 12 (HCO_3^-) mM (normal 21–28). Because of a severe anemia without an apparent etiology, bone marrow biopsy was performed, which showed ringed sideroblasts. The plasma lactic acid level was elevated. A diagnosis of Pearson syndrome was suspected based on the bone marrow findings, lactic acidosis, and abnormal fat absorption. Genetic testing using long-range polymerase chain reaction (PCR) found a deletion in the mitochondrial DNA (mtDNA), confirming the suspected diagnosis, and further delineation of this deletion was not pursued. Over her life, she required blood transfusions, which were frequent (monthly) initially, but as time moved on, these became less frequent. Filgrastim was used throughout her life. She was placed on citric acid–sodium citrate solution with some improvements in her serum bicarbonate levels. Pancreatic enzyme therapy was also instituted throughout life, and a gastrostomy tube (G-tube) was placed to assist with nutrition. Over the next several months, there was some stabilization of the weight loss, but it became clear that she was not acquiring motor milestones: her examination showed her to be hypotonic with motor weakness that appeared to be myopathic and not due to corticospinal tract dysfunction. She began walking at age 3. She was diagnosed with renal Fanconi syndrome, and over time, the renal glomerular function declined to the point that peritoneal dialysis was started when she was 4 years old, and it continued throughout her life. She developed insulin-dependent diabetes. She developed elevated transaminase enzymes without an elevation in creatine phosphokinase. Her liver ultrasound suggested fibrosis. Aside from a low albumin, there was no evidence of hepatic synthetic dysfunction.

Through this course, her language, social, and cognitive function appeared normal, but by 7 years of life, she began to talk and interact less than normal and

Mitochondrial Case Studies. http://dx.doi.org/10.1016/B978-0-12-800877-5.00004-8

lost the ability to ambulate. Her electroencephalogram was diffusely slow but did not show epileptiform activity. Her cardiac function remained normal. She was diagnosed with mixed sleep apnea but could not tolerate BiPAP therapy. Vitamin and cofactor supplements included ascorbic acid, coenzyme Q10, leucovorin, levocarnitine, and vitamin E. Numerous other dietary supplements had been tried and stopped because of lack of efficacy. Her examination at 7 years of life showed her height and weight to be less than the second percentile. She was alert with apparently normal cognition, but her speech was sparse and dysfluent, with hypophonia and dysarthria. She did not seem to be attentive, and she did fall asleep during the examination. Her vision and retinal examination was normal. She had mild ptosis but no restriction in eye movements. Drooling suggested decreased bulbar function. Her motor bulk and tone were markedly reduced, and her strength was globally in the 4/5 range. She could take a few steps with a walker. Limb ataxia was seen on a finger-to-nose examination. Muscle stretch reflexes were absent. Position sensation was normal. Plantar response was bilaterally flexor.

The girl spent many months a year hospitalized for stabilization related to central line infections, nutritional management including the intermittent need for total parental nutrition (TPN), diabetic management, and renal failure with fluid and electrolyte management. She died of acute respiratory distress syndrome related to a bacterial line infection before her ninth birthday.

DIFFERENTIAL DIAGNOSIS

Pearson syndrome should be considered when an infant presents with a sideroblastic anemia (SA). In this case, the initial laboratory evaluation led to the finding of a lactic acidosis, which suggests a mitochondrial etiology. Sideroblasts are abnormal nucleated (immature) red blood cells, with iron granules that have accumulated in the mitochondria surrounding the nucleus. However, the differential diagnosis remains extensive even when limiting the possibilities to both SA in a child with acidosis. Online Mendelian Inheritance in Man (OMIM) lists 100 entries for a search of SA during infancy, and while many of the possibilities are not relevant in this case, there are many disorders that are reasonable to consider. Many of these disorders have an unequivocal mitochondrial basis. Because of the broad nature of the illness, the diagnostic evaluation should include the mitochondrial doctor along with a hematologist and geneticist (if the mitochondrial doctor is not a geneticist).

The age of presentation with concomitant features lessens the chance of the most common cause of SA, X-linked SA. This can present at any age, but young adult onset is the most common, and the illness is primarily hematologic. X-linked SA is caused by mutations in *ALAS2*, leading to deficiency in aminolevulinic acid synthetase, the first step in heme synthesis. Diamond-Blackfan anemia (DBA) is a common cause of congenital anemia, and it is associated at times with congenital anomalies including thumb and arm anomalies, craniofacial malformation, cardiac and urogenital defects, and cleft palate. As with Pearson syndrome, failure to thrive can be part of the illness. About 20% of DBA is caused by mutations in *RPS19*, the gene encoding cytoplasmic 40S ribosomal protein S19. Mutations in *GATA1* are another common cause of DBA, a gene that encodes for erythroid transcription factor, now named GATA-binding factor 1. This protein is responsible for switching fetal hemoglobin to adult hemoglobin, and it is another cause of DBA and plays a role in some cancers and Downs syndrome myelodysplastic disorders. In addition to these two genes, there are many other genes that may be causative of DBA. There are forms of pyridoxine deficiency that can cause congenital anemia as well.

There are many mitochondrial genes associated with congenital and infantile-onset SA. Mutations in *ABCB7*, a gene also located on the X-chromosome, encodes for ATP-binding cassette sub-family B member 7, which is a mitochondrial protein that result in SA and cerebellar ataxia. The protein is involved with transporting heme from the mitochondria into the cytoplasm. This illness generally has an onset in childhood. Mutations in *SLC25A38* can cause an autosomal recessive SA. SLC25A38 handles glycine in the mitochondria and is involved in heme synthesis that can present in infancy. Mutations in *YARS2*, which encodes for tyrosyl-tRNA synthetase 2, a mitochondrial protein, are associated with infantile SA along with a myopathy and lactic acidosis. Mutations in the *YARS2* gene result in the same syndrome in both children and adults. A new disorder, labeled "*Sideroblastic anemia with immunodeficiency, fevers and developmental delay,*" without a genetic etiology at this time, was described involving an SA along with a B-cell immunodeficiency syndrome. This disorder described in a dozen cases has many clinical features (growth retardation, sensorineural deafness, seizures, cardiomyopathy, cortical magnetic resonance imaging findings, cerebellar ataxia, retinitis pigmentosa, and aminoaciduria) that mimicked a mitochondrial phenotype but without genetic findings, although these cases were from many medical centers, evaluated in different fashions, and did not have whole exome testing. There are many other disorders that can be found in OMIM and other search engines that present with the Pearson syndrome phenotype, and consideration of those disorders is necessary.

DIAGNOSTIC APPROACH

Pearson syndrome is caused by deletions or other structural defects in the mtDNA. Long-range PCR of a tissue (e.g., blood lymphocytes, bone marrow, urine sediment, buccal swab, muscle) is sensitive for demonstrating a deletion, but it will not give clarification of the size of the deletion. Other advantages of long-range PCR are costs and speed of testing. Southern blot is less sensitive but will help determine the break points, determine percent mutant heteroplasmy, and can give a more-formal estimate of the deletions size, and it is also better for detecting duplications. There is the theoretical risk of a false-negative result in lymphocyte testing because of wash out of the mutation. If the testing is not informative, consideration of expanding the evaluation to include the other disorders of SA, which includes many mitochondrial genes, should be considered. Currently, gene analysis can be ordered individually or as part of larger commercial gene panels. Many commercial panels utilize next-generation sequencing technology and deletion/duplication studies to provide both detection of point mutations, indels, small- and large-sized deletions, and duplications.

TREATMENT STRATEGIES

There are no curative therapies for Pearson syndrome. The anemia may require intensive transfusion therapy and the use of bone marrow-stimulating factors such as filgrastim. Iron overload from the infusions may require chelation therapy. Exocrine pancreatic failure can be partially treated with the use of prescription digestive enzyme therapy, and sometimes the need for supplemental nutrition that may include gastrostomy feeding and sometimes TPN. Renal tubular acidosis is common in this disorder, and buffering with bicarbonate or prescription citric acid–sodium citrate solution is necessary. Because this illness can affect vision, muscle, heart, liver, and nerve, evaluations of these organs on an ongoing basis is reasonable. Organ dysfunction using conventional therapies is advisable. As with any case having brain and muscle dysfunction, evaluation for ventilator function and sleep apnea is required. Some cases with mtDNA deletion disorders develop cerebral spinal fluid (CSF) folate deficiency, and investigation (obtaining CSF folic acid levels) and/or therapy with folinic acid can be considered. The use of vitamins and cofactors is not proven to be beneficial in this disorder but is reasonable to trial in cases.

LONG-TERM OUTCOME

The early clinical course of this illness is variable. Growth failure from birth can be present in some babies while others become ill after a few months of life. Some children have refractory anemia and acidosis, complicated by severe renal tubular acidosis, which can result in a fatal loss of tubular function. Some children have pancytopenia, in addition to the anemia, resulting in severe immune deficiency. Death in the first 2 years of life is not uncommon from complications of acidosis and nutritional deficiencies refractory to therapy, infection, or hepatic failure. Children that survive long enough begin to develop the features of Kearns-Sayre syndrome over the next few years. The cause of death is variable but includes refractory acidosis, dementia, cardiomyopathy, infection, hepatopathy, complications of refractory iron overload, and refractory epilepsy.

PATHOPHYSIOLOGY

It is not known why the bone marrow and exocrine pancreas are involved to such a severe extent as they are in this disorder, nor why the bone marrow function may improve as other features of Kerns-Sayre syndrome begin.

The illness itself is a result of various different deletions involving a portion of the mtDNA with high levels of mutant heteroplasmy, which result in inadequate translation of necessary mtDNA encoded proteins and loss of normal respiratory chain function. Although the common 4977 base pair deletion that is often associated with Kearns-Sayre syndrome may cause Pearson syndrome, many cases with Pearson syndrome have different deletions, usually larger or involving other parts of the mtDNA molecule. Deletion-duplication and other rearrangements of the mtDNA have also been seen in this syndrome. In most cases, the large-scale deletions of the mtDNA that cause Pearson syndrome are sporadic, with the deletions occurring either during oocyte formation or during early stages of embryonic development. In the rare cases of germline involvement, the deletions can be transmitted. Obviously, the reproductive fitness in Pearson syndrome is not reported, but the deletion could pose a risk in non-affected matrilineal relatives of the affected individual. For those cases with the common deletion, the best understanding as to the pathophysiology of the infantile presentation of Pearson syndrome instead of the childhood–adolescent onset of Kearns-Sayre syndrome is a high percentage of mutant heteroplasmy along with the possibly a different tissue distribution of the mutation load. For those cases with the larger size deletion, or a deletion involving more of the mitochondrial tRNAs, the understanding is the more aggressive pathophysiology is driven by the loss of more mtDNA transcript function needed for protein synthesis. There is some evidence that once the mutant heteroplasmy increases above a defined limit, the normal copies of mtDNA are degraded, resulting in the severe pathophysiology.

CLINICAL PEARLS

- Pearson syndrome is an infantile disorder that causes SA, acidosis, and exocrine pancreatic failure, resulting in poor growth and failure to thrive. Infants surviving early childhood will go on to develop features of Kearns-Sayre syndrome.
- The differential diagnosis of SA is broad, and if genetic testing of Pearson syndrome evaluation is not diagnostic, consideration for other mitochondrial and non-mitochondrial disorders is needed.
- Long-range PCR on blood lymphocytes will often detect a deletion in the mtDNA in Pearson syndrome, but if further delineation in terms of break points, structure of the deletion, and heteroplasmy is necessary, then Southern blot should be employed. Southern blot will better detect duplications or complex rearrangements in the mtDNA. If blood lymphocytes are not informative, the consideration of another tissue, such as bone marrow or muscle should be considered.
- Although Pearson syndrome is one of the most severe mitochondrial disorders, aggressive attention to basic principles of fluid and electrolyte balance, nutrition, ventilator function, and general care may improve the quality of life of the child.

FURTHER READING

[1] Alter BP. Pearson syndrome in a Diamond-Blackfan anemia cohort. Blood July 17, 2014; 124(3):312–3. http://dx.doi.org/10.1182/blood-2014-04-571687. PubMed PMID: 25035146; PubMed Central PMCID: PMC4102705.

[2] Broomfield A, Sweeney MG, Woodward CE, Fratter C, Morris AM, Leonard JV, et al. Paediatric single mitochondrial DNA deletion disorders: an overlapping spectrum of disease. J Inherit Metab Dis October 29, 2014. [Epub ahead of print] PubMed PMID: 25352051.

[3] Komulainen T, Hautakangas MR, Hinttala R, Pakanen S, Vähäsarja V, Lehenkari P, et al. Mitochondrial DNA depletion and deletions in paediatric patients with neuromuscular diseases: novel phenotypes. JIMD Rep May 5, 2015. [Epub ahead of print] PubMed PMID: 25940035.

[4] Pearson HA, Lobel JS, Kocoshis SA, Naiman JL, Windmiller J, Lammi AT, et al. A new syndrome of refractory sideroblastic anemia with vacuolization of marrow precursors and exocrine pancreatic dysfunction. J Pediatr December 1979;95(6):976–84. PubMed PMID: 501502.

[5] Rotig A, Colonna M, Bonnefont JP, Blanche S, Fischer A, Saudubray JM, Munnich A. Mitochondrial DNA deletion in Pearson's marrow/pancreas syndrome. Lancet 1989;1(8643):902–3. PubMed PMID: 2564980.

[6] van den Ouweland JM, de Klerk JB, van de Corput MP, Dirks RW, Raap AK, Scholte HR, et al. Characterization of a novel mitochondrial DNA deletion in a patient with a variant of the Pearson marrow-pancreas syndrome. Eur J Hum Genet March 2000;8(3):195–203. PubMed PMID: 10780785.

[7] Wiseman DH, May A, Jolles S, et al. A novel syndrome of congenital sideroblastic anemia, B-cell immunodeficiency, periodic fevers, and developmental delay (SIFD). Blood 2013; 122(1):112–23. http://dx.doi.org/10.1182/blood-2012-08-439083.

Chapter 5

Kearns–Sayre Syndrome

Sumit Parikh
Cleveland Clinic Lerner College of Medicine & Case Western Reserve University, Cleveland, OH, USA; Neurogenetics, Metabolism and Mitochondrial Disease Center, Cleveland Clinic, Cleveland, OH, USA

CASE PRESENTATION

Our case was first evaluated at age 12 years for complaints of weakness. He seemed to tire easily and lagged behind his peers in terms of endurance and performance of physical activities. He had developed trouble lifting objects and pitching a ball and dyspnea with exertion. His family noted difficulty climbing stairs. His symptoms were present from the morning, although he was more fatigued by the end of the day.

On his review of systems, the family remarked that he was physically smaller than others in the family as well as his peers. Others had also commented on some drooping of his eyelids over the past year or so. Pictures from early childhood showed that this was clearly a newer finding.

His past medical history was notable for a mild-to-moderate sensorineural hearing loss noted around age 10 years, and hearing aids had been recommended. He carried a diagnosis of attention-deficit hyperactivity disorder and dyslexia. His perinatal and family histories were noncontributory. He was not taking any medications.

On general examination, he was noted to have short stature with a height at the fifth percentile for age (while parents are at the 75th percentile). Other than being of small build, he was otherwise not dysmorphic, and no other concerns were identified on his general examination.

On initial neurologic examination, he was noted to have bilateral ptosis and limited extraocular movements in all directions, most prominently superiorly. Fundoscopic examination showed bilateral retinal pigmentary abnormalities. Hearing aids were in place. He had myopathic facies with a hypernasal quality to his voice. A motor examination showed normal tone but weakness with 4+ proximal strength involving his deltoids and hip flexors. Reflexes were symmetric. He was able to squat and get up from a seated position as well as sustain a straight-leg raise for 35 s bilaterally. He did have difficulty stepping up onto a chair, however, and was unable to walk in a crouched position. His sensory and cerebellar examination was normal.

Mitochondrial Case Studies. http://dx.doi.org/10.1016/B978-0-12-800877-5.00005-X

Over the past few years after the initial evaluation, our case has remained mostly stable. His ptosis has worsened leading to the need for a frontalis sling procedure. His muscle weakness had progressed slightly. His reflexes had slightly decreased responsiveness and were 1+. There was mild imbalance with sway on Romberg noted on examination.

DIFFERENTIAL DIAGNOSIS

Our case has a subacute onset of weakness, ptosis, ophthalmoplegia, retinal pigmentary abnormalities, hearing loss, and short stature.

The differential diagnosis of a child presenting with subacute onset of muscle weakness includes a variety of myopathic and large fiber neuropathic conditions. The lack of any abnormalities initially on sensory examination including intact reflexes, vibration, and proprioception sense make a primary myopathic etiology more likely. The concomitant presence of ptosis and symptom worsening at the end of the day raises the possibility of a neuromuscular junction disorder such as autoimmune myasthenia gravis.

Ophthalmoplegia may also be due to central, peripheral, or myopathic causes. However, when present bilaterally and involving multiple directions of gaze, as in our case, a central or peripheral nerve-based etiology is less likely, and a myopathic or neuromuscular condition affecting the extraocular muscles should be suspected. Findings of external ophthalmoplegia, with fluctuating severity, are seen with neuromuscular junction disorders such as myasthenia—but these findings are fixed in our case. Thus, primary myopathic causes are the more likely cause of ophthalmoplegia in this case. Add to this the case's findings of proximal muscle weakness, and a primary muscle disorder should be suspected. In an older adult with these symptoms, one may also consider a diagnosis of oculopharyngeal muscular dystrophy. External ophthalmoplegia is also seen in mitochondrial disorders such as chronic progressive external ophthalmoplegia (CPEO) and Kearns–Sayre syndrome (KSS).

Our case's associated hearing loss and short stature along with findings of pigmentary retinopathy on examination would also best fit a mitochondrial disorder. These symptoms all serve as red flags for primary mitochondrial disease. In the context of myopathy with associated ptosis, weakness, and ophthalmoplegia, a diagnosis of primary mitochondrial disease, specifically KSS should be recognized. The condition is due to a mitochondrial DNA (mtDNA) deletion. A milder presentation, CPEO, with predominantly ophthalmoplegia and ptosis and few associated or systemic findings can also occur. Cases not meeting formal diagnostic criteria for KSS, including those older than twenty at initial evaluation, but having more than just CPEO on history or examination are often labeled as having CPEO plus.

DIAGNOSTIC APPROACH AND PATHOPHYSIOLOGY

In situations where a myopathic condition is suspected, a CK level is often helpful. With the concern for a possible neuromuscular junction defect, an electromyogram

(EMG) with repetitive nerve stimulation is at times diagnostic as well. An EMG would also help identify an underlying neuropathic or myopathic disorder. The presence of neuropathy in a patient with CPEO may also indicate a higher likelihood of a nuclear DNA mutation leading to mitochondrial disease. One should also ascertain the presence of neuromuscular junction antibodies in blood when considering a diagnosis of myasthenia gravis.

If one were to recognize this case's symptoms as fitting a pattern of mitochondrial disease, biomarker testing of lactate, plasma amino acids, and urine organic acids could be undertaken to assess for signs of mitochondrial dysfunction. Such testing is now often directly followed by mitochondrial genetic studies when the clinical concern for a primary mitochondrial disorder is high.

If one recognizes the KSS phenotype, one should directly send blood for mtDNA deletion analysis. Such testing is now typically performed utilizing a next-generation sequencing approach and allows for detection and quantification of even low levels of mutation heteroplasmy.

KSS is most common due to a 1.1–10 kilobase mtDNA deletion, with the most common deletion labeled as the "common 4977 bp deletion. These deletions can be identified in peripheral blood leukocyte samples in the majority of cases with KSS. However, mtDNA heteroplasmy can lead to varying tissue distribution of mutant mtDNA. Thus, mtDNA deletion analysis in tissue such as muscle is at times necessary. This is especially true for cases with isolated CPEO. It is not yet understood as to why mtDNA deletions lead to a discrepant CPEO or KSS phenotype. Large, single mtDNA deletions can also lead to infant-onset Pearson syndrome with bone marrow suppression and pancreatic insufficiency. These cases typically progress to KSS during childhood. The size or location of the mtDNA deletion does not seem to correlate with specific symptoms or symptom severity, though the Newcastle Group in the United Kingdom has developed a risk assessment tool to ascertain whether or not symptoms may develop.

Biochemical testing in blood, urine, and spinal fluid may show markers of mitochondrial dysfunction. Cerebral spinal fluid protein levels are typically elevated. Cerebral folate deficiency can develop in KSS and may contribute to the leukoencephalopathy as well as a case's cognitive symptoms. Measurements of spinal fluid 5-methyl-tetrahydrofolate can be obtained.

Some cases develop a leukoencephalopathy, and a brain MRI will help identify this. Cases with KSS are at a high risk for cardiac conduction block and fatal cardiac arrhythmias; some develop cardiomyopathy.

Our case had a large mtDNA deletion identified in blood. He had the presence of an elevated lactate, alanine, and urine Krebs cycle intermediates. His neuroimaging did not show white matter disease. He did not receive an EMG or spinal fluid analysis.

TREATMENT

Care for cases with KSS is primarily supportive. Due to the high occurrence of cerebral folate deficiency, supplementation with prescription folinic acid is strongly recommended. Some institutions may routinely assess CSF folate levels to adjust doses, while others may make weight-based dosage adjustments.

Cognitive function might be impacted, and school-aged cases will need periodic cognitive monitoring (neuropsychological testing) and school-based modifications for optimal learning. Cases with KSS typically do not typically develop seizures. However, due to the presence of white matter disease, some may develop an alteration of tone that needs treatment with medication and splinting.

As ptosis worsens, cases typically require surgical treatment to allow them to open their eyelids. Unfortunately, there are no treatments yet for ophthalmoplegia or retinopathy. Deficits in hearing can be assisted by hearing aids, often needing regular adjustments. Cochlear implants may also be of benefit once hearing loss progresses significantly.

Due to the high risk of fatal cardiac dysrhythmias, routine and frequent cardiac monitoring is needed with periodic Holter monitor placement. There is debate as to whether prophylactic pre-symptomatic pacemaker placement is needed. These cases are also at risk for cardiomyopathy and need regular echocardiograms.

As myopathy and neuropathy worsen, a case with KSS may develop difficulties with balance, respiratory muscle function, swallowing abilities, and fatigue. Balance therapy and gait assistance can help in preventing dangerous falls. To help maintain muscle strength, cases do benefit from physical therapy and exercise. Due to respiratory muscle weakness, cases can develop carbon dioxide retention and poor airway clearance of secretions. The addition of chest physiotherapy and a cough assist device is needed. Swallowing dysfunction should be picked up pre-symptomatically as to avoid aspiration and related complications. Neuropathy can lead to discomfort and pain for which treatment may be necessary.

Mood and anxiety issues often occur in cases with chronic disease, and these should be monitored since treatment can improve quality of life.

Mitochondrial deletion disorders can also lead to associated optic nerve atrophy, thyroid dysfunction, diabetes, cortisol insufficiency, and renal tubular acidosis. The condition is progressive and additional symptoms such as cognitive difficulties, dementia, dysphagia, and gastrointestinal dysmotility may develop. Clinical monitoring for these associated conditions is needed.

LONG-TERM OUTCOME

KSS is a progressive disorder and ptosis, CPEO, myopathy, and neuropathy gradually worsen. Cognitive function typically fluctuates and declines. Cases eventually become wheelchair bound due to weakness and imbalance. If untreated, cardiac conduction defects lead to mortality. Otherwise, respiratory infections due to poor airway clearance often progress and lead to increasing morbidity and, at times, mortality.

CLINICAL PEARLS

- Weakness when combined with CPEO with ptosis has a limited differential diagnosis; KSS and CPEO+ should be high on this differential.
- mtDNA deletions cannot always be identified in blood in KSS, and especially, isolated CPEO and assessing other tissue such as muscle is at times needed.
- Cerebral folate deficiency is an eminently treatable manifestation of KSS.

FURTHER READING

[1] Mitochondrial DNA deletion syndromes at GeneReviews. http://www.ncbi.nlm.nih.gov/books/NBK1203/.

[2] Horga et al. Peripheral neuropathy predicts nuclear gene defect in patients with mitochondrial ophthalmoplegia. Brain. 2014;137(12):3200–3212.

[3] Grady JP, Campbell G, Ratnaike T, Blakely EL, Falkous G, Nesbitt V, et al. Disease progression in patients with single, large-scale mitochondrial DNA deletions. Brain 2014.

[4] Mitochondrial DNA deletion risk assessment tool. Newcastle University. http://research.ncl.ac.uk/mitoresearch/.

Chapter 6

Chronic Progressive External Ophthalmoplegia (CPEO)

Mark Tarnopolsky
Department of Pediatrics, McMaster University, Hamilton, Ontario, Canada

CASE PRESENTATION

A 55-year-old man was referred to our clinic for assessment of treatment-resistant myasthenia gravis. He presented to a community neurologist one year prior with a two-month history of ptosis and diplopia worse at the end of the day. Neurological examination showed a normal mental status, cranial nerve examination was positive for right greater than left ptosis, and ophthalmoparesis was limited to ~15° in all directions with the up-gaze being more affected, mild bilateral hearing loss, and mild (4+/5) proximal muscle weakness. The remainder of the examination was normal, and the family history was negative. A single fiber electromyography (EMG) was done and showed a significant increase in jitter. He was diagnosed with probable myasthenia gravis and started on pyridostigmine (60 mg tid) with mild subjective improvement in the diplopia. One month later, he returned to the clinic with no further clinical improvement and complained of mild dysphagia for dry foods. His anti-acetylcholine receptor antibody test returned normal, as did a computed tomography scan of the chest (looking for thymoma). He was diagnosed with antibody-negative myasthenia gravis and started on azathioprine (150 mg/d) in addition to the pyridostigmine. At follow-up six months later, there was still little improvement, and he was treated with intravenous immunoglobulin (1 g/kg/q4 weeks) for four months. When this treatment failed to improve the symptoms and he felt weaker going up stairs, he was sent for further assessment.

DIFFERENTIAL DIAGNOSIS

The three most likely diagnoses in an adult (over age 40 y) with subacute or slowly progressive ptosis and dysphagia would be myasthenia gravis, oculopharyngeal muscular dystrophy (OPMD), and chronic progressive external ophthalmoplegia (CPEO). Much less likely would be some of the congenital myopathies associated with ophthalmoplegia such as minicore myopathy (*RYR1* or *SEPN1* mutations), for they usually have an earlier onset. Although

Mitochondrial Case Studies. http://dx.doi.org/10.1016/B978-0-12-800877-5.00006-1

hyperthyroidism (Graves disease) can cause ophthalmoparesis and proximal muscle weakness, it is not associated with ptosis (often there is lid retraction), nor hearing loss, nor dysphagia.

Myasthenia gravis is an autoimmune disease caused by antibodies to some component of the neuromuscular junction. It is more common in younger women, with an equal male:female ratio in those over age 50 years. Ophthalmoplegia, ptosis, dysphagia, and dysarthria are the most common symptoms with ~10% also showing proximal muscle weakness. The symptoms tend to worsen at the end of the day and/or with repetitive activity. OPMD is an autosomal dominant muscular dystrophy caused by a trinucleotide expansion (GCG) in the *PABPN1* gene. It is far more common in the French-Canadian population due to a founder effect. The onset of symptoms usually starts in the 50 s with dysphagia and/or ptosis. Ophthalmoparesis is apparent only with up-gaze in proportion to the degree of ptosis likely due to secondary atrophy. We find that ~30% of OPMD cases also have proximal muscle weakness that worsens with age, and this can become functionally limiting. CPEO is a mitochondrial cytopathy associated with a mitochondrial DNA (mtDNA) deletion (usually a single deletion). Most cases are sporadic, yet there are rare cases of autosomal dominant inheritance due to mutations in *POLG1* or *C10orf2 (twinkle)* and even more rarely, *ANT1* or *POLG2*. All cases show external ophthalmoparesis (+/– subjective diplopia), and most have ptosis that is slowly progressive. When the ophthalmoparesis progresses without subjective diplopia, the case often has striking ophthalmoparesis not perceived by the individual at the time of diagnosis. Most cases will experience high-frequency sensori-neural hearing loss, and ~30% will have dysphagia (although not to the severity of those with OPMD). We find proximal weakness in ~50% of our cases, although it is not subjectively limiting in most cases.

DIAGNOSTIC APPROACH

In the current case, the lack of response to the immune suppression was a major clue to consider alternative diagnoses. OPMD was not felt to be likely due to the absence of a positive family history and the fact that the ophthalmoparesis was present in all directions. CPEO was considered to be a strong consideration and further testing was initiated. Unlike many other mitochondrial disorders, the plasma lactate and alanine and urine organic acids are often normal in cases with OPMD, yet the CK can be mildly elevated in a minority of cases. Of interest, the single fiber EMG often shows an increase in jitter in CPEO, and this can lead to a false-positive diagnosis of myasthenia gravis (as was the issue in the current case). Although a single mtDNA deletion is a characteristic of CPEO, it is not detectable in blood; consequently, a muscle biopsy was completed using a 5-mm modified Bergström needle. Muscle histology showed characteristic cytochrome *c* oxidase (COX) negative fibers in 4% of the total muscle fibers with rare ragged red fibers. Muscle-derived DNA was run using long-range PCR, and a single deletion was found. Quantitative mtDNA deletion analysis was done using ND1/ND4 ratio, and the deletion was found to be at 30% heteroplasmy. It is important to recognize that mtDNA deletions and COX negative muscle fibers increase with normal human aging but rarely more than 2% heteroplasmy and percentage of muscle fibers. If there was an autosomal dominant family history of any first-degree relative with ptosis and/or ophthalmoparesis and/or early onset hearing loss requiring aids (<age 55 y), we would recommend testing for mutations in *POLG1>C10orf2/twinkle>ANT1>POLG2*. If there was a maternal inheritance pattern of CPEO symptoms, the mtDNA should be sequenced. *POLG1* should also be checked if a case had Parkinsonian features or sensory neuropathy, and a history of depression would prompt assessment of the *C10orf2/twinkle* gene.

TREATMENT STRATEGY

Treatment of CPEO is mostly supportive to mitigate many of the protean manifestations of the disorder (ptosis, hearing loss, dysphagia, and muscle weakness); however, the ophthalmoplegia does not respond consistently to any therapeutic interventions. The use of prisms should be considered if there is subjective diplopia; however, over time, the eyes usually fix centrally with no dysconjugate gaze, thus no diplopia. Strabismus surgery should only be considered if the case has subjective diplopia and the ophthalmoparesis has not changed for at least one year. Most cases consider eyelid surgery when the ptosis starts to affect vision. Several options exist, but many cases are now being offered silicon sling procedures, especially if they are young. For older cases, particularly if they have redundant eyelid skin, they may be offered a simple blepharoplasty. Audiometry is performed yearly, and when appropriate, a hearing aid is usually required. In cases of profound hearing loss, some cases respond well to cochlear implants. All cases with subjective dysphagia should have a video fluoroscopic swallowing study, and if abnormalities are found, an assessment from a speech-language pathologist should be obtained. It is extremely rare for the dysphagia to progress to the extent of requiring a gastrostomy or jejunostomy tube. Nutritional therapy with coenzyme Q10 + alpha-lipoic acid + creatine monohydrate + vitamin E may improve some biochemical features of the disorder. Both endurance and resistance exercise have been shown to increase mitochondrial capacity and strength, respectively, in cases with single mtDNA deletions (most had CPEO).

LONG-TERM OUTCOME

The ophthalmoparesis usually progresses to the point of being completely paretic and centrally fixed, and the person must move their head to look in all directions. As mentioned above, the diplopia often improves as the severity of the ophthalmoparesis worsens. Once ptosis is identified, it often progresses over several years to the point of requiring surgery. The silicon sling procedure is less likely to require repeat surgery, whilst a simple blepharoplasty often requires a repeat procedure several years after the initial surgery. The problem with repeated blepharoplasty is that the person may not be able to close their eyes at night and may require eye lubrication and a night mask. High-frequency hearing loss is insidious in onset and often not noticed by the person. The hearing loss progresses over several years to the point of requiring hearing aids, and some cases progress over years to the point of needing a cochlear implant. When dysphagia is present, it also progresses very slowly, and to date, we have not had a single case out of 82 in our clinic who have required tube feeds. Muscle weakness shows a 2–3 % decline per year by objective strength testing in the 50% who have proximal weakness, but this is very rarely functionally limiting to ambulation.

PATHOPHYSIOLOGY

The pathological hallmark of CPEO is the finding of mtDNA deletion(s) at mutant levels above the normal aging-associated accumulation (aging<2% with quantitative methods). Single deletions are more commonly seen with the sporadic forms of CPEO, and multiple deletions are more often associated with mutations in genes associated with mtDNA maintenance/replication, such as *POLG1, C10orf2/twinkle, ANT*, or *POLG2*. Most of the deletions affect the region of the mtDNA between the origin of the heavy strand and the origin of the light strand and are thought to arise from errors occurring during repair of mtDNA mutations that clonally expand. Given that there are many genes that are often affected in the deletion region (i.e., ND4/3/4L/6, ATP6/8, COX1-3), the accumulation of higher levels of deletion heteroplasmy leads to the progressive loss of respiratory chain capacity. The deletion heteroplasmy in muscle homogenate is lower than when COX negative muscle fibers are specifically isolated using methods such as laser capture micro-dissection or single fiber analysis; consequently, COX–ve muscle fibers usually contain >70% mutant deletion heteroplasmy. The progressive accumulation of COX–ve fibers in the extraocular muscles leads to atrophy of the muscles and paresis.

CLINICAL PEARLS

- Cases with treatment-resistant MG should be tested for OPMD or CPEO depending on the examination and family history findings.
- Increased jitter on single fiber EMG is frequently seen in CPEO, resulting in many CPEO cases being initially diagnosed with MG.
- mtDNA deletions should not be testing in blood samples due to a very high false-negative rate in all mitochondrial disorders except Pearson bone marrow syndrome.
- High-frequency hearing loss is often not perceived by the individual even when moderately severe by audiometry, and a spouse/partner usually reports the hearing loss before the person does.

- CPEO cases with sensory neuropathy/ataxia or Parkinsonian features should be checked for *POLG1* mutations.
- CPEO cases with multiple mtDNA deletions or an autosomal dominant family history should be checked for *POLG1>C10orf2/twinkle>ANT1>POLG2* mutations.

FURTHER READING

[1] Allen RC. Genetic diseases affecting the eyelids: what should a clinician know? Curr Opin Ophthalmol 2013;24(5):463–77.

[2] Cruz-Martinez A, Arpa J, Santiago S, Pérez-Conde C, Gutiérrez-Molina M, Campos Y. Single fiber electromyography (SFEMG) in mitochondrial diseases (MD). Electromyogr Clin Neurophysiol 2004;44(1):35–8.

[3] Fawcett PR, Mastaglia FL, Mechler F. Electrophysiological findings including single fibre EMG in a family with mitochondrial myopathy. J Neurol Sci 1982;53(2):397–410.

[4] Krishnan KJ, Reeve AK, Samuels DC, Chinnery PF, Blackwood JK, et al. What causes mitochondrial DNA deletions in human cells? Nat Genet 2008;40(3):275–9.

[5] Rodriguez MC, MacDonald JR, Mahoney DJ, Parise G, Beal MF, Tarnopolsky MA. Beneficial effects of creatine, CoQ10, and lipoic acid in mitochondrial disorders. Muscle Nerve 2007;35(2):235–42.

[6] Tarnopolsky MA, Pearce E, Smith K, Lach B. Suction-modified Bergström muscle biopsy technique: experience with 13,500 procedures. Muscle Nerve 2011;43(5):717–25.

[7] Yu-Wai-Man C, Smith FE, Firbank MJ, Guthrie G, Guthrie S, et al. Extraocular muscle atrophy and central nervous system involvement in chronic progressive external ophthalmoplegia. PLoS One 2013;8(9):e75048.

Chapter 7

Leber Hereditary Optic Neuropathy

Patrick Yu-Wai-Man[1,2], Patrick F. Chinnery[3,4]
[1]Wellcome Trust Centre for Mitochondrial Research, Institute of Genetic Medicine, Newcastle University, UK; [2]Newcastle Eye Center, Royal Victoria Infirmary, Newcastle upon Tyne, UK; [3]Department of Clinical Neuroscience, University of Cambridge, Cambridge, UK; [4]MRC Mitochondrial Biology Unit, Cambridge Biomedical Campus, Cambridge, UK

CASE PRESENTATION

Clinical Case

A 26-year-old healthy adult man developed painless visual blurring affecting his left eye that was associated with subjective color desaturation. He experienced rapid visual deterioration, and when seen 1 week later by his local optometrist, his visual acuities were recorded at 20/20 in his right eye and 20/200 in his left eye. Fundoscopic examination was thought to be normal, and he was referred for a neurological opinion for suspected optic neuritis. Three weeks later, he developed similar visual symptoms in his right eye, and he was seen on an urgent basis.

At his baseline clinic visit, he denied any history of head trauma, recent drug abuse, or antecedent illnesses. His visual acuity was recorded at counting fingers in both eyes, and there was pronounced bilateral dyschromatopsia on color plate testing. The pupillary light reflexes were brisk with no relative afferent pupillary defect (RAPD) noted. Goldman visual field perimetry showed dense centrocecal scotomas bilaterally with a sharply demarcated border. Dilated fundoscopic examination revealed optic disc hyperemia with tortuous retinal vessels and segmental swelling of the peripapillary retinal nerve fiber layer (Figure 1(a)). The remainder of the neurological examination was normal.

Routine hematological and biochemical blood tests were normal. A cranial magnetic resonance imaging (MRI) scan with intravenous gadolinium contrast did not show any areas of enhancement along the visual pathways. A spinal tap confirmed normal opening pressures, and his cerebrospinal fluid was acellular with normal protein and glucose contents. No unmatched oligoclonal bands were detected. On visual electrophysiological testing, there was marked attenuation of the visual evoked potentials bilaterally pointing toward a primary

Mitochondrial Case Studies. http://dx.doi.org/10.1016/B978-0-12-800877-5.00007-3

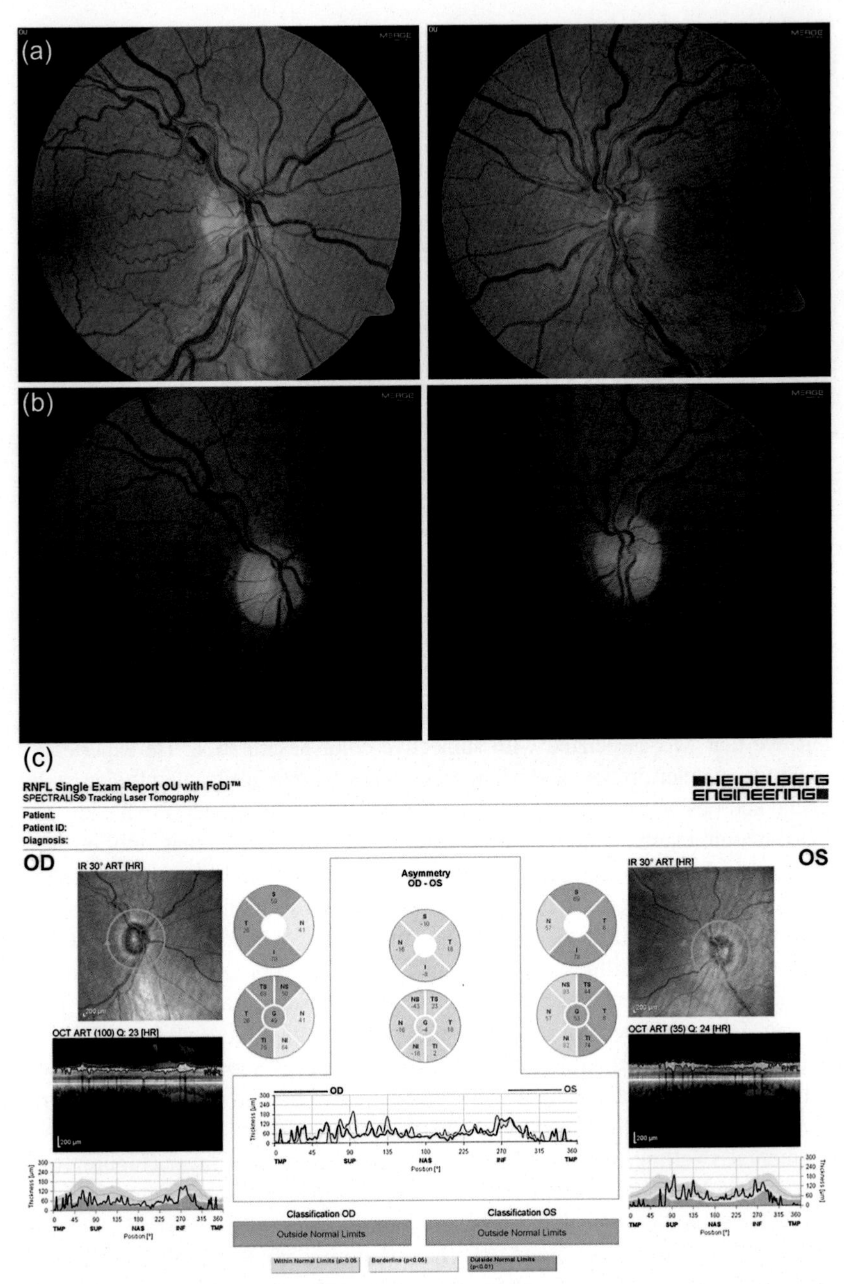

FIGURE 1 Disease progression from the acute to the chronic stage. (a) Classical fundoscopic appearance in acute Leber hereditary optic neuropathy (LHON) with optic disc hyperemia, swelling of the peripapillary retinal nerve fiber layer, prominent vascular tortuosity, and fine telangiectatic vessels. (b) Established optic atrophy with marked pallor of the neuroretinal rim. (c) Spectralis™ optical coherence tomography imaging in the chronic phase of LHON showing marked retinal nerve fiber layer thinning with relative preservation of the nasal bundle.

pathological optic nerve process. There was no clinically significant response to a 3-day course of high-dose intravenous methylprednisolone.

On follow-up 2 months later, the man had not experienced any visual improvement, and there was now overt bilateral optic atrophy (Figure 1(b)). High-resolution spectral-domain optic coherence tomography (OCT) imaging showed marked thinning of the retinal nerve fiber layer with relative preservation of the nasal quadrant (Figure 1(c)). The case was registered as severely sight impaired as he met the legal criteria for blind registration. Shortly afterward, genetic testing confirmed that he harbored the m.11778A>G mitochondrial DNA (mtDNA) mutation at near homoplasmic level, providing a definitive molecular diagnosis of Leber hereditary optic neuropathy (LHON).

The man was the eldest of three brothers and one sister. His parents were fit and healthy with no medical problems. There was no relevant medical history on his father's side, but his mother had two brothers and a sister. One of her two brothers was registered blind after he developed subacute bilateral visual failure in his early 30s. He was known to be a heavy smoker. The man himself was a nonsmoker, and he had a modest alcohol consumption. Two years after the diagnosis was made, his visual acuity had remained unchanged at counting fingers in both eyes.

Discussion

Disease conversion in LHON is characterized by acute or subacute central visual loss in one eye, followed 2–4 months later by the fellow eye. Unilateral optic nerve involvement in LHON is exceptionally rare, and another underlying pathological process should be actively excluded in these atypical cases. Bilateral simultaneous onset probably occurs in about 25% of cases, although it can be difficult for some individuals to accurately report whether visual loss had been ongoing in one eye prior to the fellow eye being affected. The peak age of onset is in the second and third decades of life, and it is unusual for symptoms to develop beyond 50 years of age. The initial visual loss in LHON is severe, and it usually plateaus over the next 6 months, with most cases achieving visual acuities of 20/200 or worse.

Occasionally, cases with LHON have a more complex phenotype with additional features such as cardiac conduction defects, peripheral neuropathy, dystonia, and myopathy. There is a well-reported association between the three primary mtDNA LHON mutations and a multiple sclerosis-like illness, especially among female carriers (so-called Harding's disease). Rarer pathogenic mtDNA variants have been linked with more atypical LHON plus syndromes, where the optic neuropathy segregated with prominent neurological features including spastic dystonia, ataxia, juvenile-onset encephalopathy, and psychiatric disturbances.

DIFFERENTIAL DIAGNOSIS

Cases with subacute visual failure present as a neuro-ophthalmological emergency, and it is critically important to exclude an inflammatory or structural basis for the visual loss. Urgent MR neuroimaging of the brain and orbit imaging is essential, ideally with intravenous gadolinium contrast, to identify any compressive or infiltrative lesions or areas of enhancement suggestive of an inflammatory process. Depending on the clinical history, other investigations such as a spinal tap, autoantibody testing, and an infectious or vasculitic screen need to be tailored appropriately to exclude potentially reversible causes of a bilateral optic neuropathy and the initiation of prompt treatment before visual loss becomes irreversible.

LHON needs to be distinguished from acute optic neuritis, either as an isolated event or as part of multiple sclerosis, since both disease entities can present similarly. However, periorbital discomfort and pain on eye movements are characteristic features of an inflammatory optic neuropathy that usually allow for an early distinction to be made. Furthermore, there is rapid visual recovery without treatment, and bilateral involvement is rare in typical demyelinating optic neuritis. In cases with a rapidly progressive bilateral optic neuropathy associated with poor visual function, it is important to consider the possibility of neuromyelitis optica and other potentially treatable causes associated with systemic diseases such as sarcoidosis and systemic lupus erythematosus. The majority of cases with LHON are young adults, but in cases over the age of 50 years, erythrocyte sedimentation rate and C-reactive protein levels are mandatory to exclude the possibility of giant cell arteritis.

DIAGNOSTIC APPROACH

The minimum prevalence of LHON has been estimated at about 1 in 35,000 of the population with a much higher prevalence in men at about 1 in 14,000. It is, therefore, a relatively rare disorder, and except for known mutation carriers coming from established families, the onus is on the clinician to adopt a systematic diagnostic approach to exclude other acquired optic nerve pathologies. An accurate chronological history of the presenting complaint and probing questions regarding the extended family history are essential. The neuro-ophthalmological examination includes documentation of the following: (1) color vision, which is invariably severely depressed; (2) assessment of pupillary light reflexes for the presence or absence of an RAPD; and (3) visual field perimetry to document the pattern of field defect, which is typically a dense central or centrocecal scotoma; and OCT imaging of the retinal nerve fiber layer. In the acute stage, OCT usually shows swelling of the nerve fiber layer around the optic disc secondary to axonal stasis, and this is followed by marked thinning as optic atrophy ensues. In keeping with postmortem histopathological studies, there is characteristic sparing of the nasal nerve fiber layer in LHON in contrast to the temporal papillomacular bundle, which is affected early in the disease process, and much more severely (Figure 1(c)).

LHON molecular genetic testing is now widely available in most diagnostic laboratories worldwide. Blood or buccal swab DNA samples are usually sufficient because the mutation is usually homoplasmic. Three mtDNA point mutations, m.3460G>A (*MTND1*), m.11778G>A (*MTND4*), and m.14484T>C (*MTND6*), account for the majority (90%) of LHON cases, and for this reason, they are often referred to as primary. If these common mutations are not present and there is a high index of clinical suspicion, further mitochondrial genome sequencing should be requested when this service is available, with a particular focus on genes encoding complex I subunits of the mitochondrial respiratory chain (Table 1).

TABLE 1 Mitochondrial DNA Variants Identified in Cases with Leber Hereditary Optic Neuropathy

	Mitochondrial Gene	Nucleotide Change
Common mutations (~90%)	*MTND1*	m.3460G>A[a]
	MTND4	m.11778G>A[a]
	MTND6	m.14484T>C[a]
Rare Rh variants (~10%)	*MTND1*	m.3376G>A, m.3635G>A[a], m.3697G>A, m.3700G>A[a], m.3733G>A[a], m.4025C>T, m.4160T>C, m.4171C>A[a]
	MTND2	m.4640C>A, m.5244G>A
	MTND3	m.10237T>C
	MTND4	m.11696G>A, m.11253T>C
	MTND4L	m.10663T>C[a]
	MTND5	m.12811T>C, m.12848C>T, m.13637A>G, m.13730G>A
	MTND6	m.14325T>C, m.14568C>T, m.14459G>A[a], m.14729G>A, m.14482C>A[a], m.14482C>G[a], m.14495A>G[a], m.14498C>T, m.14568C>T[a], m.14596A>T
	MTATP6	m.9101T>C
	MTCO3	m.9804G>A
	MTCYB	m.14831G>A

[a]*These mtDNA variants are definitely pathogenic. They have been identified in two or more independent LHON pedigrees and show segregation with affected disease status. The remaining putative LHON mutations have been found in singleton cases or in a single family, and additional evidence is required before pathogenicity can be irrefutably ascribed.*

PATHOPHYSIOLOGY

The majority of pathogenic mtDNA LHON mutations are found in genes encoding for complex I subunits of the mitochondrial respiratory chain. Cases harboring these mutations may have a measurable defect of complex I activity in skin fibroblasts, blood platelets, or in a muscle biopsy. Impaired mitochondrial oxidative phosphorylation has also been documented in vivo with functional phosphorus magnetic resonance spectroscopy (^{31}P-MRS) imaging. The most severe biochemical defects are seen in cases harboring the m.3460A>G mutation and the least severe in cases with the m.14484T>C mutation, in keeping with the relatively better visual prognosis of this specific *MTND6* mutation.

The downstream consequences of this impairment in mitochondrial respiratory chain function are complicated, and they include the increased production of reactive oxygen species (ROS), disordered calcium signaling, increased sensitivity to excitotoxic stimuli, and eventually apoptosis leading to retinal ganglion cell (RGC) loss and optic nerve degeneration. Limited postmortem histopathological studies have shown a stereotypical pattern of neurodegeneration with the preferential loss of RGCs within the papillomacular bundle—an anatomical feature that has been linked to the limited mitochondrial energy reserve of their smaller axons. Severe dysfunction of the papillomacular bundle in the early stage of LHON accounts for the dense centrocecal scotoma on visual field testing and the marked dyschromatopsia. Post-ganglionic volume loss has also been documented affecting the optic radiation on diffusion tensor MRI, and this is thought to be a secondary post-synaptic phenomenon caused by primary RGC loss. Intriguingly, a special class of melanopsin-expressing RGCs are preferentially spared in LHON, providing an explanation for the relatively preserved pupillary reflexes, which are somewhat atypical in the context of such a severe optic neuropathy.

LHON pedigrees display strict maternal inheritance, but not all mutation carriers will experience visual loss during their lifetime. The conversion rate for male carriers is about 50% compared with about 10% for female carriers. LHON is, therefore, characterized by both marked incomplete penetrance and a striking male bias for visual loss. The pathological consequences of the mtDNA LHON mutations are clearly being influenced by other disease modifiers, and these can be classified into four main groups: (1) secondary mtDNA factors, (2) nuclear susceptibility genes, (3) hormonal differences, and (4) environmental triggers.

Secondary mtDNA Factors

Approximately 10% of cases presenting with LHON harbor a mixture of mutated and wild-type mtDNA—a feature known as heteroplasmy. A case series has shown a level of heteroplasmy greater than 60% in blood samples taken from affected LHON individuals, suggesting a minimum threshold effect.

Lower levels of heteroplasmy in some mutation carriers could therefore account for the characteristic incomplete penetrance seen in LHON pedigrees.

The clinical penetrance of primary mtDNA LHON mutations is also thought to be influenced by the specific mitochondrial genetic background on which they occur. These mtDNA haplogroups are defined by a number of common genetic polymorphisms that have clustered together during human evolution and population migrations. The haplogroup J background, which is found in about 10% of people of European extraction, increases the clinical penetrance in LHON pedigrees harboring the m.11778G>A and m.14484T>C mutations. On the other hand, carriers of the m.3460G>A mutation are more likely to become visually affected on a haplogroup K background.

Nuclear Susceptibility Genes

The predilection for males to lose vision in LHON cannot be explained by mitochondrial genetics, and a visual loss susceptibility gene on the X-chromosome has long been suspected, the so-called two-locus disease model. Three independent linkage studies have revealed overlapping candidate regions on the X-chromosome, but the actual modifier gene has still not yet been identified, which further complicates genetic counseling.

Hormonal Differences

Another obvious factor to account for the observed male bias in LHON is a protective influence conferred by female sex hormones. This hypothesis was recently investigated using LHON cybrid cell lines, and interestingly, treatment with 17β-oestradiol resulted in reduced ROS levels, increased activity of the antioxidant enzyme superoxide dismutase, and more efficient mitochondrial biogenesis. Interestingly, unaffected LHON carriers tend to have higher levels of mtDNA in their circulating blood leucocytes compared with affected cases, consistent with a compensatory mechanism linked to increased mitochondrial biogenesis. Since RGC bodies have a high concentration of estrogen β receptors, supplementation with estrogen-like compounds is currently being explored as a possible therapeutic option in LHON.

Environmental Triggers

Various environmental factors have been implicated in precipitating visual loss among LHON carriers including head trauma, industrial toxins, and drugs with mitochondrial toxic effects. The evidence for these environmental triggers is largely anecdotal, but there is relatively strong evidence for an increased risk of visual failure among smokers, and to a lesser extent, heavy drinkers.

CASE MANAGEMENT

Supportive Measures

LHON causes severe visual impairment, and the sudden onset of visual loss in otherwise healthy individuals carries a significant psychological and socioeconomic burden. Clinicians therefore have an important role to play in facilitating access to rehabilitative services such as low visual aids and occupational therapy. Genetic counseling in this mitochondrial disorder is complicated by the sex- and age-dependent penetrance of the underlying pathogenic mtDNA mutations. As a rule of thumb, the lifetime risk of disease conversion is about 50% and 10% in male and female carriers, respectively, and the majority (95%) of carriers who experience visual loss will do so before the age of 50 years. Several LHON support networks have been established recently in North America and Europe, and these organizations can provide tremendous support to recently affected cases and their families (http://www.lhon.org/ and www.lhonsociety.org/, accessed January 13, 2015).

Preventative Measures

Cases with LHON should be strongly advised not to smoke and to moderate their alcohol intake. This is especially the case for unaffected LHON carriers as smoking, and to a lesser extent excessive binge drinking, have been linked with an increased risk of disease conversion. It also seems sensible to avoid exposure to other putative environmental triggers for visual loss in LHON, in particular industrial toxins and drugs with mitochondrial toxic effects, for example, ethambutol. With regard to contact sports, unaffected LHON carriers should be advised that there is anecdotal evidence of head trauma precipitating disease conversion in some instances.

Mitochondrial Cocktails

Over the years, various combinations of vitamins (B2, B3, B12, C, E, and folic acid) and other compounds with putative mitochondrial antioxidant and bioenergetics properties have been used to treat cases with LHON, but none with any convincing proof of efficacy. The list of supplements that are frequently promoted on the Internet includes alpha-lipoic acid, carnitine, creatine, and L-arginine, again without clear evidence of clinical efficacy.

Neuroprotective Agents

Idebenone is a short-chain synthetic analogue of ubiquinone that promotes mitochondrial ATP synthesis in addition to having antioxidant properties. Recently published studies on the use of this compound in LHON have shown promise, with a greater proportion of cases receiving idebenone recovering vision compared with the untreated natural history, and a very favorable safety profile. Although based on retrospective analysis, the most consistent prognostic factors associated with visual recovery were the early initiation of treatment and a more prolonged course of treatment with idebenone. Although idebenone is currently not licensed for clinical use, individuals with LHON frequently opt to gain access to it at their own expense from various Internet sources. Another newer generation ubiquinone analogue is EPI-743, and preliminary data from an open-labeled study of five cases with LHON who were treated within 90 days of disease conversion are encouraging. The preliminary results will need to be substantiated in a properly designed clinical trial to further investigate the efficacy of this experimental drug in halting disease progression and improving the final visual outcome in LHON. At this point in time, there are currently no proven agents that can prevent disease conversion in asymptomatic LHON carriers.

Gene Therapy

Gene replacement therapy for LHON is an attractive strategy given the easy anatomical accessibility of the RGC layer for direct manipulation. Clinical trials will soon be launched in North America and in Europe to assess the benefit of this novel therapeutic approach. It is therefore important to emphasize to individuals that gene therapy for LHON is still at an early stage of development, and stringent safety and efficacy data are still needed.

Unproven Treatments

There is highly anecdotal Internet evidence of cases with LHON benefiting from hyperbaric oxygen therapy. The purported rationale for this treatment is to provide increased levels of oxygen to RGCs during the acute phase of LHON with the aim of improving mitochondrial biogenesis. Various poorly monitored stem cell institutes worldwide are also promoting the use of non-validated experimental protocols, and individuals with LHON need to be carefully advised before embarking on multiple expensive courses of treatment with the possible associated biological risks.

CLINICAL PEARLS

- LHON classically presents in young adult males with a peak age of onset in the second and third decades of life. However, this diagnosis should be considered in all cases of bilateral simultaneous or sequential optic neuropathy, irrespective of age and sex, in particular when visual function is severely affected with little or no recovery.
- There is frequently no clear history of other affected family members having experienced early-onset visual loss because of the reduced clinical penetrance masking the maternal inheritance pattern of LHON, or simply lack of extended family contacts.

- Relative preservation of the pupillary light reflexes in the context of severe visual loss is a characteristic feature of LHON.
- In about 20% of cases of LHON, the fundus looks entirely normal in the acute stage, and these individuals, especially children, are frequently labeled as having functional visual loss. It should also be stressed that optic atrophy takes about 6–8 weeks before it becomes established.
- Useful ancillary tests in suspected cases of LHON are OCT, which can identify areas of segmental retinal nerve fiber layer swelling, and visual electrophysiology, which can provide objective evidence of primary RGC pathology.
- If an individual is only seen in the chronic stage of LHON, it can be difficult to exclude other possible causes of optic atrophy, especially if there is no clear maternal family history. In these cases, MR neuroimaging of the anterior visual pathways is mandatory whilst awaiting the results of molecular genetic testing.

FURTHER READING

[1] Barboni P, Savini G, Valentino ML, et al. Retinal nerve fiber layer evaluation by optical coherence tomography in Leber's hereditary optic neuropathy. Ophthalmology 2005;112(1):120–6.
[2] Barboni P, Savini G, Valentino ML, et al. Leber's hereditary optic neuropathy with childhood onset. Invest Ophthalmol Vis Sci 2006;47(12):5303–9.
[3] Kirkman MA, Yu Wai Man P, Korsten A, et al. Gene-environment interactions in Leber hereditary optic neuropathy. Brain 2009;132(9):2317–26.
[4] Klopstock K, Yu Wai Man P, Dimitriadis K, et al. A randomized placebo-controlled trial of idebenone in Leber's hereditary optic neuropathy. Brain 2011;134(9):2677–86.
[5] Moura AL, Nagy BV, La Morgia C, et al. The pupil light reflex in Leber's hereditary optic neuropathy: evidence for preservation of melanopsin-expressing retinal ganglion cells. Invest Ophthalmol Vis Sci 2013;54(7):4471–7.
[6] Pfeffer G, Burke A, Yu-Wai-Man P, Compston DA, Chinnery PF. Clinical features of MS associated with Leber hereditary optic neuropathy mtDNA mutations. Neurology 2013;81(24): 2073–81.
[7] Sadun AA, Chicani CF, Ross-Cisneros FN, et al. Effect of EPI-743 on the clinical course of the mitochondrial disease Leber hereditary optic neuropathy. Arch Neurol 2012;69(3):331–8.
[8] Yu Wai Man P, Chinnery PF. Leber hereditary optic neuropathy. In: Pagon RA, Bird TC, Dolan CR, Stephens K, editors. Gene reviews. 2013. Available online at: http://www.ncbi.nlm.nih.gov/books/NBK1174/. [accessed 13.01.15].
[9] Yu Wai Man P, Votruba M, Moore AT, Chinnery PF. Treatment strategies for inherited optic neuropathies—past, present and future. Eye 2014;28(5):521–37.

Chapter 8

Leigh Syndrome

Sumit Parikh[1,2]
[1]*Cleveland Clinic Lerner College of Medicine & Case Western Reserve University, Cleveland, OH, USA;*
[2]*Neurogenetics, Metabolism and Mitochondrial Disease Center, Cleveland Clinic, Cleveland, OH, USA*

CASE PRESENTATION

Our case presented at age 13 months with a slight alteration in gait and deviation of her left eye. She had been in her usual state of health until a month before when her mother noted that she was falling more and also unsteady when walking. This was a clear change in function since she had learned to ambulate independently around age 10 months and had been quite surefooted by 11–12 months. In addition, her mother had noted outward deviation of her left eye over the week before her initial assessment associated with her tilting her head to the left.

There was no history of intercurrent or recent illness, infections, or immunizations. She had been developing typically to date. Her review of systems and past medical and family history were noncontributory.

Her neurologic examination showed an alert and interactive girl with a clear head tilt to the left with an outward deviation of the left eye, which was worse when her head was held in the midline position. There was a slight facial asymmetry with drooping of the left half of the face. Her gait was wide-based and ataxic. There was a slight tremor noted in the hands on intention and mild dysmetria. Her general examination was normal.

Over the subsequent few years, our case developed periodic regression, often presenting with apnea or an alteration of consciousness. After each episode, she would be left with new and residual neurologic deficits. At this point, she is several years old, bright with dysarthric speech, with strabismus, facial palsy, dysphagia, and she is nonambulatory due to imbalance and hemiparesis with spasticity.

DIFFERENTIAL DIAGNOSIS

New and rapid onset strabismus in a young child is typically a concerning finding, raising the concern of a new intracranial process impacting the function of nerves controlling extraocular movements. The association of additional cranial nerve abnormalities on neurologic examination (facial asymmetry) and ataxia localize the condition to the brain stem, specifically the lower midbrain, pontine, and cerebellar region. The cause of multiple cranial neuropathies in pediatric cases is relatively

Mitochondrial Case Studies. http://dx.doi.org/10.1016/B978-0-12-800877-5.00008-5

short and is primarily a new lesion in the brain stem traversing the course of several cranial nerve nuclei. The common pediatric sources of such a lesion include neoplasm (brain-stem glioma or central nervous system lymphoma), demyelinating disease, infectious/inflammatory disease, or a disorder of energy metabolism. Our case did not have a recent history of infection or immunization—thus our initial concerns were more focused on a neoplasm or disorder of energy metabolism.

DIAGNOSTIC APPROACH AND PATHOPHYSIOLOGY

The initial diagnostic test for a case with these symptoms is an urgent brain magnetic resonance imaging (MRI), with and without contrast. This study will often be diagnostic and quickly help narrow the differential. Our case's brain MRI showed bilateral symmetric lesions involving the periaqueductal gray and dentate nuclei of the cerebellum. Such findings, especially when involving other deep gray matter nuclei such as the basal ganglia and/or thalami, are classified under the classification of Leigh syndrome (LS). Diagnostic criteria for LS were defined by Rahman et al. in 1996 and include progressive neurologic disease with motor and intellectual developmental delay, signs and symptoms of brain-stem and/or basal ganglia disease, raised lactate concentration in blood and/or cerebrospinal fluid, and typical neuroradiologic changes or neuropathologic changes in the case or affected siblings [1]. When all diagnostic criteria are not met, a diagnosis of Leigh-like syndrome is often used.

LS is most typically seen in disorders of energy metabolism, whether due to pyruvate dehydrogenase complex deficiency or disruption of electron transport chain function. A rare disorder of thiamine transport can lead to LS-like findings and must be kept in the differential since symptoms and neuroimaging abnormalities improve with supplementation of thiamine and biotin.

Mitochondrial disease-related LS can occur due to an mtDNA or mitochondrially-directed nDNA mutation. When due to an mtDNA mutation, pathogenic variants at m.8993T>G or m.8993T>C account for 10% of cases. Another 10–20% of cases have variants in other mtDNA genes. The mothers of these individuals are often asymptomatic or may have milder features including neurogenic muscle weakness, ataxia, and retinitis pigmentosa (NARP).

Once neuroimaging raised the possibility of LS in our case, additional mitochondrial markers were assessed in our case's blood and urine, showing a lactate of 3.2 mM (normal <2.2), elevation in alanine of 650 μM on plasma amino acid analysis (normal <450), and elevations in tricarboxylic cycle intermediates on urine organic acid analysis.

Mitochondrial DNA genome sequencing was pursued that showed an m.3688G>A mutation at 90% heteroplasmy in blood. The mutation was assessed in her asymptomatic mother and found to be present at less than 10% heteroplasmy.

Most cases with mtDNA-associated LS present with symptoms between the ages of 3 and 12 months, although later childhood and adult disease onset can also occur [2]. Exacerbations are often heralded by the presence of a viral infection. The condition is progressive, and decompensation during periods of illness or over time often lead to exacerbation of the cranial lesions with associated worsening neurologic function. Due to involvement of the brain stem, cases may often present with worsening respiratory or cardiovascular function, alteration of consciousness, and apnea. Recovery after each decompensation is limited with an accumulation of symptoms over time. Cases gradually develop worsening spasticity, dystonia and movement disorders from the basal ganglia lesions, epilepsy from cortical disease, and dysphagia and respiratory insufficiency from brain-stem disease. Some cases also develop ophthalmologic disease including optic atrophy, retinopathy, strabismus, and dysconjugate gaze. Non-neurologic manifestations of mitochondrial disease such as cardiomyopathy, hepatopathy, endocrinopathy and nephropathy may also occur. Cognitive function can be spared in some patients, especially early in the course of the illness, a factor often overlooked in children that are not able to communicate because of severe pyramidal and extrapyramidal involvement. Death often occurs due to respiratory insufficiency, complications of aspiration, or cardiovascular deterioration. The use of invasive ventilation has affected the natural history or the disorder by extending life, often by years.

TREATMENT

Treatment of LS is currently supportive. Brain-stem dysfunction often leads to an alteration in respiratory drive, heart rate, and blood pressure, therefore close monitoring of these variables during an illness is crucial. Non-invasive ventilatory support such as nasal BiPap may obviate the need for more invasive treatment. Depending on a family's wishes, respiratory support via a tracheostomy and home ventilation are options that can often improve quality of life, especially in cases with a later onset or otherwise milder constellation of symptoms. The neurologic treatment of spasticity, dystonia, and movement disorders is necessary.

Monitoring of cardiac, hepatic, endocrine, ophthalmic, and renal functioning is reasonable, especially to ensure that treatable problems such as hypothyroidism or dysrhythmias are not overlooked. As swallowing function may worsen with progression of brain-stem disease, the individual's diet should be adjusted to prevent aspiration. The early use of a gastrostomy tube may help prevent aspiration as well. Therapy services are crucial to allow the individual to maintain mobility and balance.

Mood and anxiety issues often occur in cases with chronic disease, and these should be monitored because treatment can improve quality of life.

CLINICAL PEARLS

- Mitochondrial disease can present with new-onset multiple cranial neuropathies at any age.
- LS due to either mtDNA or nDNA mutations has very similar clinical symptoms to mitochondrial disease; when due to mtDNA mutations, the individual's mother is often asymptomatic.
- Treatment is primarily supportive though the early use of a gastrostomy tube, or selectively, a tracheostomy and home ventilator can improve quality of life for many cases.

REFERENCES

[1] Rahman S, Blok RB, Dahl HH, Danks DM, Kirby DM, Chow CW, et al. Leigh syndrome: clinical features and biochemical and DNA abnormalities. Ann Neurol 1996;39:343–51.

[2] GeneReviews. mtDNA-associated Leigh Syndrome, http://www.ncbi.nlm.nih.gov/books/NBK1173/.

Chapter 9

Neuropathy, Ataxia, and Retinitis Pigmentosa

Mary Kay Koenig[1], Leon Grant[2]

[1]University of Texas Medical School at Houston, Department of Pediatrics, Division of Child and Adolescent Neurology, Endowed Chair of Mitochondrial Medicine, Houston, TX, USA; [2]University of Hawaii, Kapi'olani Medical Center for Women and Children, Honolulu, HI, USA

CASE PRESENTATION

A 34-year-old man was seen for evaluation of progressive neurological decline. His family had noted worsening ataxia for the last 12 months. Medical history revealed that he was born at term following an uncomplicated pregnancy. His parents had no concerns until 18 months of age when his pediatrician noted that he was not yet walking. Magnetic resonance imaging studies of the brain and spinal cord were normal. Screening laboratory studies were unremarkable, and they included the following: complete metabolic profile, complete blood count, creatine kinase level, liver function tests, thyroid studies, plasma and urine amino acids and organic acids, plasma free and total acyl-carnitine profile, and plasma levels of very long chain fatty acids. Karyotype was 46XY. The case struggled with fine and gross motor skills, and received physical therapy throughout childhood. Language milestones were reached appropriately; however, he was always difficult to understand. He attended special education classes through his public school system and ultimately received a high school diploma. At 8 years of age, the case was diagnosed with retinitis pigmentosa (RP), and at 14 years of age, he was declared legally blind. At 20 years of age, he began to lose his hearing. Testing demonstrated bilateral sensorineural hearing loss, and the case currently uses bilateral hearing assist devices.

The case has a healthy 32-year-old brother. His 82-year-old father has hypertension. His mother is 62 years old and has short stature and migraine headaches. There is a maternal uncle with short stature, who is described by the family as sickly with a learning disability. His maternal grandmother is described by the family as slow, lazy, and sickly. She died from presumptive amyotrophic lateral sclerosis in her 70s. His maternal grandmother had a brother who died at birth. His maternal great-grandmother had a brother and sister who were also described as sickly and slow. She also had a brother who died at 6 months of age.

Mitochondrial Case Studies. http://dx.doi.org/10.1016/B978-0-12-800877-5.00009-7

The individual lives at home with parents and is able to perform his own activities of daily living. He is able to stay alone for several days when his parents travel. He uses a cane to assist with ambulation as he is legally blind. He is an avid golfer through his participation in Special Olympics.

On examination, the case was awake, alert, and oriented to person and place. His speech was fluent but moderately dysarthric, and his father often had to clarify what he said. His cognition was impaired. The case was able to spell three-letter words (e.g., cat, dog) but unable to read or write. He was able to perform simple addition (5+5) but unable to perform even simple multiplication (2×2). He had full ophthalmoplegia, and visual acuity was poor. He was able to see light but unable to count fingers. Sensory examination was unremarkable with normal response to fine-touch, proprioception, pain, and temperature. Motor examination demonstrated normal and equal muscle bulk with diffusely decreased muscle tone and strength. His reflexes were absent in the upper and lower extremities. His feet were grossly deformed with bilateral pes cavus. His toes were down-going bilaterally. The case was steady when sitting; however, he demonstrated bilateral intention tremor. Upon rising from a chair, he became markedly ataxic requiring assistance to maintain his balance. Once secure in a standing position, he was able to stand independently with a wide stance and everted feet. While his gait was unsteady, he was able to walk independently without falling for at least 20 feet.

DIFFERENTIAL DIAGNOSIS AND DIAGNOSTIC APPROACH

The differential diagnosis of a child presenting at 18 months of age with delayed motor milestones is diverse. Considerations include a wide range of myopathies, muscular dystrophies, neuropathies, chromosomal disorders, metabolic diseases, and structural malformations. Current reasoning leads to an approach centered on screening studies and typically includes brain and spine imaging, muscle enzymes, electromyography, nerve conduction studies, and basic chromosomal analysis. If other organ involvement exists, metabolic testing may be performed as well. Although for our case testing was unremarkable at the time of presentation, the development of RP led to a secondary evaluation for his symptoms.

RP is an inherited eye disease that results in progressive visual impairment from degeneration of the retinal pigment epithelium. Mutations in over 50 different genes have been described with nonsyndromic RP. As our case presented with syndromic findings, including developmental delay and hearing loss, syndromic causes of RP such as Usher syndrome, Waardenburg syndrome, Alport syndrome, Refsum disease, mitochondrial diseases, abetalipoproteinemia, mucopolysaccharidoses, Bardet-Biedl syndrome, and neuronal ceroid lipofuscinosis were considered. The maternal family history of short stature, migraines, and developmental disability increased the concern for a mitochondrial disease. Mitochondrial DNA sequencing was performed and found a heteroplasmic mutation (90%) at position 8993 (A>G), confirming clinical suspicions for neuropathy, ataxia, and retinitis pigmentosa (NARP).

CLINICAL PRESENTATION

NARP is a genetic disorder chiefly affecting the nervous system. The syndrome was first described by Holt et al. in 1990 [1]. Symptoms typically arise in childhood or early adulthood and consist of proximal neurogenic muscle weakness, sensory neuropathy, ataxia, and RP. Affected individuals may also develop seizures, sensorineural hearing loss, ophthalmoplegia, short stature, and cardiac conduction defects [2,3]. Learning disabilities and developmental delays are common, and cases may develop dementia later in the disease course [4,5].

Electromyography and nerve conduction studies may demonstrate sensorimotor axonal neuropathy. Magnetic resonance imaging of the brain may identify cerebral or cerebellar atrophy or show midline abnormalities. Ophthalmologic examination may reveal RP with or without optic atrophy [6]. Lactate elevation may be seen in blood or urine samples, but this elevation is more consistent in cerebral spinal fluid [7]. Diagnosis is confirmed via detection of the heteroplasmic change at position 8993 in the mitochondrial DNA (mtDNA).

PATHOPHYSIOLOGY

NARP arises from the heteroplasmic transversion m.8993T>G or the transition m.8993T>C in the ATP6 gene [4]. The clinical expression of a mtDNA pathogenic variant such as this is influenced by several factors including the variant pathogenicity, the relative amount of mutant and wild-type mtDNA (the heteroplasmic mutant load), the variation in mutant load among different tissues, and the energy requirements of the brain and other tissues, which may vary with age. Biochemical experiments in platelets and lymphocytes from cases with NARP demonstrated impairment of ATP synthase resulting in unidirectional impairment of proton flow and a close relationship between tissue heteroplasmy levels and clinical expression [8]. Both mutations result in energy deprivation and overproduction of reactive oxidative species, though studies have identified a pathomechanism difference in the two mutations that may, in addition to heteroplasmy levels, explain the difference in phenotypes. The 8993T>G transversion mainly results in energy deficiency, while the 8993T>C mutation mainly favors increased production of reactive oxidative species [9]. Individuals with heteroplasmy levels below 60% are generally asymptomatic, while moderate heteroplasmy levels (~70–90%) of the m.8993T>G mutation present with the NARP phenotype, and heteroplasmy rates exceeding 95% manifest as maternally inherited Leigh syndrome. Maternal relatives of individuals with NARP can also present with individual or multiple symptoms associated with Leigh syndrome or NARP. These include mild learning difficulties, muscle weakness, night blindness, deafness, diabetes mellitus, migraine, or sudden unexpected death. Individuals with the m.8993T>C are generally symptomatic only when mutant loads exceed 90% [4,10–12].

TREATMENT

There are no specific treatments for NARP. Supportive and symptomatic care are essential because NARP is often the less severe phenotype seen in family members with Leigh syndrome; those recommendations provided for Leigh syndrome may apply.

CLINICAL PEARLS

- Most mtDNA inherited disorders are progressive with continued neurologic deterioration throughout a person's lifetime.
- The combination of neuropathy, ataxia, and visual loss should lead to a suspicion of NARP.
- Cognitive findings may be an early feature of NARP.
- Many mtDNA inherited disorders are not associated with lactic acidosis.
- Although gross motor delay presents a wide differential, following a child over time may lead to new diagnostic clues.

REFERENCES

[1] Holt IJ, Harding AE, Petty RK, Morgan-Hughes JA. A new mitochondrial disease associated with mitochondrial DNA heteroplasmy. Am J Hum Genet 1990;46:428–33.

[2] Santorelli FM, Tanji K, Sano M, Shanske S, El-Shahawi M, Kranz-Eble P, et al. Maternally inherited encephalopathy associated with a single-base insertion in the mitochondrial tRNATrp gene. Ann Neurol 1997;42:256–60.

[3] Sembrano E, Barthlen GM, Wallace S, Lamm C. Polysomnographic findings in a patient with the mitochondrial encephalomyopathy NARP. Neurology 1997;49:1714–7.

[4] Thorburn DR, Rahman S. Mitochondrial DNA-associated Leigh syndrome and NARP. Gene Review. [Online] April 2014. http://www.ncbi.nlm.nih.gov/pubmed/20301352.

[5] Rahman S, Blok RB, Dahl HH, Danks DM, Kirby DM, Chow CW, et al. Leigh syndrome: clinical features and biochemical and DNA abnormalities. Ann Neurol 1996;39:343–51.

[6] Ortiz RG, Newman NJ, Shoffner JM, Kaufman AE, Koontz DA, Wallace DC. Variable retinal and neurologic manifestations in patients harboring the mitochondrial DNA 8993 mutation. Archiv Ophthalmol 1993;111:1525–30.

[7] Rabier D, Diry C, Rotig A, Rustin P, Heron B, Bardet J, et al. Persistent hypocitrullinaemia as a marker for mtDNA NARP T 8993 G mutation? J Inherit Metab Dis 1998;21:216–9.

[8] Lenaz G, Baracca A, Carelli V, D'Aurelio M, Sgarbi G, Solaini G. Bioenergetics of mitochondrial diseases associated with mtDNA mutations. Biochim Biophys Acta 2004;1658:89–94.

[9] Debray FG, Lambert M, Lortie A, Vanasse M, Mitchelle GA. Long-term outcome of leigh syndrome caused by the NARP T8993C mtDNA mutation. Am J Med Genet 2007;143A:2046–51.

[10] Baracca A, Sgarbi G, Mattiazzi M, Casalena G, Pagnotta E, Valentino ML, et al. Biochemical phenotypes associated with the mitochondrial ATP6 gene mutations at nt8993. Biochim Biophys Acta 2007;1767:913–9.
[11] Sgarbi G, Baracca A, Lenza G, Valentin LM, Carelli V, Solaini G. Inefficient coupling between proton transport and ATP synthesis may be the pathogenic mechanism for NARP and Leigh syndrome resulting from the T8993G mutation in mtDNA. Biochem J 2006;395:493–500.
[12] Carelli V, Baracca A, Barogi S, Pallotti F, Valentino ML, Montagna P, et al. Biochemical-clinical correlation in patients with different loads of the mitochondrial DNA T8993G mutation. Archiv Neurol 2002;59:264–70.

FURTHER READING

[1] Mitochondrial DNA-associated Leigh syndrome and NARP. Thorburn DR, Rahman S. Gene Reviews. Seattle (WA): University of Washington, Seattle; 1993–2014. October 30, 2003 [updated April 17, 2014].
[2] Neuropathy, ataxia, and retinitis pigmentosa. Online Mendelian inheritance in man, OMIM®. Johns Hopkins University, Baltimore, MD. MIM Number: 551500:11/10/2009. World Wide Web URL: http://omim.org/.
[3] DiMauro S, Schon EA. Mitochondrial disorders in the nervous system. Annu Rev Neurosci 2008;31:91–123.

Chapter 10

Maternally Inherited (Mitochondrial) Diabetes

Amy Goldstein
Neurogenetics & Metabolism, Division of Child Neurology, Children's Hospital of Pittsburgh, Pittsburgh, PA, USA

CASE PRESENTATION

Two cases with different mitochondrial disorders will be discussed: the first case has Kearns–Sayre syndrome, and the second case is a mother with maternally inherited diabetes and deafness (MIDD) whose son has mitochondrial encephalomyopathy and lactic acidosis with stroke-like episodes (MELAS). Both of these mitochondrial disorders show an increased prevalence of diabetes mellitus [1].

Our first case was a 2.5-year old at initial presentation for failure to thrive. Her evaluation at that time revealed normal thyroid function studies, basic metabolic profile, complete blood count, celiac antibodies, and a sweat chloride test. She had a vitamin D deficiency, which was treated with supplemental vitamin D. Failure to thrive was attributed to psychosocial causes. At 3.2 years old, she was referred to Endocrinology for continued failure to thrive, again thought to be psychosocial. She continued to be a picky eater and had difficulty gaining weight. She presented again at the age of 7 years with tetanic hand spasms during an episode of viral gastroenteritis. Her calcium was 7.5 mg/dL (normal range 8.8–10.8 mg/dL), her parathyroid hormone level was 3 pg/mL (normal range 9–59 pg/mL) and she was diagnosed with hypoparathyroidism. She was admitted for further evaluation at that time. Additional laboratory studies revealed a mildly elevated lactate of 3.2 mMol/L (normal range 0.5–2.2 mMol/L, and an elevated alanine of 87 umol/dL (normal range 9–60 umol/dL), with an elevated alanine:lysine ratio of 4.35 (normal less than 3:1). At that time, a single glucose measurement was 161 mg/dL (normal range 70–100 mg/dL). Neurology was consulted and noted a history of developmental delay and hypotonia and ptosis on examination. A magnetic resonance imaging (MRI) of her brain was ordered and revealed bilateral symmetric lesions in her basal ganglia (see Figure 1). Mitochondrial DNA (mtDNA) studies were ordered, and they eventually revealed a large (8 kb) mtDNA deletion. While we were still awaiting her mtDNA studies, within 2 months of presenting with hypoparathyroidism, she presented with polyuria, polydipsia, and glucosuria. Her

Mitochondrial Case Studies. http://dx.doi.org/10.1016/B978-0-12-800877-5.00010-3

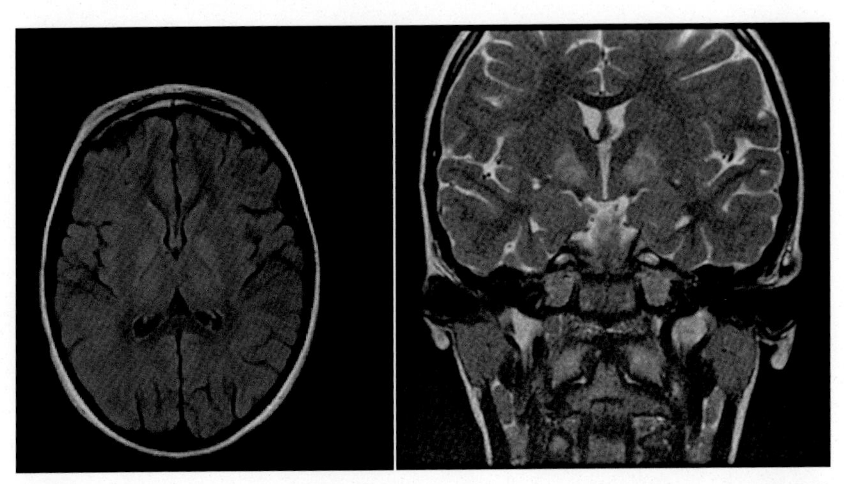

FIGURE 1 Case 1 MRI brain; on the left is an axial fluid attenuation inversion recovery (FLAIR) showing increased signal of basal ganglia; on the right is a coronal T2 demonstrating the same.

serum glucose was 629 mg/dL (normal range 70–105 mg/dL) with a hemoglobin A1C of 9.6% (normal range 4.3–6.1%). She was not in diabetic ketoacidosis, and her DM1antibody was negative. She was given insulin and developed significant hypoglycemia (glucoses in the 30–50s), compounded by her inconsistent meals. Her total daily dose of insulin was approximately 0.25 units/kg/day; short-acting insulin was dosed post-prandially based on her oral intake. Given her long-standing short stature with failure to thrive and emerging endocrinopathies, clinically, she appeared to have Kearns–Sayre syndrome, as opposed to the other clinical phenotypes seen in mtDNA deletion disorders (Pearson syndrome, Leigh syndrome, and progressive external ophthalmoplegia (PEO)) [2]. Over the next year, she developed a pigmentary retinopathy with progressive night blindness (electroretinogram consistent with retinal dystrophy). Her electrocardiogram (EKG) initially showed sinus tachycardia with nonspecific intraventricular conduction delay and with intermittent aberrant ventricular conduction as well as T-wave inversion in the inferolateral leads. Her EKGs were followed closely, and within a year, she developed a progressive arrhythmia with bifascicular block with intermittent bundle branch block, especially at higher heart rates (detected via Holter monitoring). She was under the care of a pediatric cardiologist familiar with Kearns–Sayre syndrome and the incidence of conduction block, which can lead to complete heart block if untreated with a prophylactic pacemaker [3]. A prophylactic pacemaker was placed to prevent sudden death. Her diabetes continues to be difficult to control on a day-to-day basis, with unpredictable insulin needs even with a steady dietary intake.

Our second case is a 19-year-old male, who was referred to Endocrinology at 7.5 years for poor growth. He had always been a picky eater but also had difficulty chewing and swallowing. At 9.5 years old, he was referred to Gastroenterology for continued failure to thrive and constipation with intermittent diarrhea. At that time, he was also noted to have exercise intolerance, intellectual disability, and progressive sensorineural hearing loss requiring hearing aids. The family history was notable for a mother with deafness and diabetes mellitus (type II) with a healthy diet and low BMI. His thyroid function studies had been followed yearly since age 7, and by age 13, he had developed hypothyroidism requiring levothyroxine. He had a possible seizure at age 13 with normal routine and ambulatory electroencephalograms. A brain MRI was ordered to evaluate his pituitary, which was abnormal for global volume loss. At age 19, he presented with seizures, 5 days of left arm and leg paresthesias, and dysarthria. He had had recent onset increased fatigue and mood lability. He was admitted and found to have elevated serum lactates (3.2–3.5 mMol/L), elevated cerebrospinal fluid lactate of 5.7 mEq/L, as well as elevated lactate level in the brain parenchyma on MR spectroscopy, presenting as 1.33 ppm peak in both the area of the stroke as well as unaffected brain tissue. (see Figure 2 for images at age 13, and follow-up at age 19 years.) Due to his MRI findings of stroke-like episode (ischemia not following an arterial territory), as well as his history and maternal family history, MELAS was strongly suspected. Gene sequencing of

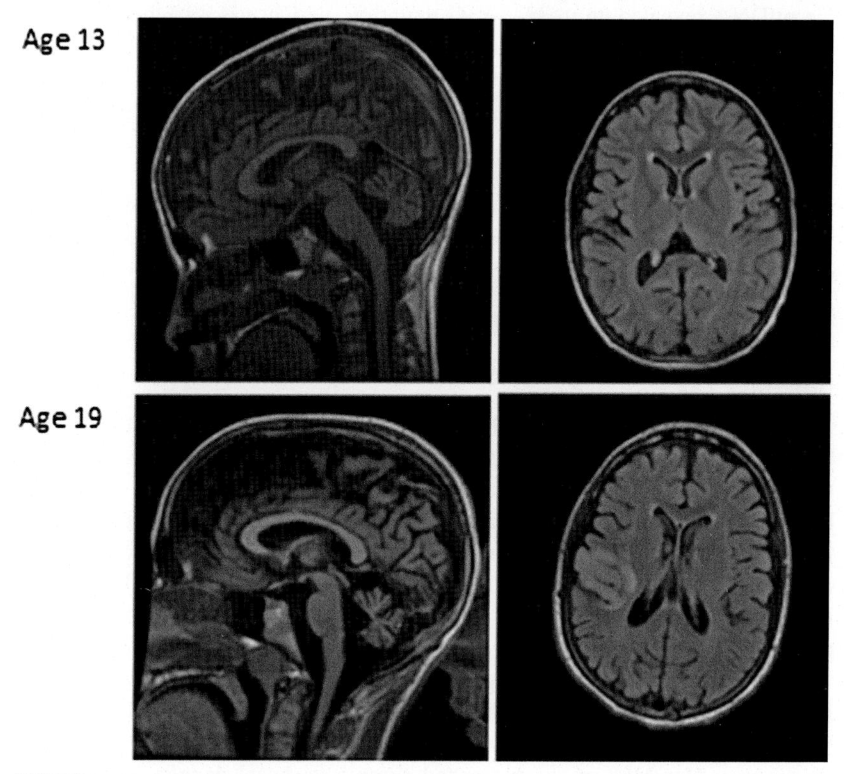

FIGURE 2 Case 2 MRI brain at ages 13 and 19. Images on left are T1 sagittal; images on right are axial fluid attenuation inversion recovery (FLAIR) sequences.

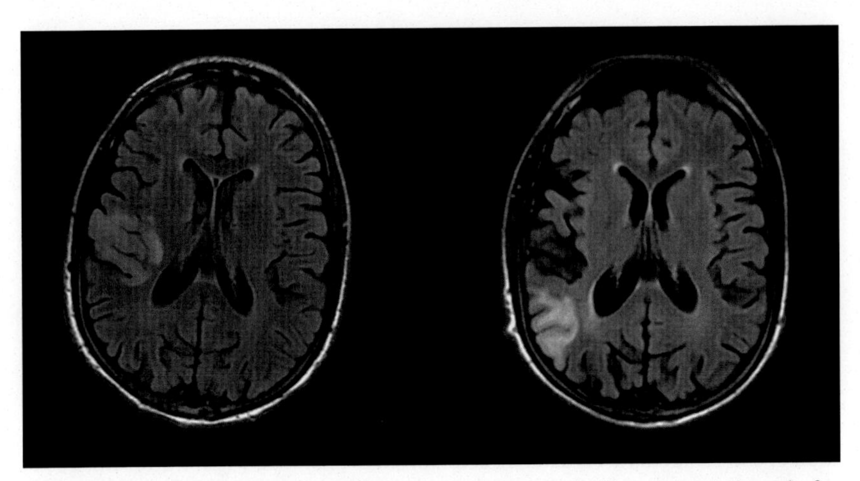

FIGURE 3 MRI brain axial FLAIR taken at time of recurrent stroke-like symptoms, 1 month after his first symptoms. The old stroke can be seen (in left panel), as well as a new stroke of higher signal intensity (in the right panel).

his mtDNA revealed an m.3243A>G mutation, which was 89% heteroplasmy (blood). His mother was found to have a heteroplasmy level of 39% (blood). After explaining her diagnosis of MIDD to her endocrinologist, her diabetes is now under slightly better control. About 1 month after his diagnosis, he returned with similar symptoms.

Figure 3 shows an MRI performed at that time, with stroke-like symptoms involving paresthesias and weakness in the left arm. He was treated with arginine IV 0.5 g/kg/day until symptoms resolved (about 72 h later). He was discharged on oral arginine 0.3 g/kg/day indefinitely. No further stroke-like events or seizures to date.

DIFFERENTIAL DIAGNOSIS

Pediatric cases who present with diabetes mellitus typically have type 1 diabetes mellitus (T1DM). This disorder is due to an insulin deficiency due to the destruction of pancreatic beta cells of the islets of Langerhans, and it may be associated with positive disease specific antibodies. Type 2 diabetes mellitus (T2DM) is due to insulin resistance, and it is typically associated with obesity and presents in adult life. Mitochondrial diabetes typically presents in adulthood, and symptoms are usually insidious, especially when seen with m.3243A>G mutation. It is rare to see T2DM present in childhood [4].

In our first case, her difficulty to gain weight and short stature was her presenting symptom in early childhood. She then had developmental delay and hypotonia when she presented with tetany and was found to have hypoparathyroidism. Her MRI of the brain had symmetric basal ganglia lesions. She then presented with increased thirst and urination, typically seen in diabetes.

Her other multisystemic features of mitochondrial disease, especially failure to thrive and developmental delays, should alert the clinician to the possibility of mitochondrial diabetes.

In our second case, a long history of failure to thrive, intellectual disability, and hearing loss followed by stroke-like episodes was very suggestible for MELAS. His mother has MIDD, and other maternal relatives are, therefore, at risk for MIDD and from the m.3243A>G mutation, which was found at a lower level heteroplasmy.

DIAGNOSTIC APPROACH

Several of the mitochondrial disorders are associated with a higher prevalence of diabetes mellitus, including mitochondrial DNA point mutations m.3243A>G, m.8296A>G, m.14577T>C, and m.14709T>C. Cases with mtDNA deletions leading to Kearn–Sayre syndrome develop DM 11–14% of the time. Several nuclear DNA mutations, including *POLG*, *RRM2B*, *OPA1* and others, also have DM as part of their phenotype. Based on other organ symptoms and family history, testing for mtDNA and/or nuclear DNA mutations is indicated. Unless the phenotype is typical for a specific syndrome, one should begin with mtDNA genome sequencing followed by an analysis of mitochondrial nuclear genes via a gene-panel or whole exome sequencing [5].

TREATMENT STRATEGY

Cases with diabetes should be under the care of an endocrinologist. There are no specific management recommendations for mitochondrial diabetes mellitus. Insulin therapy should be utilized when needed and titrated based on case needs. Some cases may not require insulin at the onset, but they still need to be managed with an oral agent. Caution should be taken with the use of metformin and thiazolidinediones, due to the inhibition of mitochondria complex I and lactic acidosis (which already preexists in many MELAS cases), although metformin has been used without problems in this case group. The preferred treatment is a short-acting sulphonylurea, with close observation for hypoglycemia [6].

In a recent study by El-Hattab et al., glucose metabolism in MELAS cases with and without diabetes (as compared to healthy controls) was studied using stable isotope infusion techniques [6]. They found that cases harboring the m.3243A>G mutation had increased glucose production due to increased gluconeogenesis, and those with diabetes had both insulin resistance and insulin deficiency.

Once diabetes is diagnosed, deafness is usually already present but may not have been yet discovered; these cases need audiograms and hearing augmentation if necessary. Other end-organ involvement also needs to be monitored and should include the following: an ophthalmology evaluation, an echocardiogram, EKG/Holter monitoring, kidney function, gastrointestinal motility studies, and a thorough neurological examination [5].

Cases with MELAS typically have maternal relatives who are asymptomatic carriers, but 30% will develop symptoms over the next 5 years and need to be followed closely for symptoms of diabetes mellitus [5].

LONG-TERM OUTCOME

Mitochondrial diabetes, especially when seen in MELAS or m.3243A>G cases, is different than type 1 or type 2 diabetes mellitus. The onset of diabetes is earlier in life than typical T2DM and cases have a lower body mass index (BMI). Cases typically have a higher hemoglobin A1C than type 1 or type 2 and an earlier insulin requirement than typical for type 2. Diabetes in MELAS typically begins in mid-life (late 30s). Twenty percent of people will have an acute onset of symptoms, yet only 8% present in diabetic ketoacidosis. Only 13% of MELAS cases (from the m.3243A>G mutation) require insulin at diagnosis. The remaining 80% have a more insidious onset, with 45.2% progressing to an insulin requirement within 2–4 years. In terms of end organ involvement, mitochondrial diabetes has higher rates of neuropathy and nephropathy compared to T1DM or T2DM, but lower rates of retinopathy or cataracts [7].

PATHOPHYSIOLOGY/NEUROBIOLOGY OF DISEASE

Mitochondrial cytopathy in the pancreas leads to slow destruction of the β cells, a decrease in insulin production (as opposed to insulin resistance), and the development of mitochondrial diabetes (MIDD). The m. 3243A>G mutation can cause MELAS or MIDD, depending on heteroplasmy level. While the newest technology allows for testing of mtDNA mutations down to very low levels of heteroplasmy, MELAS can be difficult to detect in blood, as the amount of the mutation in blood declines at a rate of 1.4% per year [8]. Urine epithelial cells (from a noninvasive urine collection) are a more specific tissue to test, and heteroplasmy correlates well with disease progression [9].

CLINICAL PEARLS

- MELAS should be suspected in a case with diabetes who also has the following: pre-dementia, sensorineural hearing loss/deafness, neuromuscular disease, end-stage renal disease, short stature/thin body habitus (non-obese), gastrointestinal dysmotility, and a maternal family history of deafness and diabetes.
- Kearns–Sayre syndrome should be suspected in a case with diabetes who also has the following: cardiomyopathy or cardiac arrhythmia, neuropathy, skeletal muscle weakness, gastrointestinal dysmotility, pigmentary retinopathy, ptosis, ataxia, or short stature/failure to thrive. Mitochondrial disease is more likely if a diabetic case has a low BMI, short stature, or a younger age of onset than expected for T2DM.

REFERENCES

[1] Karaa A, Goldstein A. The spectrum of clinical presentation, diagnosis, and management of mitochondrial forms of diabetes. Pediatr Diabetes October 20, 2014.

[2] DiMauro S, Hirano M. Mitochondrial DNA deletion syndromes. In: Pagon RA, Adam MP, Ardinger HH, et al., editors. GeneReviews® [Internet]. Seattle (WA): University of Washington, Seattle; December 17, 2003. 1993–2015. Available from: http://www.ncbi.nlm.nih.gov/books/NBK1203/. [Updated May 3, 2011].

[3] Aure K, Ogier de Baulny H, Laforet P, Jardel C, Eymard B, Lombes A. Chronic progressive ophthalmoplegia with large-scale mtDNA rearrangement: can we predict progression? Brain 2007;130:1516–24.

[4] Guillausseau PJ, Dubois-Laforge D, Massin P, Laloi-Michelin M, Bellanne-Chantelot C, Gin H. Heterogeneity of diabetes phenotype in patients with 3243-bp mutation of mitochondrial DNA (maternally inherited diabetes and deafness or MIDD). Diabet Metab 2004;30:181–6.

[5] Schaefer AM, Walker M, Turnbull DM, Taylor RW. Endocrine disorders in mitochondrial disease. Mol Cell Endocrinol 2013;379(1–2):2–11.

[6] El-Hattab AW, Emrick LT, Hsu JW, Chanprasert S, Jahoor F, Scaglia F, et al. Glucose metabolism derangements in adults with the MELAS m.3243A>G mutation. Mitochondrion September 2014;18:63–9. http://dx.doi.org/10.1016/j.mito.2014.07.008. Epub July 30, 2014.

[7] Holmes-Walker D, Mitchell P, Boyages S. Does mitochondrial genome mutation in subjects with maternally inherited diabetes and deafness decrease severity of diabetic retinopathy. Diabet Med 1998;15:946–52.

[8] Rahman S, Poulton J, Marchington D, Suomalainen A. Decrease of 3243A–>G mtDNA mutation from blood in MELAS syndrome: a longitudinal study. Am J Hum Genet 2001;68(1): 238–40.

[9] Whittaker RG, Blackwood JK, Alston CL, Blakely EL, Elson JL, McFarland R, et al. Urine heteroplasmy is the best predictor of clinical outcome in the m.3243A>G mtDNA mutation. Neurology 2009;72(6):568–9.

FURTHER READING

[1] Whittaker RG, Schaefer AM, McFarland R, Taylor RW, Walker M, Turnbull DM. Prevalence and progression of diabetes in mitochondrial disease. Diabetologia 2007;50(10):2085–9.

[2] Laloi-Michelin M, Meas T, Ambonville C, Bellanne-Chantelot C, Beaufils S, Massin P, et al. The clinical variability of maternally inherited diabetes and deafness is associated with the degree of heteroplasmy in blood leukocytes. J Clin Endocrinol Metab 2009;94(8):3025–30.

[3] Whittaker RG, Schaefer AM, McFarland R, Taylor RW, Walker M, Turnbull DM. Diabetes and deafness: is it sufficient to screen for the mitochondrial 3243A>G mutation alone? Diabetes Care 2007;30(9):2238–9.

Chapter 11

Sporadic Myopathy

Salvatore DiMauro, Kurenai Tanji
Columbia University Medical Center, New York, NY, USA

CASE PRESENTATIONS

Case 1 [1]

A 38-year-old man complained of lifelong intolerance to exercise. He described premature fatigue and myalgia even after moderate exercise, such as walking on level ground for 20 min. Walking uphill or upstairs, he could only tolerate for a few minutes. If he forced himself to exercise, he felt exhausted on the following day, and the muscles he used felt sore. He never noted pigmenturia. He believed his legs were more affected than his arms. A cardiovascular consultation revealed no cardiac dysfunction. The neurological examination was normal, including muscle bulk and strength. Resting serum CK was normal. Lactic acid was 5.2 mEq/L (normal values, 0.5–2.2 mEq/L) and rose excessively after a standardized aerobic exercise (9.2 mEq/L after 1 min). Electromyography (EMG) findings were normal. His family history was negative for neuromuscular disorders; specifically, his mother and two sisters did not complain of exercise intolerance. Biochemical analysis showed a distinct complex I deficiency in his frozen muscle biopsy, and histochemistry revealed scattered hyper-SDH stained and COX-positive (ragged blue) fibers (Figure 1); Sanger sequencing identified a pathogenic mutation (m.11832G>A) in *MTND4* encoding the ND4 subunit of complex I.

Case 2 [2]

A 61-year-old Filipino woman presented with a 5-year history of progressive weakness and atrophy involving the proximal and distal muscles of the arms and legs symmetrically. She had fallen repeatedly in the past 3 years, resulting in elbow and ankle fractures. She could not walk more than 100 m without resting. There was no family history of neuromuscular disease. Two of her sisters were examined and found to be neurologically normal. By report, her adult daughter had no neurological impairments.

Physical examination revealed a thin woman with marked muscle atrophy, worst in the intrinsic muscles of the hands and in the forearms bilaterally. She

Mitochondrial Case Studies. http://dx.doi.org/10.1016/B978-0-12-800877-5.00011-5

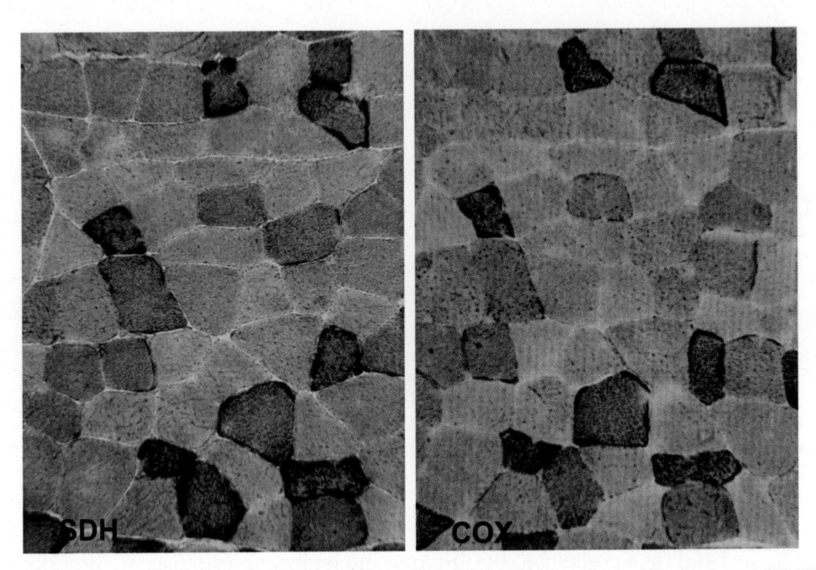

FIGURE 1 Cross-sections of frozen muscle specimen stained for succinate dehydrogenase (SDH) reaction and for cytochrome *c* oxidase (COX) reaction. Scattered fibers stained intensely with SDH, a sign of excessive mitochondrial proliferation (ragged blue fibers are good substitutes for traditional ragged red fibers [RRF], shown with the modified Gomori trichrome stain). Ragged blue and COX-positive fibers indicate a defect of specific mtDNA protein synthesis (ND4 of complex I, in this case, or a defect of cytochrome *b* in complex III [6]), whereas presence of ragged blue and COX-negative fibers is a sure sign of defective mtDNA COX-subunits or mtDNA-controlled protein synthesis in toto.

was able to move against gravity in all muscles tested, but could not resist the examiner in the distal muscles of the arms and legs. There was no evidence of spasticity, clinical myotonia, or involuntary movements. She had mild left ptosis, but no diplopia, nystagmus, hearing impairment, dysphagia, or dysphonia. Gait was wide-based and waddling. Deep tendon reflexes were diminished in the arms, but normal at the knees and ankles. Sensory examination and coordination were normal.

Serum creatine phosphokinase (CK) ranged between 600 and 700 U/L (normal, 39–238 U/L). Venous lactate was elevated at 4.5 mmol/L (normal < 1.8 mmol/L). Nerve conduction studies were normal. Electromyography (EMG) revealed myopathic changes, with positive waves, fibrillation potentials, and myotonic discharges in all muscles tested.

Muscle histochemistry showed numerous atrophic fibers, dispersed among normal-sized or hypertrophic fibers. Generally moderate, but focally severe endomysial fibrosis was present, and perimysial adipose tissue was significantly increased. Roughly 40% of the fibers stained intensely with the succinate dehydrogenase (SDH) reaction (ragged blue fibers), but not all ragged blue fibers were COX-deficient. Immunohistochemical staining for dystrophin, caveolin 3, α-sarcoglycan, merosin, and dysferlin showed no specific deficiency.

Biochemical analysis of respiratory chain complexes was essentially normal, except for mildly increased activities of citrate synthase and SDH and slightly decreased activity of complex I.

Sequencing of the entire muscle mtDNA revealed a heteroplasmic single base mutation, m.4403G>A in the *MTTM* gene. This change is not reported in the URLs of MITOMAP and mtDBHuman Mitochondrial Genome Database and was not found in 100 mtDNAs from normal or disease controls. Pathogenic mutations in the *MTTM* gene are extremely rare, but a similar mutation was reported in a similarly affected sporadic myopathic woman, who also had a mixed muscle biopsy, showing mitochondrial changes together with dystrophic features [3].

Case 3 [4]

A 24-year-old woman had been in good health until the age of 10, when she complained of exercise intolerance and aching pains in her legs, worse in cold weather. During her early teenage tears, she also complained of weakness in her arms. These symptoms were relatively stable, and she was not investigated until her 3-year-old daughter was found to have high serum CK. The mother also had increased CK values (368 U/L, normal < 155), and muscle biopsy of the right deltoid showed cytochrome *c* oxidase (COX)-positive ragged red fibers with increased subsarcolemmal SDH activity. Biochemical analysis showed normal activities of respiratory chain complexes I–V. Molecular analysis revealed a novel point mutation (m.3288A>G) in $tRNA^{Leu(UUR)}$, a common hot spot gene often associated with multisystem disorders (*in primis* mitochondrial encephalomyopathy, lactic acidosis and stroke-like symptoms: MELAS) but sometimes associated only with myopathy or cardiopathy.

The 26-year-old brother of the woman had been in good health until 5 years of age, when he developed slowly progressive proximal limb weakness. When he walked a medium distance, he had sharp pain in the legs. He also had poor weight gain and morning vomiting. The nausea and vomiting were due to respiratory acidosis, which developed in the recumbent position during sleep. A muscle biopsy, performed at age 9, revealed accumulation of subsarcolemmal mitochondria. An EMG performed at age 10 was consistent with myopathy. At age 14, surgery for strabismus was done, and a spinal fusion with Harrington rod placement was performed at age 15 due to severe scoliosis. Because of respiratory muscle weakness, he suffered respiratory arrests and required permanent tracheostomy at age 18. During the second decade, he developed progressive bilateral ptosis (without limitation of ocular movements) requiring surgery at age 26. Motor examination showed symmetric weakness of both flexor and extensor neck muscles and mild shoulder and pelvic girdle weakness. Muscle stretch reflexes were absent in the arms and present in the legs. Except for bilateral ptosis, cranial nerve function was intact. Electrocardiogram (ECG) and funduscopic examination were normal. This individual could walk up steps and play half-court basketball for a couple of hours. However, his respiratory function was stable for the last few years with forced vital capacity approximately 25% of normal, and he used nocturnal positive pressure ventilation.

DIFFERENTIAL DIAGNOSIS

In all three cases, a myopathy was the main diagnostic problem, and a myopathy characterized by exercise intolerance in case 1 and a slowly progressive weakness in cases 2 and 3.

Case 1 has severe episodic exercise intolerance not followed by pigmenturia, but it was reminiscent of metabolic myopathies, either due to disorders of glycogen breakdown (GSD V, McArdle disease) or to defects of long-chain fatty acids beta-oxidation (CPTII deficiency). However, the differential clinical diagnosis distinguishes the type of strenuous exercise causing painful cramps (McArdle disease) and the long-term moderate exertion often triggered by fasting (CPTII deficiency). The situation of case 1 is due to shortage of energy (ATP) generation from glycogen or lipids, explaining the exercise intolerance often culminating in muscle breakdown (rhabdomyolysis) and myoglobinuria. The situation of case 2 is much more indefinite than due to a defect of acute energy crisis but a rather ill-described chronic myopathy (possibly limb-girdle or distal myopathy). The situation in case 3 is dominated by a slowly progressive myopathy revealing itself in a young woman but also in her brother and her affected child, strongly suggesting a maternal inheritance, thus ruling out a sporadic myopathy. For completeness, we would like to add an unusual differential diagnosis that could explain a mitochondrial myopathy occurring as a consequence of a primary glycogenosis type zero, that is, the opposite of a glycogen storage disease but rather a loss of glycogen (or "*aglycogenosis*") due to mutations in the gene (*GYS1*) encoding glycogen synthetase [5]. Glycogen synthetase deficiency is associated in children with exercise intolerance and sudden cardiac death, and compensatory accumulation of mitochondria in skeletal muscle may distract attention from lack of glycogen and lead to a mitochondrial myopathy.

DIAGNOSTIC APPROACH

A question became obvious to students of metabolic myopathies: what would happen to defects in the mitochondrial respiratory chain, the quintessential final ATP-generating pathway? Exercise intolerance soon became a common presentation in many mitochondrial encephalomyopathies due to either genetic mitochondrial DNA (mtDNA) or to various Mendelian defects. However, many mutations in mtDNA presented peculiarly as sporadic myopathies, which contradict most principles of mitochondrial genetics, tissue-specificity, and maternal inheritance.

Typically, pathogenic mtDNA mutations were heteroplasmic and often associated with maternally inherited multisystem disorders, and various tissues contained different mutation loads. Thus, exercise intolerance was a typical symptom of MELAS (mitochondrial encephalomyopathy, lactic acidosis, stroke-like episodes) and of MERRF (myoclonus epilepsy with ragged red fibers). However, we were deceived to find some cases with isolated severe exercise intolerance or chronic weakness and no maternal inheritance, which led us first to remove from consideration an mtDNA mutation. Nonetheless, a simple test such as increased blood lactate led us to trust mitochondrial dysfunction and to perform a muscle biopsy, which revealed ragged red fibers, and biochemical analysis showed isolated defects in complex I (case 1) but no specific biochemical defect (case 2 and case 3). It was easy to sequence a pathogenic mutation in one of the six genes encoding complex I subunits (*MTND4*) and to sequence the whole mtDNA in cases 2 and 3. To follow our suspicion that these mutations were present only in skeletal muscle, we found them absent in other more accessible tissues (blood, buccal smear, urinary sediment, cultured skin fibroblasts), but we found them present in the same tissues in case 3. To follow our second clue, lack of maternal inheritance, in case 2, we found no mutation in any of the non-muscle-accessible tissues from the mother or non-affected siblings, but in case 3, the mutation was obviously present in all accessible tissues from both affected maternal relatives. The sporadic nature of the myopathies was explained in cases 1 and 2 by de novo mutations in myogenic stem cells after germ-layer differentiation, excluding maternal inheritance and involvement of any other tissues than skeletal muscle [6]. However, the maternal inheritance of the myopathy in case 3 became apparent, and the isolated involvement of skeletal muscle requires a trick of mitochondrial genetics (see below).

The first two cases are exemplary genuine sporadic mitochondrial myopathy, but the third case qualifies with caveats, with emphasis on pseudo-sporadic and pseudo-myopathy.

PATHOPHYSIOLOGY

The family in case 3 presented a mild myopathy affecting two of four siblings and one affected maternal child, clearly excluding a sporadic but documenting a maternally inherited myopathy. As the mutant gene (*MTTL-UUR*), mutation was heteroplasmic with very low mutation loads in blood from two affected and two unaffected children, but it had a very high mutation load (97%, almost homoplasmic) in the muscle from the affected woman. This indicated how a heteroplasmic multisystemic mutation could cause a tissue-specific phenotype in some cases because, in a given tissue like muscle, a mitotic segregation occurs, skewing heteroplasmy to surpass the threshold effect [7].

Still another example of a typically sporadic ocular myopathy (progressive external ophthalmoplegia, PEO) is due to another generalized defect of mitochondrial protein synthesis, defined by large-scale single mtDNA deletions that can cause three different syndromes: Pearson syndrome (PS), Kearns-Sayre syndrome (KSS), or PEO [8]. Although maternal inheritance is extremely rare in these conditions, it is believed that a few mtDNAs with giant deletions in the blastocyst can enter all three germ layers resulting in KSS, a multisystem disorder, segregate to the hematopoietic lineage causing PS, or segregate to muscle resulting in PEO.

A last question: are there sporadic mitochondrial myopathies due to nuclear gene defects? As a matter of fact, in 1991 and 1993, Haller and coworkers described a young Swedish man [9,10] with lifelong exercise intolerance, dyspnea, and episodes of myoglobinuria. This syndrome was called mitochondrial myopathy with succinate dehydrogenase and aconitase deficiency and attributed to altered metabolism of iron-sulfur (Fe-S) cluster proteins, which are the prosthetic groups in complexes I, II, and III, and in the Krebs cycle enzyme aconitase. Accordingly, muscle histochemistry showed SDH deficiency, and biochemical analysis showed deficiencies of complex II, complex III, and aconitase. In 2008, homozygosity mapping revealed a single pathogenic mutation in the *ISCU* gene in three Swedish families [11]. This is one of a few autosomal recessive mitochondrial myopathies with the single involvement of skeletal muscle.

CLINICAL PEARLS

- Sporadic mitochondrial myopathies are more often due to mtDNA mutations than to Mendelian heredity.
- Although mtDNA-related diseases are transmitted maternally, most sporadic mitochondrial myopathies contradict the first rule of mitochondrial genetics and often fool a diagnostic clue.
- Most sporadic mitochondrial myopathies are due to mutations in protein-coding genes (e.g., ND4 mutation, Case 1) but a few are due to genes involved in protein synthesis (e.g., $tRNA^{Met}$ [MTTP] mutation in Case 2).
- Myopathies due to mtDNA protein-coding genes deceive diagnostic criteria, such as high serum CK, altered EMG, mild/absent weakness, but a crucial diagnostic criterion is high blood lactic acid (e.g., Case 1).
- Maternally inherited, mitochondrial myopathies are usually parts of multisystem disorders.
- Mendelian mitochondrial myopathies are rarely muscle-specific but a few involve skeletal and cardiac muscles (and leukopenia), such as X-linked Barth disease.

REFERENCES

[1] Andreu AL, Tanji K, Bruno C, et al. Exercise intolerance due to a nonsense mutation in the mtDNA ND4 gene. Ann Neurol 1999;45:820–3.

[2] Peverelli L, Gold CA, Naini AB, et al. Mitochondrial myopathy with dystrophic features due to a novel mutation in the MTTM gene. Muscle Nerve 2014;50:292–5.

[3] Vissing J, Salamon MB, Arlien-Soborg P, et al. A new mitochondrial tRNAMet gene mutation in a patient with dystrophic muscle and exercise intolerance. Neurology 1998;50:1875–8.

[4] Hadjigeorgiou GM, Kim SH, Fischbeck KH, et al. A new mitochondrial DNA mutation (A3288G) in the TRNALeu(UUR) gene associated with familial myopathy. J Neurol Sci 1999;164:153–7.

[5] Kollberg G, Tulinius M, Gilljam T, et al. Cardiomyopathy and exercise intolerance in muscle glycogen storage disease 0. New Engl J Med 2007;357:1507–14.

[6] Andreu AL, Hanna MG, Reichmann H, et al. Exercise intolerance due to mutations in the cytochrome b gene of mitochondrial DNA. New Engl J Med 1999;341:1037–44.

[7] DiMauro S, Bonilla E, Mancuso M, et al. Mitochondrial myopathies. Basic Appl Myol 2003;13:145–55.

[8] Moraes CT, DiMauro S, Zeviani M, et al. Mitochondrial DNA deletions in progressive external ophthalmoplegia and Kearns-Sayre syndrome. New Engl J Med 1989;320:1293–9.

[9] Haller RG, Henriksson KG, Jorfeldt L, et al. Deficiency of skeletal muscle succinate dehydrogenase and aconitase. J Clin Invest 1991;88:1197–206.

[10] Hall RE, Henriksson KG, Lewis SF, Haller RG, Kennaway NG. Mitochondrial myopathy with succinate dehydrogenase and aconitase deficiency. Abnormalities of several iron-sulfur proteins. J Clin Invest 1993;92:2660–6.

[11] Mochel F, Knight MA, Tong W-H, et al. Splice mutation in the iron-sulfur cluster scaffold protein ISCU causes myopathy with exercise intolerance. Am J Hum Genet 2008;82:652–60.

Part II

Nuclear Encoded Diseases

Chapter 12

Pyruvate Dehydrogenase Complex Deficiency

Suzanne D. DeBrosse[1], Douglas S. Kerr[2]
[1]Departments of Genetics and Genome Sciences, Pediatrics, and Neurology, Case Western Reserve University School of Medicine, University Hospitals Case Medical Center, Cleveland, OH, USA;
[2]Departments of Pediatrics, Biochemistry, Nutrition and Pathology, Case Western Reserve University School of Medicine, University Hospitals Case Medical Center, Cleveland, OH, USA

CASE PRESENTATION

A 2-year-old girl presents with globally delayed developmental milestones and a seizure disorder. She had an unremarkable birth history, but she had been noted to be a moderately floppy infant. She appeared to be healthy in early infancy, although her pediatrician noted that her head circumference was tracking just below the 3rd percentile for age. She did not sit until 14 months and now crawls but does not yet walk. She babbles and says "mama" and "dada" nonspecifically. She is developing a mature pincer grasp. She had her first seizure, described as a generalized motor seizure, several months ago during a febrile illness, followed by a second seizure the previous week with focal features in the absence of fever. Her examination was notable for developmental delay, axial and appendicular hypotonia, preserved strength, and slightly brisk muscle stretch reflexes. Her weight, length, and systemic examination, including heart and abdomen, were normal.

Despite microcephaly and developmental delay, a magnetic resonance imaging (MRI) of the brain had been deferred because of parental reluctance to sedate the child, who was not a candidate for routine pediatric sedation because of her neurological issues. After the seizure with focal features and an electroencephalogram supporting a diagnosis of focal epilepsy, the MRI was pursued. The pediatric neurologist decided to draw basic metabolic laboratory studies first to decide if special studies, such as magnetic resonance spectroscopy (MRS), would be indicated. She ordered the following: basic chemistry, ammonia, lactate, plasma amino acids, plasma acylcarnitine profile with free and total carnitine, and urine organic acids. She did not order pyruvate because doing so typically necessitates the use of a special tube (with perchloric acid), not easily procured at the outpatient laboratory, and because it would have a longer turnaround time. In contrast, blood lactate measurement could be obtained the same day.

Mitochondrial Case Studies. http://dx.doi.org/10.1016/B978-0-12-800877-5.00012-7

The child's parents took their daughter to the outpatient laboratory. They were deliberately instructed not to fast the child, who had some pureed fruit about an hour before the blood draw. She became agitated and struggled during the blood draw. Later that day, the ordering physician received the results: Blood lactate was elevated at 6.0 mmol/L (reference range 1.00–2.40 mM/L), and ammonia was slightly elevated at 45 µmol/L (reference range < 25 µM/L). The modest elevation in ammonia was attributed to delay in specimen processing. The chemistry panel was normal except for a slightly low bicarbonate level (18 mM). The pediatric neurologist decided to wait for the rest of the laboratory results before deciding which labs to repeat.

Over the following week, the remainder of the laboratory results came back, and they were notable for elevated alanine (650 µM) in the plasma amino acid profile, which was otherwise normal, basically normal plasma acylcarnitine profile, and urine organic acids that included elevated levels of both lactate and pyruvate. The neurologist decided to repeat the blood lactate, this time also obtaining blood pyruvate for the lactate-to-pyruvate ratio. She noted that both lactate (4.6 mM, reference range 0.80–2.40 mM) and pyruvate (0.38 mM, reference range 0.03–0.12 mM) were elevated, as she suspected based on increased alanine (which comes from pyruvate) and the urine organic acid results. The ratio of lactate to pyruvate was normal at 12 (range 10–20). She decided to add an MRS to her MRI order.

Brain magnetic resonance imaging (MRI) without contrast and proton magnetic resonance spectroscopy (MRS) were performed. The MRI was abnormal, with moderate ventriculomegaly and a thin and slightly dysmorphic corpus callosum. The MRS was notable for a lactate peak in the basal ganglia. Repeat blood lactate and pyruvate levels were drawn under anesthesia, and both were elevated with a normal lactate-to-pyruvate ratio (L/P) ratio.

Based on these laboratory results and imaging findings, the physician ordered a pyruvate dehydrogenase complex (PDC) enzyme assay in blood lymphocytes. Both PDC complex activity and E3 activity, the two components of the test, were normal. Still suspecting pyruvate dehydrogenase complex deficiency or another disorder of pyruvate metabolism, she performed a skin biopsy and sent cultured fibroblasts for PDC and pyruvate carboxylase (PC) enzyme studies. This time, the E3 and PC activities were normal, but overall PDC activity was definitely decreased compared to the control (25% of the reference mean).

The child's parents had read online about the whole exome sequencing test and asked if this would be an appropriate test for their daughter. The neurologist consulted with a metabolic geneticist regarding the enzyme test results and was advised to perform focused gene testing first before ordering the exome test. A small panel of gene tests related to PDC deficiency was ordered, to start with *PDHA1* and reflex to other PDC genes if *PDHA1* was negative. The results returned with a previously reported pathogenic missense mutation in the *PDHA1* gene. When the biochemical geneticist was notified, he said this made sense given her discordant blood and fibroblast enzyme results. The patient's healthy mother was then tested for the mutation seen in her daughter and, to her surprise, was found to carry it.

DIFFERENTIAL DIAGNOSIS

By Clinical Presentation

There are many disorders that present with clinical features overlapping those seen in PDC deficiency. Developmental delay is a very frequent nonspecific finding, with hundreds of potential explanations, including genetic, metabolic, acquired, and multifactorial. Hypotonia (seen in the vast majority of children with PDC deficiency) can have both central and neuromuscular causes. The combination of delay and hypotonia may make the clinician consider chromosomal disorders (aneuploidies and microdeletion or microduplication syndromes), Prader-Willi syndrome (especially when combined with poor feeding), or even neuromuscular disorders such as spinal muscular atrophy. Agenesis or dysgenesis of the corpus callosum is also a nonspecific finding, and agenesis can be seen in *ARX*-related disorders, Aicardi syndrome, and many other genetic syndromes. Thinning of the corpus callosum can be seen in many disorders that result in decreased white matter volume. Microcephaly is not a consistent finding in PDC deficiency (approximately half of patients). The differential diagnosis of microcephaly also includes chromosomal disorders, primary microcephaly syndromes, fetal alcohol syndrome, and especially when combined with seizures and delay, Rett syndrome, Angelman syndrome, and similar disorders. Seizures, seen in about half of cases of PDC deficiency with highly variable type, severity, and treatment response, likewise have a broad differential diagnosis including metabolic, genetic, and acquired causes. The constellation of developmental delay, hypotonia, microcephaly, and seizures should prompt a search for an inborn error of metabolism. Leigh syndrome (lesions in the brain stem, basal ganglia, and/or cerebellum with corresponding neurological symptoms) can be seen in some patients with PDC deficiency or with a variety of mitochondrial disorders associated with either mitochondrial DNA mutations or nuclear DNA mutations. Older children with PDC deficiency may present with axonal neuropathy and intermittent ataxia mimicking Guillain-Barre syndrome.

By Laboratory Features

Lactic acidemia is a nonspecific finding that can be associated with severe illness associated with respiratory or circulatory insufficiency. If these are not part of the clinical scenario, this should prompt a search for an inborn error of metabolism, particularly if lactic acidemia is persistent and accompanied by neurological symptoms. It is true that the squirming child (exercise-induced oxygen deficit) may cause a temporary increase in blood lactate, but these are usually mild-to-moderate elevations and not usually reproducible under resting conditions. The presence of elevated blood alanine or pyruvate, or lactate or pyruvate in the urine, is a red flag that the lactic acidosis may be clinically significant. There is a separate differential diagnosis for elevated lactate based on whether pyruvate is also elevated, and consequently, whether the lactate-to-pyruvate ratio is elevated or not.

1. Differential diagnosis of increased lactate with a normal L/P ratio: disorders of gluconeogenesis (e.g., PC deficiency), other disorders of pyruvate oxidation (e.g., pyruvate transporter deficiency), and recovery from transient hypoxemia may also increase both lactate and pyruvate with a normal lactate-to-pyruvate ratio.
2. Differential diagnosis of increased lactate with an elevated ratio: systemic or focal hypoxemia or other disorders of oxidative metabolism, including mitochondrial disorders affecting the electron transport chain. These generally cause impaired oxidation of nicatinamide adenine dehydrogenase (NADH), shifting the equilibrium between lactate and pyruvate toward lactate, resulting in an increased lactate-to-pyruvate ratio. There are multiple disorders of the electron transport chain that can lead to a Leigh syndrome phenotype, similar to PDC deficiency.

DIAGNOSTIC APPROACH

Initial biochemical studies include lactate and pyruvate measurements, and lactate-to-pyruvate ratio, as well as plasma amino acid and urine organic acid profiles. Due to the metabolic block at the PDC, pyruvate increases. Since pyruvate is in equilibrium with lactate and NADH oxidation is not impaired, much of the pyruvate will be converted to lactate, but the ratio of lactate to pyruvate will remain normal, at 10–20. Blood lactate is sometimes normal in PDC deficiency, especially if the patient is fasting, so measurement after a carbohydrate-rich meal (in an individual not already on a carbohydrate-restricted diet) may be needed. (In a person already on a carbohydrate-restricted diet, the sudden reintroduction of a carbohydrate load is a metabolic stressor that should be avoided or approached with caution.) Some patients will have lactic acid elevations only seen in CSF or by MRS, although these lactate levels also can sometimes be normal in PDC deficiency. Plasma amino acid profile may reveal increased alanine, due to transamination of pyruvate to alanine. Urine organic acid analysis may reveal lactate and pyruvate elevations, and the absence of metabolite elevations that may characterize other inborn errors.

Once PDC deficiency is suspected, the next step is often measurement of PDC enzyme activity in blood lymphocytes or skin fibroblasts and/or sequencing of the *PDHA1* gene. Cultured skin fibroblasts for assay of PDC activity are sometimes preferred because they are more stable for shipping and can be used for repeated or additional testing. In females with the X-linked form of PDC deficiency, enzyme activities may vary between tissues, requiring analysis of more than one tissue type. Abnormally low PDC activity is diagnostic, and the test may distinguish abnormal function of the E3 portion of the enzyme complex (dihydrolipoamide dehydrogenase) from other PDC component dysfunction

(E1, E2, etc.). Molecular confirmation with genetic testing can then be sought, or if suspicion is very high, this step is sometimes pursued without performing enzyme analysis. Typically, testing begins with the *PDHA1* gene (sequencing and deletion/duplication testing to detect larger deletions that cannot be identified by sequencing), and then includes *PDHB*, *PHDX*, *DLAT*, *DLD*, and *PDP1*. The X-linked form (*PDHA1* gene) accounts for the majority of cases that can be molecularly confirmed, and these other genes encode catalytic and regulatory enzymes of the PDC complex. Defects in genes encoding other aspects of pyruvate oxidation are increasingly recognized as causes of PDC deficiency. With the discovery of these other genes encoding cofactors, regulators, and the mitochondrial pyruvate carrier, genetic testing options are increasing. However, not all individuals with biochemically confirmed PDC deficiency obtain a DNA diagnosis with the currently available sequencing of PDC-related genes, thus a negative result on genetic testing does not rule out PDC deficiency. Combined assaying of PDC activity and *PDHA1*sequencing should exclude almost all cases of PDC deficiency. A molecular diagnosis is useful not only for confirming the diagnosis, but for understanding the mode of inheritance. In about 25% of cases of PDC deficiency due to a *PDHA1* mutation, the mother of the affected child will be a carrier of this mutation despite typically being asymptomatic. Determining whether the mutation(s) is X-linked vs. autosomal recessive, and de novo versus inherited in the X-linked form, has major implications for recurrence risk.

TREATMENT STRATEGY

Several supplements have been proposed to treat patients with PDC deficiency. The most widely used is thiamine (vitamin B1), from which the cofactor thiamine pyrophosphate (TPP) is derived. TPP is a cofactor for the E1 catalytic enzyme, pyruvate dehydrogenase. Accordingly, it has been proposed that some cases of PDC deficiency may be more thiamine responsive than others, dependent on the specific location of the metabolic block or even the site of the mutation within the *PDHA1* gene, which encodes a subunit of E1. However, dramatically responsive cases appear to be rare and not well-clinically characterized. Doses of thiamine ranging from 25 to at least 2000 mg/day have been employed in patients with PDC deficiencies.

The ketogenic diet (low carbohydrate or nearly carbohydrate-free diet) is frequently employed, to bypass the metabolic block in carbohydrate metabolism. By significantly reducing glycolysis, production of pyruvate declines dramatically, and lactic acidemia improves. Fatty acid oxidation supplies acetyl-CoA to the TCA cycle and electron transport chain, improving ATP production, and increased circulating ketones (β-hydroxybutyrate, βOHB; and acetoacetate, AcAc) that are efficiently used by the brain as an alternative energy source. Use of the diet is not universal, and not all patients tolerate the diet. The ideal formulation of the ketogenic diet has not been established in clinical trials (and unlikely to be so), but a zebrafish model of PDC deficiency demonstrates benefit, comparative data from males with the same mutations indicate that strict reduction of carbohydrate is beneficial, and it is the most beneficial treatment by parental reports. The diet may be particularly helpful in patients without devastating prior brain damage, or with both PDC deficiency and epilepsy, in whom it may also help control otherwise intractable seizures. Monitoring of blood βOHB and AcAc is recommended. L-carnitine supplementation should be used in conjunction with ketogenic diets, which deplete free carnitine due to increased excretion of acylcarnitines.

Dichloroacetate (DCA) has been proposed as a potential treatment of PDC deficiency. It is an inhibitor of the major kinase that inactivates E1 (pyruvate dehydrogenase). Giving DCA may optimize residual E1 activity in patients with *PDHA1* or *PDHB* mutations, or may help provide substrate for those with mutations in downstream catalytic enzymes. However, clinical trials so far have not yet proven benefit in PDC deficiency, and this drug has been associated with peripheral neuropathy, at least in adults with MELAS. Phenylbutyrate has a similar effect by inhibiting pyruvate dehydrogenase kinases, but at this time, neither DCA nor phenylbutyrate have been proven effective and safe in clinical trials and are not approved for this use by the US FDA.

Supportive therapies are a mainstay of treatment, particularly antiepileptic drugs for seizures. While valproic acid may be relatively contraindicated/used with caution because of its effects on mitochondria, no specific drugs have yet been shown to be favorable in PDC deficiency, and a variety of seizure types observed in different affected individuals may necessitate different therapies for different patients.

LONG-TERM OUTCOME

Long-term outcome in PDC deficiency is extremely variable, and it generally correlates with age of symptom onset. Patients who present with the severe neonatal presentation of intractable lactic acidosis and accompanying respiratory failure frequently do not survive the neonatal period. Patients who present as toddlers with developmental delay, with or without microcephaly, seizures, and CNS malformations, frequently show a more static encephalopathy with intellectual disability. Those who manifest Leigh syndrome, mainly males, may have early death or a later onset, milder, intermittent course of lesions and symptoms. Patients who experience normal development prior to the development of peripheral neuropathy or ataxia in the school-age years often have normal cognition. Survival curves show the steepest decline in early infancy, owing to the severity of the neonatal presentation, and survival to age 4 years is a good predictor of long-term survival. Females have a higher rate of survival than males, but do so with greater average intellectual disability. This paradox is likely due to the high prevalence of X-linked disease, with females able to survive more deleterious mutations than males, but at the cost of increased disability. Some females may asymptomatically carry a *PDHA1* mutation, likely due to a favorable pattern of X-inactivation within the nervous system. There is great genotype–phenotype variability, and outcomes cannot be predicted by mutation alone.

PATHOPHYSIOLOGY/NEUROBIOLOGY OF DISEASE

PDC is a multiple enzyme complex that catalyzes the production of acetyl-CoA from pyruvate produced by glycolysis. PDC contains three catalytic enzymes, two regulatory enzymes, and a binding protein. It also requires the cofactors TPP, lipoic acid, and flavin adenine dinucleotide (FAD). The first enzyme of the PDC complex is pyruvate dehydrogenase (E1). In this step, one carbon is removed from pyruvate, and the resulting 2-carbon molecule is hydrohyethyl bound to TPP. The activity of E1 is regulated by phosphatases (activators) and kinases (inhibitors). Thus, E1 is the rate-limiting step of PDC. The next step in converting pyruvate to acetyl-CoA is catalyzed by dihydrolipoamide S-acetyltransferase (E2). E2 transfers the hydroxyethyl group from TPP to an oxidized form of covalently bound lipoamide, and the resulting acetyl group is then transferred to free coenzyme A to form acetyl-CoA and reduced dihydrolipoamide-E2. Finally, the flavoprotein dihydrolipoamide dehydrogenase (E3) re-oxidizes the lipoyl group of dihydrolipoamide-E2 to form lipoamide-E2 and NADH. In the process, FAD bound to E3 accepts the electrons (reduction to $FADH_2$) then donates them to NAD+, producing NADH + H^+ (Figure 1). E3 is

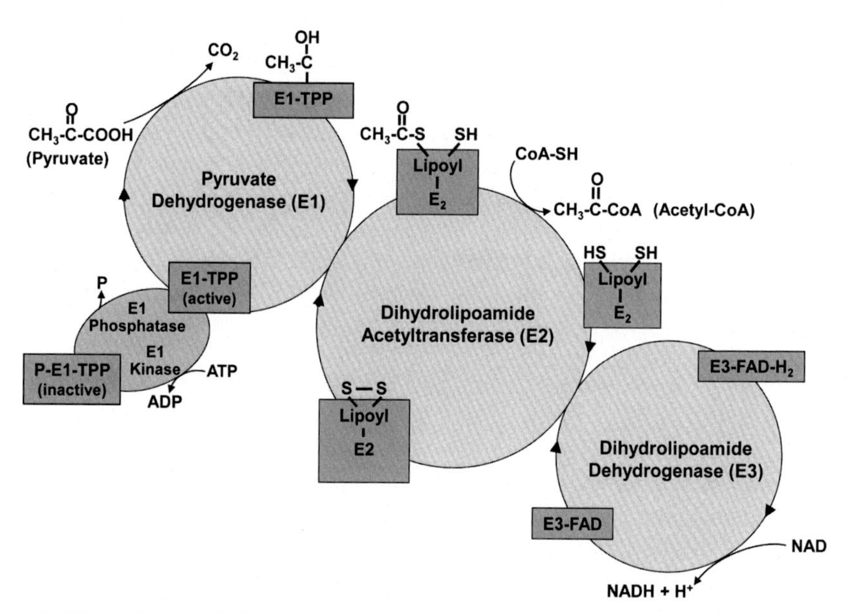

FIGURE 1 The overall biochemical reactions of the pyruvate dehydrogenase complex, including its regulation by phosphorylation/dephosphorylation [3]. *With kind permission from Springer Science and Business Media.*

shared by PDC as well as alpha-ketoglutarate and branched-chain 2-ketoacid dehydrogenase complexes and the glycine cleavage enzyme. The E3 binding protein, which is specific to PDC, binds E3 to E2 in the core of PDC. There are six genes that encode components of this pathway (*PDHA1, PDHB, DLD, DLAT, PDHX, PDP1*) that are associated with PDC deficiencies. Mutations in one of these genes are responsible for about three-fourths of all mutations identified in persons with PDC deficiency, with mutations in *PDHA1* the single most common genetic cause. PDC deficiency associated with these genes is inherited in an X-linked manner (*PDHA1*) or an autosomal recessive manner. As stated above, it is increasingly recognized that mutations in other genes related to this pathway can also be implicated in PDC deficiency.

EFFECTS ON BRAIN PATHOLOGY PRE- AND POSTNATALLY

Brain energy metabolism is highly dependent on glycolysis and subsequent pyruvate oxidation. The resulting energy adenine triphosphate (ATP) deficit can injure tissue, particularly that with a high energy demand, such as the basal ganglia. Lactic acidosis may further impair energy metabolism and the synthesis of proteins, lipids, and neurotransmitters in the brain. As in other disorders with impaired ATP production, there may be excitotoxicity due to reduced uptake of glutamate by astrocytes, and excessive glutamate and free radicals may harm oligodendrocytes.

The resulting brain abnormalities that may be seen with PDC deficiency are highly variable. These include both congenital malformations, postulated to be related to damage to proliferating and migrating cells, and destructive lesions. Congenital malformations include agenesis or dysgenesis of the corpus callosum, colpocephaly, subependymal cysts, and other defects of neuronal migration (e.g., ectopia or dysplasia of brain stem or cerebellum, heterotopia, and pachygyria). Destructive lesions include widespread injury to white matter, which may result in cystic lesions. Leigh syndrome (subacute necrotizing encephalopathy, including radiographically visualized lesions of the basal ganglia, thalamus, and brain stem) is sometimes observed. Cerebral atrophy and/or ventriculomegaly is frequently seen, with frank hydrocephalus an occasional finding.

CLINICAL PEARLS

- There are a variety of presentations, which tend to correlate with the age at symptom onset.
- The lactate-to-pyruvate ratio is typically normal in PDC deficiency, due to the increase in both lactate and pyruvate, which contrasts with defects of the mitochondrial respiratory chain in which this ratio is elevated.
- A combination of PDC enzyme testing (skin fibroblasts, blood lymphocytes, or muscle) and PDHA1 gene testing can identify most cases of PDC deficiency.

- Even the X-linked form of PDC deficiency can affect females, who sometimes are paradoxically more impaired than males due to their ability to survive more severe mutations compared to males. Females may also carry the X-linked form asymptomatically.
- New genetic causes are still being discovered. These genes encode proteins not part of the PDC complex itself, but that are important for proper functioning of the complex.

FURTHER READING

[1] Barnerias C, Saudubray JM, Touati G, et al. Pyruvate dehydrogenase complex deficiency: four neurological phenotypes with differing pathogenesis. Dev Med Child Neurol 2010;52:e1–9.

[2] DeBrosse S, Okajima K, Schmotzer C, Frohnapfel M, Kerr DS. Spectrum of neurological outcomes in pyruvate dehydrogenase complex deficiencies. Mol Genet Metab November 2012;107(3):394–402.

[3] DeBrosse SD, Kerr DS. Pyruvate dehydrogenase complex deficiencies. In: Wong L-JC, editor. Mitochondrial disorders caused by nuclear genes. New York: Springer Science+Business Media, LLC; 2013. p. 301–17.

[4] De Meirleir L, Brivet M, Garcia-Cazorla A. Disorders of pyruvate metabolism and the tricarboxylic acid cycle. In: Fernandes J, Saudubray J-M, van den Berg G, Walter JH, editors. Inborn metabolic diseases: diagnosis and treatment. 5th ed. Heidelberg: Springer-Verlag; 2012. p. 187–200.

[5] Giribaldi G, Doria-Lamba L, Biancheri R, Severino M, Rossi A, Santorelli FM, et al. Intermittent-relapsing pyruvate dehydrogenase complex deficiency: a case with clinical, biochemical, and neuroradiological reversibility. Dev Med Child Neurol May 2012;54(5):472–6.

[6] Imbard A, Boutron A, Vequaud C, et al. Molecular characterization of 82 patients with pyruvate dehydrogenase complex deficiency. Structural implications of novel amino acid substitutions in E1 protein. Mol Genet Metab 2011;104:507–16.

[7] Lissens W, De Meirleir L, Seneca S, et al. Mutations in the X-linked pyruvate dehydrogenase (E1) alpha subunit gene (PDHA1) in patients with a pyruvate dehydrogenase complex deficiency. Hum Mutat 2000;15:209–19.

[8] Robinson BH, MacMillan H, Petrova-Benedict R, Sherwood WG. Variable clinical presentation in patients with defective E1 component of pyruvate dehydrogenase complex. J Pediatr 1987;111:525–33.

[9] Patel KP, O'Brien TW, Subramony SH, Shuster J, Stacpoole PW. The spectrum of pyruvate dehydrogenase complex deficiency: clinical, biochemical and genetic features in 371 patients. Mol Genet Metab July 2012;106(3):385–94.

[10] Soares-Fernandes JP, Teixeira-Gomes R, Cruz R, et al. Neonatal pyruvate dehydrogenase deficiency due to a R302H mutation in the PDHA1 gene: MRI findings. Pediatr Radiol 2008;38:559–62.

[11] Sperl W, Fleuren L, Freisinger P, et al. The spectrum of pyruvate oxidation defects in the diagnosis of mitochondrial disorders. J Inherit Metab Dis December 20, 2014. [E-pub ahead of print, DOI 10.1007/s10545-014-9787-3].

[12] Wexler ID, Hemalatha SG, McConnell J, et al. Outcome of pyruvate dehydrogenase deficiency treated with ketogenic diets. Studies in patients with identical mutations. Neurology 1997;49:1655–61.

Chapter 13

Friedreich Ataxia

Mary Kay Koenig
University of Texas Medical School at Houston, Department of Pediatrics, Division of Child and Adolescent Neurology, Endowed Chair of Mitochondrial Medicine, Houston, TX, USA

CASE PRESENTATION

A 7-year-old boy was seen for evaluation of tripping and falling. His mother had noted worsening clumsiness for the past two years along with progressive loss of fine motor control, changes in his voice, and difficulty chewing. Medical history revealed that he was born at term following an uncomplicated pregnancy. There were no concerns until 5 years of age when he began falling.

The case had a healthy 8-year-old paternal half-brother with attention-deficit disorder and a healthy 1-year-old maternal half-sister. His father died at the age of 25 years from advanced cardiac disease of unknown etiology. There was a paternal cousin who died from unknown causes at 5 years of age. The child's mother had hypothyroidism, and his maternal grandmother had childhood onset hearing loss.

The patient lives at home with his mother and younger sister and is able to perform his own activities of daily living. He attends regular first grade classes and has no difficulty with the course work.

On examination, the patient was awake, alert, and oriented to person and place. He was small for his age (height and weight <10% for age). Digital clubbing was present on both hands. His speech was fluent but moderately slurred. His cognition appeared intact. Cranial nerves were grossly intact. Sensory examination was unremarkable with normal response to fine touch, proprioception, pain, and temperature. Motor examination demonstrated normal and equal muscle bulk, tone, and strength. His muscle stretch reflexes were absent in the upper and lower extremities. Toes were up-going bilaterally. The patient was steady when sitting; however, he was unable to perform finger-nose-finger or rapid, alternating movements. Upon rising from the chair, he became ataxic, requiring assistance to maintain his balance. Once secure in a standing position, he was able to stand independently with a wide stance. While his gait was unsteady, he was able to walk independently without falling.

Testing of the *FXN* gene demonstrated homozygous GAA repeat expansion within the first intron with >1000 repeats on each allele. The disease progressed with the child becoming wheelchair bound by the age of 9 years and developing

Mitochondrial Case Studies. http://dx.doi.org/10.1016/B978-0-12-800877-5.00013-9

concentric biventricular hypertrophic cardiomyopathy. He remains alive at 10 years of age with aggressive medical management.

DIFFERENTIAL DIAGNOSIS AND DIAGNOSTIC APPROACH

The differential diagnosis of a child presenting at 5 years of age with progressive ataxia is broad, and clinical features for many disorders overlap. Although Friedreich ataxia (FRDA) is the most common cause of hereditary ataxia in the Caucasian population, accounting for up to half of all genetic ataxia and 75% of inherited ataxia in individuals under the age of 25 years, its prevalence is still less than 1:30,000 [8,10,25,34]. Considerations include ataxia with vitamin E deficiency, ataxia with CoQ10 deficiency, abetalipoproteinemia, Refsum's disease, late-onset GM2 gangliosidosis, congenital disorder of glycosylation type 1a, cerebrotendinous xanthomatosis, Wilson disease, ataxia with apraxia, ataxia telangiectasia, ataxia due to mitochondrial DNA mutations, hereditary motor and sensory neuropathy, hereditary spastic paraparesis, spastic ataxia of Charlevoix-Saguenay, Marinesco-Sjögren syndrome, the spinocerebellar ataxias, giant axonal neuropathy, Charcot-Marie tooth, and CACNA1A-related disorders [8,10,11,34,35]. Treatable conditions such as vitamin E deficiency are especially crucial to diagnose as early treatment can improve the long-term outcome. Determination of inheritance patterns, age of onset, imaging findings, and comorbid features such as involvement of the eyes, heart, gastrointestinal tract, and cerebellum will help guide the diagnostic evaluation [8]. Although magnetic resonance imaging studies can show atrophy to the posterior columns of the cervical spinal cord and medulla, significant cerebellar atrophy is unusual in FRDA, and its presence should lead to other considerations [8,25,34].

CLINICAL PRESENTATION

Nikolaus Friedreich first described FRDA in a series of five papers published from 1863 through 1877. Dr Friedreich described six patients in two families with adolescent onset of poor balance, leg weakness, decreased ambulation, impaired coordination, dysarthria, nystagmus, impaired sensation, kyphoscoliosis, foot deformity, and fatty degeneration of the heart [10,13–17,26,37]. He recognized that the condition arose from degeneration of the posterior columns of the spinal cord [37].

Over the next 100 years, the clinical spectrum of FRDA expanded [10]. Onset is typically in the first or second decade of life with slow progression [8,10,34,45], although early onset has been described [35]. Initial symptoms are gait ataxia that progresses to truncal ataxia with impairment of limb coordination, titubation, dysarthria, dysphagia, eye movement abnormalities, areflexia, extensor plantar response, sensory loss, and weakness of the legs, feet, hands, and arms [8,10,25,29,34,35,45]. Ambulation is typically lost within 10–15 years of symptom onset [8,10,34].

Common comorbid features include the following:

- Central sleep apnea [8]
- Depression [34]
- Executive dysfunction: Cognition is spared, although speech can become ataxic and patients develop executive dysfunction with impaired verbal fluency, attention, and working memory. Information processing speed is reduced as are visuoconstructive, visuoperceptive, and visuospatial reasoning. [8,10,25,34]
- Ophthalmologic dysfunction: Square wave jerks with fixation, slow extraocular movements, prolonged saccadic latency, horizontal nystagmus, visual field restriction, and optic atrophy with decreased visual acuity [8,10,18,25,34,36]
- Hearing loss along with impaired auditory processing resulting in poor perception of speech and auditory signals [8,10,18,25,34,36]
- Posterior column sensory neuropathy with impaired vibratory sense and proprioception [8,10,25,34]
- Kyphoscoliosis [8,10,29,34,35,45]
- Hypertrophic cardiomyopathy, concentric cardiomyopathy, or dilated cardiomyopathy. EKG may show T-wave inversion, left-axis deviation, or repolarization abnormalities [4,8,10,25,29,35,45]
- Glucose intolerance or diabetes mellitus [8,10,25,34–36,45]
- Urinary incontinence [8,10,34]
- Foot deformities: pes cavus or talipes equinovarus [8,10,29,34,35]
- Cold, cyanosed distal lower limbs due to autonomic dysfunction [8,10,34]

Survival is typically three to four decades with an average age of death between 30 and 40 years [8,10]. The most common cause of death is congestive heart failure and/or cardiac arrhythmia [4,8,10,34,45]. Other causes of death include aspiration pneumonia, diabetes mellitus, myocardial infarction, and bronchopneumonia [8,34].

PATHOPHYSIOLOGY

The first clinical descriptions of FRDA by Friedreich appeared in a series of articles published between 1863 and 1877 [13–17]. Friedreich's descriptions detailed the clinical features and basic pathophysiology of posterior spinal cord degeneration [10,13–17]. Our understanding of both the clinical and pathophysiologic features of FRDA has expanded tremendously in the last century. In 1907, Mott expanded on Friedreich's neuropathologic description, and in 1957, Urich emphasized the existence of suprasegmental lesions [32]. In 1980, Lamarche discovered iron granules within the cardiomyocytes of FRDA patients [27]. A major breakthrough came in 1996 when Campuzano et al. [7] identified the genetic defect in FRDA and recognized its role in iron metabolism [7]. In 1997, Rotig discovered the iron-sulfur protein deficiency affecting complexes I, II, and III of the electron transport chain along with reduced activity of aconitase in endocardial biopsies of FRDA patients [52]. Currently, iron–sulfur cluster deficiency is the accepted critical factor in the pathogenesis of FRDA [25].

FRDA affects both the central and peripheral nervous systems along with the heart, the skeleton, and the endocrine pancreas [8,19,20,25,29]. Within the central and peripheral nervous systems, there is a selective vulnerability of specific types of motor and sensory neurons [8]. Early in the disease, efferent components of the large, myelinated sensory neurons of the dorsal columns degenerate, producing an axonal neuropathy with impairment of proprioceptive, pressure, and vibratory senses [8,20,29,34]. Other sensory neurons including auditory and visual neurons may also degenerate along with the motor neurons within the lateral corticospinal tracts, producing the upper motor neuron weakness seen in most patients [8,29]. This combined atrophy of the dorsal columns and the corticospinal tracts is characteristic of FRDA [8]. Atrophy of the dentate nucleus of the cerebellum may also produce a mild cerebellar component to the disease [8,20,29].

Iron deposits within the myocardial cells of the heart produce hypertrophy of the individual cells, particularly within the left ventricular wall and septum, leading to progressive replacement by connective tissue and dilation of the heart with congestive heart failure [8,29].

A major breakthrough in our understanding of the pathophysiology of FRDA came in 1996 when Campuzano et al. [7] discovered the unstable GAA expansion in intron 1 of the *X25* gene located on chromosome 9q13 encoding the protein frataxin [7]. The majority of FRDA patients (>96%) are homozygous for this expansion [3,7,8,10,29,34,36] with 1–4% of cases being compound heterozygotes with a single GAA expansion and a point mutation or deletion on their other allele [8,10,25,31,34]. The frataxin gene (*X25*) contains nine exons that encode a 210 amino acid mitochondrial matrix protein [8,20]. The GAA triplet repeat within the first intron of this gene is polymorphic with normal alleles containing eight to 38 repeats. The critical threshold for pathogenicity is approximately 90 repeats with FRDA patients typically displaying between 70 and 1500 repeats on each allele [8,29,34]. Transmission of the repeat is unstable, resulting in either contraction or expansion when passed from a parent to a child [8]. Although the mechanisms underlying the instability of this repeat are not clear, problems with DNA replication, recombination events, and aberrant DNA repair have all been postulated [8,31].

The size of the smaller repeat correlates positively with the presence and number of clinical features and negatively with age at onset; however, repeat size cannot be used accurately to predict prognosis in an individual patient [3,10,29,34,36,48]. The presence of the repeat expansion results in reduced quantity of structurally and functionally normal frataxin protein (10–25% of normal) with a higher number of repeats producing less functional protein [7,8,19,20,51]. In individuals with expanded alleles of different sizes, the smaller allele determines the amount of residual frataxin production and the allele associated with the age of onset of disease [8]. Although they produce significantly less frataxin than normal controls, heterozygous carriers of a single repeat expansion are typically not affected [8,34].

Although the exact function of frataxin remains unknown, it is known to be vital for cellular function and to have a role in iron metabolism [3,8,31]. Frataxin is important in the biogenesis of iron–sulfur clusters that are necessary for heme synthesis, function of complexes I, II, and III of the electron transport chain, and function of the Krebs cycle enzyme aconitase [7,8,19,20,29,31,36,46]. Deficiency of frataxin not only disrupts the Krebs cycle and oxidative phosphorylation but also disturbs iron homeostasis. Excess iron enters the mitochondria and reacts with hydrogen peroxide to form insoluble iron [8,20,49]. This iron precipitates generating hydroxyl radicals and produces increased oxidative stress. The dysfunctional electron transport chain complexes also allow electrons to leak across the inner mitochondrial membrane. These electrons both uncouple the electron transport chain and react with oxygen to form hydrogen peroxide, generating even more hydroxyl radicals. Free radicals damage intracellular proteins, lipids, and DNA, stimulate stress response cell signaling pathways, disrupt cellular cytoskeleton structure, impair axonal transport, and produce cell injury and death through apoptosis [8,19,20,35,36,50–52].

TREATMENT

Despite extensive ongoing research, there remains no disease-modifying therapy or cure for FRDA [8,45]. Current treatment options are supportive. Institution of aggressive symptomatic medical therapy early in the course of the disease, along with screening for known comorbid conditions, may improve the quality and quantity of life for individual patients [4].

Although no definitive treatment has emerged, our improved understanding of disease mechanisms and identification of candidate drugs based on pathophysiology of the genetic defect has moved ahead steadily since the identification of the *FXN* gene in 1996 [36,45].

There are three biochemical defects that guide most current approaches to therapeutic development:

- Increase cellular frataxin
- Reduce mitochondrial iron accumulation
- Reduce cellular oxidative stress [45]

Increase Cellular Frataxin

Studies have established that the repeats on exon 1 of the *FXN* gene reduce its transcription with a concomitant decrease in frataxin protein production [21]. The size of the expansion correlates inversely with age of onset and directly with rate of disease progression [45]. As the repeats do not alter the FXN protein structure, one therapeutic target is to increase the transcription of the defective *FXN* genes [8,21,45].

Candidate drugs showing in vitro success include resveratrol, PPAR gamma agonists, and interferon gamma [8,45]. High-throughput drug screens using a frataxin reporter system to identify compounds that increase frataxin expression are also in progress [45].

Fusion of frataxin with the protein transduction domain of the TAT protein has successfully delivered frataxin to mitochondria in cultured cells from both affected humans and in a mouse model. Studies have found that this process can improve the lifespan in mouse models of FRDA [45].

One mechanism by which GAA repeat expansions decrease the expression of *FXN* is through hydroacetylation of histones, leading to the possibility that histone deacetylase inhibitors might increase frataxin expression. This approach was pioneered by Gottesfeld et al., who demonstrated the ability of histone deacetylase inhibitors to counteract the chromatin-condensing effect of the GAA repeat expansions, thus increasing frataxin expression in human neuronal cells derived from patient-induced pluripotent stem cells and in mouse models of FRDA [8,21,31,45].

Human recombinant erythropoietin is also under investigation secondary to its ability to stabilize the frataxin transcript and increase protein levels in cell lines. An open label study demonstrated an improvement of ataxia scores with reduced oxidative stress markers and increased frataxin protein levels in leukocytes of patients [8,29,45].

Reduce Mitochondrial Iron Accumulation

Frataxin localizes to the mitochondrial matrix where it binds iron and is involved in iron-sulfur cluster assembly. FRDA leads to an increased mitochondrial iron uptake and disturbed iron homeostasis [33]. Deferiprone is an iron chelator that effectively shuttles iron between subcellular compartments. It can also transfer iron from iron-overloaded cells to extracellular apotransferrin and pre-erythroid cells [33]. In FRDA mice, deferiprone increased mitochondrial membrane redox potential, ATP production, and resistance to cellular apoptosis [45]. Clinical studies on the efficacy of deferiprone suggest a beneficial effect of mild iron chelation with low doses of deferiprone on cardiac function, whereas higher doses produce worsening of the condition [33,45]. However, deferiprone carries a significant risk of agranulocytosis, and further studies are needed to assess if ataxia can be improved [33].

Reduce Cellular Oxidative Stress

Frataxin-deficient cells produce more free radicals and are more sensitive to oxidative damage suggesting that antioxidants should help [29,36]. Multiple studies involving the use either alone or in combination of vitamin E, coenzyme Q10, and Idebenone have been undertaken. Although several studies have shown possible benefit, to date there are no consistent, statistically significant benefits in the use of antioxidant therapy in patients with symptomatic FRDA [1,2,5,6,8,9,12,22–24,29,30,32,33,38–42,43–47,49]. In addition to the previously studies agents, other antioxidants are in preclinical and early clinical development, though little has been published regarding efficacy of these agents (EPI-A0001, EPI-743, EgB-761, OX1, mitochondrial radical quenchers, and deuterated polyunsaturated fatty acids) [8,45]. The lack of clinical evidence for efficacy does not exclude these agents as potential therapeutic targets. As FRDA is a very slowly progressive disease, minimal changes over short time periods are difficult to capture. Clinical studies need to be long, and clinical assessment tools need to be defined. More sensitive biomarkers that monitor smaller changes would also improve ability to detect significant changes in shorter time frames [35,36].

CLINICAL PEARLS

- FRDA should be suspected in any adolescent presenting with progressive ataxia.
- FRDA is caused by a triplet repeat expansion on the *FXN (X25)* gene and will be missed on whole exome sequencing. Directed analysis is required.
- Monitoring for comorbid conditions is crucial to improve outcomes.
- Improved understanding of the pathophysiology of FRDA is leading to abundant research toward development of treatments.

REFERENCES

[1] Arnold P, Boulat O, Mairec R, Kuntzer T. Expanding view of phenotype and oxidative stress in Friedreich's ataxia patients with and without idebenone. Schweizer Archiv Neurol Psychiatr 2006;157:169–76.

[2] Artuch R, Aracil A, Mas A, Colome C, Rissech M, Monros E, et al. Friedreich'as ataxia: idebenone treatment in early stage patients. Neuropediatrics 2002;33:190–3.

[3] Bevans-Galea MV, Lockhart PJ, Galea CA, Hannan AJ, Delatycki M. Beyond loss of frataxin: the complex molecular pathology of Friedreich ataxia. Discov Med 2014;17:25–35.

[4] Bourke T, Keane D. Friedreich's ataxia: a review from a cardiology perspective. Irish J Med Sci 2011;180:799–805.

[5] Brandsema JF, Stephens D, Hartley J, Yoon G. Intermediate-dose idebenone and quality of life in Friedreich ataxia. Pediatr Neurol 2010;42:338–42.

[6] Buyse G, Mertens L, Di Salvo G, Matthijs I, Weidemann F, Eyskens B, et al. Idebenone treatment in Friedreich's ataxia: neurological, cardiac, and biochemical monitoring. Neurology 2002;60:1679–81.

[7] Campuzano V, Montermini L, Lutz Y, Cova L, Hindelang C, Jiralerspong S, et al. Frataxin is reduced in Friedreich ataxia patients and is associated with mitochondrial membranes. Hum Mol Genet 1997;6:1771–80.

[8] Collins A. Clinical Neurogenetics friedreich ataxia. Neurol Clin 2013;31:1095–120.

[9] Cooper JM, Korlipara LVP, Hart PE, Bradley JL, Schapira AHV. Coenzyme Q10 and vitamin E deficiency in Friedreich's ataxia: predictor of efficacy of vitamin E and coenzyme Q10 therapy. Eur J Neurol 2008;15:1371–9.

[10] Delatycki MB, Corben LA. Clinical features of Friedreich ataxia. J Child Neurol 2012;27:1133–7.

[11] Delatycki M, Corben L, Pandolfo M, Lynch D, Schulz J. Consensus clinical management guidelines for Friedreich's ataxia. November 2014. www.curefa.org.

[12] Di Prospero NA, Sumner CJ, Penzak SR, Ravina B, Fischbeck KH, Taylor JP. Safety, tolerability, and pharmacokinetics of high-dose idebenone in patients with Friedreich ataxia. Arch Neurol 2007;64:803–8.

[13] Friedreich N. Ueber degenerative Atrophie der spinalen Hinterstrange. Virchows Arch Pathol Anat Physiol Klin Med 1963;26:391–419.

[14] Friedreich N. Ueber degenerative Atrophie der spinalen Hinterstrange. Virchows Arch Pathol Anat Physiol Klin Med 1963;26:433–59.

[15] Friedreich N. Ueber degenerative Atrophie der spinalen Hinterstrange. Virchows Arch Pathol Anat Physiol Klin Med 1963;27:1–26.

[16] Friedreich N. Ueber Ataxie mit besonderer Berucksichtingung der hereditaren Formen. Virchows Arch Pathol Anat Physiol Klin Med 1876;68:145–245.

[17] Friedreich N. Ueber Ataxie mit besonderer Berucksichtigung der hereditaren Formen. Virchows Arch Pathol Anat Physiol Klin Med 1877;70:140–52.

[18] Geoffroy G, Barbeau A, Breton G, Lemieux B, Aube M, Leger C, et al. Clinical description and roentgenologic evaluation of patients with Friedreich's ataxia. Can J Neurol Sci 1976;3:279–86.

[19] Gomes CM, Snatos R. Neurodegeneration in Friedreich's ataxia: from defective frataxin to oxidative stress. Oxid Med Cell Longev 2013.

[20] Gonzalez-Cabo P, Palau F. Mitochondrial pathophysiology in Friedreich's ataxia. J Neurochem 2013;126:53–64.

[21] Gottesfeld JM, Rusche JR, Pandolfo M. Increasing frataxin gene expression with histone deacetylase inhibitors as a therapeutic approach for Friedreich's ataxia. J Neurochem 2013;126:147–54.

[22] Hart PE, Lodi R, Rajagopalan, et al. Antioxidant treatment of patients with Friedreich ataxia: four-year follow-up. Archiv Neurol 2005;62:621–6.

[23] Hausse AO, Aggoun Y, Bonnet D, Sidi D, Munnich A, Rotig A, et al. Idebenone and reduced cardiac hypertrophy in Friedreich's ataxia. Heart 2002;87:346–9.

[24] Kearney M, Orrell RW, Fahey M, Pandolfo M. Antioxidants and other pharmacological treatment for Friedreich ataxia (Review). Cochrane Collab 2012. Issue 4.

[25] Koeppen AH. Friedreich's ataxia: pathology, pathogenesis, and molecular genetics. J Neurol Sci 2011;303:1–12.

[26] Koeppen AH. Nikolaus Friedreich and degenerative atrophy of the dorsal columns of the spinal cord. J Neurochem 2013;126:4–10.

[27] Lamarche JB, Cote M, Lemieux B. the cardiomyopathy of Friedreich's ataxia. Morphological observations in 3 cases. Can J Neurol Sci 1980;7:389–96.

[28] Lodi R, Hart PE, Rajagoalan B, et al. Antioxidant treatment improves in vivo cardiac and skeletal muscle bioenergetics in patients with Friedreich's ataxia. Ann Neurol 2001;49:590–6.

[29] Pandolfo M. Friedreich ataxia: the clinical picture. J Neurol 2009;256:S3–8.

[30] Mariotti C, Solari A, Torta D, Marano L, Fiorentini C, Di Donato S. Idebenone treatment in Friedreich patients: one-year-long randomized placebo-controlled trial. Neurology 2002;60:1676–9.

[31] Meier T, Buyse G. Idebenone: an emerging therapy for Friedreich ataxia. J Neurol 2009;256(Suppl. 1):25–30.
[32] Mott FW. Case of Friedreich's disease, with autopsy and systematic microscopical examination of the nervous system. Arch Neurol Psychiat Lond 1907;3:180–200.
[33] Pandolfo M, Hausmann L. Deferiprone for the treatment of Friedreich's ataxia. J Neurochem 2013;126:142–6.
[34] Parkinson MH, Boesch S, Nachbauer W, Mariotti C, Giunti P. Clinical features of Friedreich's ataxia: classical and atypical phenotypes. J Neurochem 2013;126:103–77.
[35] Parkinson MH, Schulz JB, Giunti P. Co-enzyme Q10 and idebenone use in Friedreich's ataxia. J Neurochem 2013;126:125–41.
[36] Perlman SL. A review of Friedreich ataxia clinical trial results. J Child Neurol 2012;27: 1217–22.
[37] Pineda M, Arpa J, Montero R, et al. Idebenone treatment in paediatric and adult patients with Friedreich ataxia: long-term follow-up. Eur J Paediatr Neurol 2008;12:470–5.
[38] Ribai P, Pousset F, Tanguy ML, et al. Neurological, cardiological, and oculomotor progression in 104 patients with Friedreich ataxia during long-term follow-up. Archiv Neurol 2007;64:558–64.
[39] Rustin P, Bonnet D, Rotig A, Munnich A, Sidi D. Idebenone treatment in Friedreich patients: one-year-long randomized placebo-controlled trial (letter). Neurology 2004;62:524–5.
[40] Rustin P, Rotig A, Munnich A, Sidi D. Heart hypertrophy and function are improved by idenbenone in Friedreich's ataxia. Free Radic Res 2002;36:477–9.
[41] Rustin P, von Kleist-Retzow J-C, Chantrel-Groussard K, Sidi D, Munnich A, R€otig A. Effect of idebenone on cardiomyopathy in Friedreich's ataxia: a preliminary study. Lancet 1999;354:477–9.
[42] Schols L, Vorgerd M, Schillings M, Skipka G, Zange J. Idebenone in patients with Friedreich ataxia. Neurosci. Lett 2001;306:169–72.
[43] Schulz JB, Dehmer T, Sch€ols L, et al. Oxidative stress in patients with Friedreich ataxia. Neurology 2000;55:1719–21.
[44] Velasco-Sanchez D, Aracil A, Montero R, et al. Combined therapy with idebenone and deferiprone in patients with Friedreich's ataxia. Cerebellum 2011;10:1–8.
[45] Wilson RB. Therapeutic developments in Friedreich ataxia. J Child Neurol 2012;27: 1212–6.
[46] Rotig A, de Lonlay P, Chretien D, Foury F, Koenig M, Sidi D, Munnich A, Rustin P. Aconitase and mitochondrial iron-sulphur protein deficiency in Friedreich ataxia. Nat Genet 1997;17:215–7.
[47] Lodi R, Cooper JM, Bradley JL, Manners D, Styles P, Taylor DJ, Schapira AHV. Deficit of in vivo mitochondrial ATP production in patients with Friedreich ataxia. Proceedings of the National Academy of Science 1999;96:11492–5.
[48] Bradley JL, Blake JC, Chamberlain S, Thomas PK, Cooper JM, Schapira AHV. Clinical, biochemical and molecular genetic correlations in Friedreich's ataxia. Hum Mol Genet 2000;9:275–82.
[49] Delatycki MB, Camakaris J, Brooks H, Evans-Whipp T, Thorburn DR, Williamson R, Forrest SM. Direct evidence that mitochondrial iron accumulation occurs in Friedreich ataxia. Ann Neurol 1999;45:673–5.
[50] Piemonte F, Pastore A, Tozzi G, Tagliacozzi D, Santorelli FM, Carrozzo R, Casali C, Damiano M, Federici G, Bertini E. Glutathione in blood of patients with Friedreich's ataxia. Eur J Clin Invest 2001;31:1007–11.

[51] Houshmand M, Panahi MS, Nafisi S, Soltanzadeh A, Alkandari FM. Identification and sizing of GAA trinucleotide repeat expansion, investigation for D-loop variations and mitochondrial deletion in Iranian patients with Friedreich's ataxia. Mitochondrion 2006;6:82–8.
[52] Heidari HM, Houshmand M, Hosseinkhani S, Nafissi S, Khatami M. Complex I and ATP content deficiency in lymphocytes from Friedreich's ataxia. Can J Neurol Sci 2009;36:26–31.

FURTHER READING

[1] Consensus clinical management guidelines for Friedreich's ataxia. www.curefa.org/clinical-care-guidelines.
[2] Kearney M, Orrell RW, Fahey M, Pandolfo M. Antioxidants and other pharmacological treatment for Friedreich ataxia (Review). Cochrane Collab 2012. Issue 4.

Chapter 14

Nuclear Genetic Causes of Leigh and Leigh-Like Syndrome

James E. Davison[1], Shamima Rahman[1,2]
[1]*Metabolic Medicine, Great Ormond Street Hospital for Children NHS Foundation Trust, London, UK;* [2]*Mitochondrial Research Group, Genetics and Genomic Medicine, UCL Institute of Child Health, London, UK*

INTRODUCTION

Leigh syndrome and Leigh-like syndrome are mitochondrial encephalopathies with widely heterogeneous genetic etiologies that usually arise from cerebral mitochondrial respiratory chain enzyme (RCE) dysfunction and subsequent adenosine triphosphate (ATP) depletion. Cellular energy failure is thought to be the underlying cause of the characteristic neuropathology first described by Leigh [1]. Classical Leigh syndrome is associated with focal and bilateral lesions, particularly involving the basal ganglia, thalamus, and brain stem regions, neuropathologically characterized by spongiform changes with vacuolation of the neuropil and relative preservation of neurons, associated with demyelination, gliosis, necrosis, and capillary proliferation [1]. Although the original description by Leigh was from postmortem pathological specimens, typical *in vivo* magnetic resonance neuroimaging (MRI) patterns are now the cornerstone of diagnosis of the disease, in the context of compatible clinical features and (usually) elevated lactate levels in blood and/or cerebrospinal fluid (CSF) [2].

Clinically, Leigh syndrome usually presents in the first year of life, often with developmental regression and ophthalmic signs including early nystagmus, retinal dysfunction, and ophthalmoplegia. Nonspecific early features include feeding difficulties, poor weight gain, and vomiting, which may be attributed to gastroesophageal reflux. Progressive neurological involvement with ataxia, dystonia, and seizures is typical, with death usually ensuing within the first 5 years of life, often from brain stem failure. Leigh-like syndrome shares similar clinical presentations, but affected cases may not have elevated lactate levels and may have different neuroanatomical patterns of involvement and subsequent variation in the clinical features and course [2].

The genetic basis of Leigh syndrome is very diverse, and it includes both nuclear [15] and mitochondrial DNA-encoded mutations [3] responsible for

Mitochondrial Case Studies. http://dx.doi.org/10.1016/B978-0-12-800877-5.00014-0

abnormal assembly and function of any of the respiratory chain complexes, associated enzymes including the pyruvate dehydrogenase complex (PDHc), or generalized mitochondrial function or turnover (Tables 1 and 2). Delineating the specific genetic cause of a child's Leigh syndrome is important in establishing a concrete diagnosis, may provide further prognostic information, and is necessary for providing accurate genetic counseling about risks in future pregnancies and for facilitating prenatal and preimplantation genetic diagnosis and wider genetic testing within the family if indicated.

DIAGNOSTIC PIPELINE

The initial clinical presentation may lead to consideration of a possible diagnosis of Leigh syndrome. Presentation may be in the neonatal period with a generally unwell infant, with poor feeding and hypotonia. Congenital lactic acidosis may be present. There may be early indications of brain involvement with abnormal eye signs, seizures, and abnormal respiratory pattern. The presentation is, however, often later at a few months of age, typically preceded by a relatively minor viral intercurrent illness seeming to precipitate catastrophic neurological sequelae. While the neuropathology dominates the clinical picture of Leigh syndrome, evidence of multisystem involvement is also suggestive of mitochondrial disease in general. Multisystem features should be sought actively, to help confirm the diagnosis, but also so treatable complications can be addressed (including cardiomyopathy, hormone deficiencies, and electrolyte imbalance arising from renal tubular dysfunction).

Following clinical suspicion, initial biochemical testing may suggest the diagnosis. In particular, the demonstration of persistent lactic acidosis is suggestive, although neither fully sensitive nor specific for Leigh syndrome. Associated with this, plasma amino acid analysis may demonstrate elevated alanine, also seen in urine amino acids, while urine organic acids may reveal elevated lactate, pyruvate, and other markers of mitochondrial dysfunction. In occasional cases, plasma and/or urinary metabolites may suggest a specific genetic diagnosis; for example, methylmalonic aciduria and elevated plasma succinyl and propionylcarnitine might indicate *SUCLA2* or *SUCLG1* mutations, while 3-methylglutaconic aciduria in a child with Leigh syndrome is suspicious of *SERAC1* mutations (see case 4 below). CSF studies may demonstrate elevated lactate and alterations in neurotransmitters or reduced levels of 5-methyltetrahydrofolate, the major transport folate.

Although Leigh syndrome is predominantly neurological, there may be markers of other system involvement including renal tubular dysfunction (e.g., elevated N-Acetylglucosaminidase/creatinine ratio), myopathy (mildly elevated creatine kinase), or cardiomyopathy (abnormal echocardiography or electrocardiogram).

Neuroimaging plays a crucial role in diagnosing Leigh syndrome, although as the case examples that follow demonstrate, the pattern seen can be variable, and it is rare that the imaging appearance suggests a specific genetic diagnosis.

TABLE 1 Nuclear genes associated with Leigh and Leigh-like syndrome with specific respiratory chain enzyme complex deficiencies

Muscle RCE Deficiency	Gene	Neuroimaging	Other Features
Isolated complex I	*NDUFS1*	Cystic leukoencephalopathy	
	NDUFS2		HCM
	NDUFS3		
	NDUFS4	Leukoencephalopathy	HCM
	NDUFS7	Cystic leukoencephalopathy	
	NDUFS8	Cystic leukoencephalopathy	
	NDUFV1		
	NDUFV2		
	NDUFA1 (X-linked)		
	NDUFA2		HCM
	NDUFA9		
	NDFUA10	Symmetric lesions in mamillo-thalamic tracts, substantia nigra/medial lemniscus, medial longitudinal fasciculus, and spinothalamic tracts	HCM
	NDUFA12		Hypertrichosis
	NDUFAF2		
	NDUFAF5		FILA
	NDUFAF6		
	FOXRED1		
Isolated complex II	*SDHA*	+/− Succinate peak on MRS	HCM
	SDHB		
	SDHAF1	+/− Succinate peak on MRS; leukoencephalopathy	
Isolated complex III	*UQCRQ*		
	TTC19	Olivopontocerebellar atrophy	
	BCS1L		SNHL, pili torti, renal tubulopathy, liver disease

Continued

TABLE 1 Nuclear genes associated with Leigh and Leigh-like syndrome with specific respiratory chain enzyme complex deficiencies—cont'd

Muscle RCE Deficiency	Gene	Neuroimaging	Other Features
Isolated complex IV	*NDUFA4*		Sensory neuropathy
	SURF1		Hypertrichosis
	COX10		Anemia, SNHL, HCM
	COX15		
	SCO2		HCM
	LRPPRC		Metabolic crises, stroke-like episodes
	TACO1		Seizures
	PET100		
Multiple deficiencies: mitochondrial DNA depletion syndrome (MDDS)	*POLG*		Seizures, liver disease
	SUCLA2		SNHL
	SUCLG1		Liver disease
	FBXL4		Facial dysmorphism, skeletal problems, seizures, renal tubulopathy
Multiple deficiencies: defects of mitochondrial translation	*TRMU*		Acute liver failure (reversible)
	MTFMT	Cystic leukoencephalopathy	
	FARS2		Alpers-like
	EARS2	LTBL	
	IARS2		Multisystem disease
	GFM1		Liver disease
	C12orf65		Sensory neuropathy

FILA, fatal infantile lactic acidosis; HCM, hypertrophic cardiomyopathy; LTBL, leukoencephalopathy with thalamus and brain stem involvement and high lactate; RCE, respiratory chain enzyme(s); SNHL, sensorineural hearing loss.
Adapted from Ref. [15].

TABLE 2 Other nuclear genes associated with Leigh and Leigh-like syndrome not due to primary OXPHOS deficiency

Enzyme Deficiency	Gene	Neuroimaging	Other Features
PDHc deficiency	*PDHA(X-linked)*	CC agenesis/hypoplasia	
	PDHB	CC agenesis/hypoplasia	
	DLD (PDH E3)		Hypoglycemia, ketoacidosis, elevated plasma branched chain amino acids, liver failure
	PDHX (PDH E3 binding)	Thin CC/CC agenesis	
Biotin and thiamine related enzyme and transporter deficiencies	*BTD (biotinidase)*	Leukoencephalopathy	Alopecia, eczema, optic atrophy, SNHL, organic aciduria, responsive to biotin
	SLC25A19	Bilateral striatal necrosis	Peripheral neuropathy
	TPK1		2-Ketoglutaric aciduria
	SCL19A3	Biotin and thiamine responsive basal ganglia disease	Responsive to *high dose* biotin and thiamine
Polyribonucleotide nucleotidyltransferase deficiency	*PNPT1*		SNHL
Coenzyme Q_{10} biosynthesis deficiency	*PDSS2*		Nephrotic syndrome

Continued

TABLE 2 Other nuclear genes associated with Leigh and Leigh-like syndrome not due to primary OXPHOS deficiency—cont'd

Enzyme Deficiency	Gene	Neuroimaging	Other Features
Lipoic acid synthesis deficiency	*LIAS*		Combined deficiency of PDH + glycine cleavage enzyme, elevated plasma glycine, deficient lipoylated proteins on western blot, seizures + burst suppression on EEG
	LIPT1		Combined deficiency of PDH, α-KGDH + BCKDH, increased glutamine + proline, low levels of lysine and branched chain amino acids, normal glycine
SERAC1 lipid remodeling deficiency	*SERAC1*		MEGDEL: 3-Methylglutaconic Aciduria, SNHL, liver disease in infancy
Other	*ETHE1*		Acrocyanosis, petechiae, diarrhea, ethylmalonic aciduria
	HICBH		Elevated plasma 4-hydroxybutyrylcarnitine levels (+/– deficiencies of RCEs and PDHc)
	ECHS1		Increased urinary excretion of S-(2-carboxypropyl) cysteine

CC, corpus callosum; MEGDEL, 3-methylglutaconic aciduria, deafness, encephalopathy (Leigh-like); PDHc, pyruvate dehydrogenase complex; RCE, respiratory chain enzyme(s); SNHL, sensorineural hearing loss.
Adapted from Ref. [15].

Magnetic resonance spectroscopy (MRS) may demonstrate elevated brain lactate, but it is again neither sensitive nor specific for Leigh syndrome. MRS can be useful in identifying a succinate peak, which is suggestive of an underlying deficiency of succinate dehydrogenase (SDH) [4], but this is an extremely rare cause of Leigh syndrome (see case 2 below).

Tissue samples may be obtained for skin fibroblast culture for assays of PDHc, and a muscle biopsy for standard histopathologic and electron microscopic examination and for assays of RCE activities. DNA may be extracted from muscle for molecular genetic analysis of the mitochondrial DNA (mtDNA), including whole mtDNA sequencing and quantitation. DNA from blood is sufficient for analysis of nuclear-encoded genes. DNA from the parents is also required to verify genetic findings in the proband.

The cumulative results from these initial investigations may then direct analysis of potential candidate genes, by pinpointing either typical clinical or neuroimaging phenotypes, biochemical clues, or specific involvement of individual RCE complexes.

The family history may also provide diagnostic pointers as to whether a nuclear or mtDNA mutation is the likely cause. A consanguineous family with multiple affected family members on both sides of the pedigree raises the suspicion of nuclear (i.e., autosomal recessive) inheritance, whereas a matrilineal pattern of inheritance is more typical of mtDNA mutations. Of course, in many families, the child presenting may be the first affected; in this case, any mode of inheritance is possible, including nuclear (recessive or X-linked) or mtDNA mutations, or occasionally there may be a *de novo* mutation.

CASE STUDIES

Case 1: Complex I Deficiency

A male infant presented at 6 months of age with developmental regression, irritability, and poor visual fixation. Feeding had deteriorated, necessitating nasogastric nutrition. Of note, his parents were consanguineous, and he had an elder sister who had presented at a similar age with developmental regression following a febrile illness. Initial biochemical testing demonstrated a lactic acidosis (4.9 mmol/L, reference range < 2.0 mmol/L) with elevated plasma and urinary alanine. Furthermore, the lactate/pyruvate ratio was elevated at 92 (normal < 25). Liver function tests were normal. MRI brain demonstrated a symmetrical leukoencephalopathy (see Figure 1). A muscle biopsy in the elder sibling had found an isolated deficiency of complex I activity. Because of the large number of candidate genes, including 44 subunits of complex I, and numerous assembly factors [5], whole exome sequencing (WES) was performed and revealed a novel homozygous missense variant in the nuclear subunit gene *NDUFV2*. Both parents were shown to be heterozygous carriers, and studies in cultured skin fibroblasts demonstrated a reduction in NDFUV2 protein.

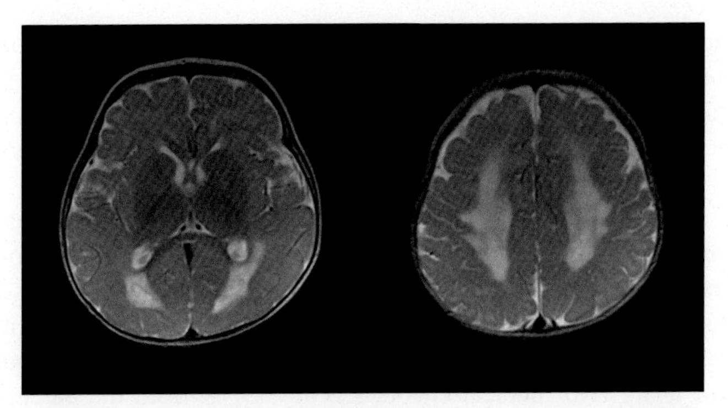

FIGURE 1 MRI brain scan obtained at 8 months of age from case 1 (complex I deficiency due to *NDFUV2* mutation). Axial T2 weighted MR images demonstrating widespread symmetrical increased signal in white matter (leukoencephalopathy).

The first complex in the respiratory chain, complex I (NADH:ubiquinone oxidoreductase, EC1.6.5.3) oxidizes NADH and reduces coenzyme Q_{10} (ubiquinone). Complex I consists of 37 nuclear-encoded subunits and seven encoded by mtDNA [5]. *NDUFV2* encodes the 24kD subunit and together with the nuclear-encoded subunits NDUFV1 and NDUFV3 forms the flavoprotein domain of the enzyme. Cases with *NDUFV1* mutations have been reported to have a similar cystic leukoencephalopathy to that seen in this case [6].

Case 2: Complex II Deficiency

A male infant was born at term after an uncomplicated pregnancy. Early neurodevelopment was normal, with the child sitting at 5 months and walking at 11 months. From 2 years of age, he deteriorated and lost ability to walk, and he became increasingly hypotonic. An MRI brain scan demonstrated bilateral symmetrical signal abnormalities in the posterior and medial thalami, periaqueductal gray matter, anterior pons, medulla, and upper cervical cord suggestive of Leigh syndrome (see Figure 2).

Plasma lactate levels were elevated (3.7–5.0 mmol/L). His clinical course continued to be variable, with fluctuating motor skills. A clinical trial of coenzyme Q_{10} had no obvious benefit. Owing to fluctuating neurology with dystonia and rigidity, he was treated with levodopa. At age 15 years, he had a significant residual movement disorder, with progressive thoracic kyphosis.

Muscle biopsy demonstrated deficiency of complexes II+III, with subsequent testing demonstrating isolated reduction of complex II activity.

Complex II (succinate dehydrogenase complex, EC1.3.5.1) is entirely encoded by nuclear genes, including the flavoprotein subunit SDHA, together with an iron-sulfur containing unit (SDHB) and membrane-bound subunits SDHC and SDHD. Molecular genetic testing targeted the *SDHA* gene in this case, since this is most frequently mutated in complex II deficient Leigh

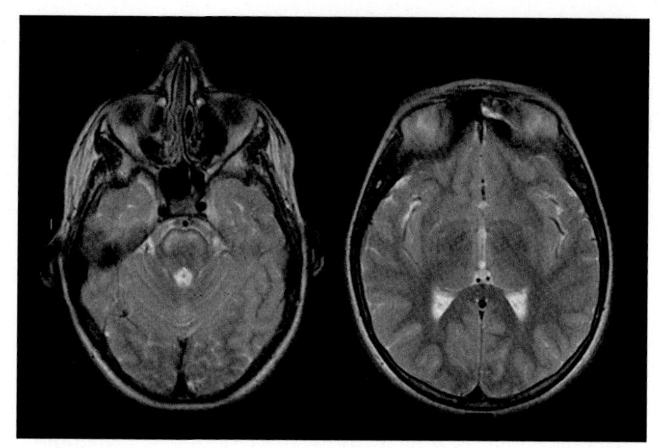

FIGURE 2 MRI brain scan from case 3 (complex II deficiency due to *SDHA* mutation). Axial T2 weighted images showing signal abnormalities in medial and posterior thalami and brain stem.

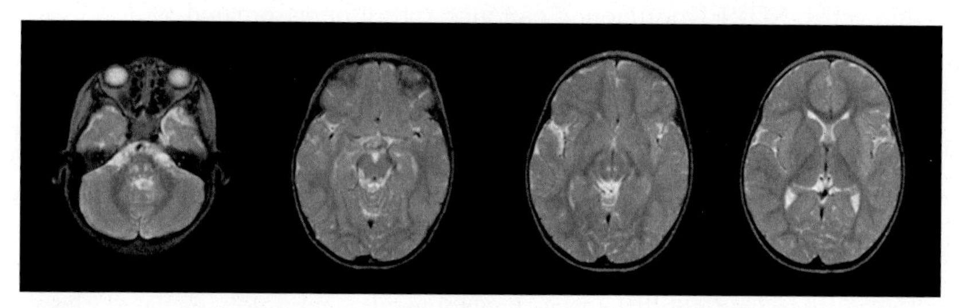

FIGURE 3 Axial T2 weighted images demonstrating symmetrical lesions in anterior medulla and inferior colliculi, and in bilateral lateral putamina.

syndrome, and it identified a homozygous mutation [7]. Both *SDHA* and *SDHB* mutations have been reported to cause a leukoencephalopathy [4]. In contrast, dominant mutations affecting the SDHC and SDHD subunits have been linked to tumorigenesis.

Case 3: Complex IV Deficiency

A Caucasian female born to nonconsanguineous parents presented at 23 months of age with a 4-month history of ataxic gait. She was found to be ataxic with intention tremor in the upper limbs, and she also had intermittent horizontal nystagmus. An MRI brain scan demonstrated lesions within the brain stem (anterior medulla and inferior colliculi), and a subsequent repeat scan also demonstrated bilateral lateral putaminal changes (see Figure 3) consistent with Leigh syndrome.

Plasma amino acids showed a mildly elevated alanine (665 μmol/L, reference range 150–450). Muscle and skin biopsies were performed. Muscle histopathological examination showed fiber type disproportion with type 1 fibers significantly smaller than type 2. Muscle RCE analysis demonstrated deficiency of cytochrome oxidase (activity ratio to citrate synthase 0.003, reference range 0.014–0.034). Examination of cultured fibroblasts revealed a systemic cytochrome oxidase deficiency (activity of 5.6 nmol/mg protein/min, reference range 30–90).

Mutations in the complex IV assembly factor SURF1 are the most frequent cause of cytochrome oxidase deficient Leigh syndrome, responsible for up to 75% of cases [8,9]. Genetic investigations in this case, therefore, targeted the *SURF1* gene and identified compound heterozygous mutations [9].

The individual was treated with coenzyme Q_{10} with apparent benefit. Her clinical course varied with exacerbations (worsening ataxia and fatigue) after intercurrent illness including varicella, and after vaccinations. At 14 years, she was in mainstream education, and she was well other than continuing ophthalmoplegia and ataxia and possible absence seizures. This represents a much milder course than is typical for SURF1 deficiency; median survival is 5.4 years, although occasional cases with survival into the second decade have been reported [9]. Like Leigh syndrome caused by complex I and complex II deficiencies, SURF1 deficiency can also rarely be associated with a cystic leukoencephalopathy [10].

Case 4: MEGDEL Syndrome

A male infant born to non-consanguineous Bengali parents presented unwell on day three of life, with hyperammonemia and lactic acidosis requiring hemodialysis. After an initial recovery from the acute metabolic decompensation, he had continuing feeding difficulties, hepatomegaly, and mildly elevated plasma lactate (2.3 mmol/L). He developed acquired microcephaly. At 1.5 years, he was diagnosed with hearing impairment requiring hearing aids. By 20 months, developmental progress was noted to be static. An MRI of his brain showed an abnormal signal in the striatum bilaterally, with atrophy especially of the caudate heads. He developed a movement disorder with dystonia and self-injurious behaviors. No further developmental progress was seen. Subsequently, he was treated with coenzyme Q_{10} and riboflavin, with apparent good clinical response evidenced by stabilization of his general condition and improved alertness.

Investigations from the neonatal period included a muscle biopsy, which showed normal histological features with no ragged red fibers or cytochrome oxidase negative fibers, but an isolated complex III deficiency was seen on enzyme assay. Urine organic acid analysis showed grossly raised lactate and pyruvate suggestive of disturbed lactate metabolism. There was also strongly elevated 3-methylglutaconate with moderately raised 3-methylglutarate consistent with mitochondrial dysfunction.

The initial molecular genetic approach focused on complex III genes, and it did not reveal any pathogenic mutations in cytochrome b (the only mtDNA-encoded subunit of complex III), or in the complex III assembly gene *BCS1L*, the gene most frequently affected in isolated complex III deficiency [11]. Analysis of a repeat muscle biopsy at 12 years revealed normal light microscopy

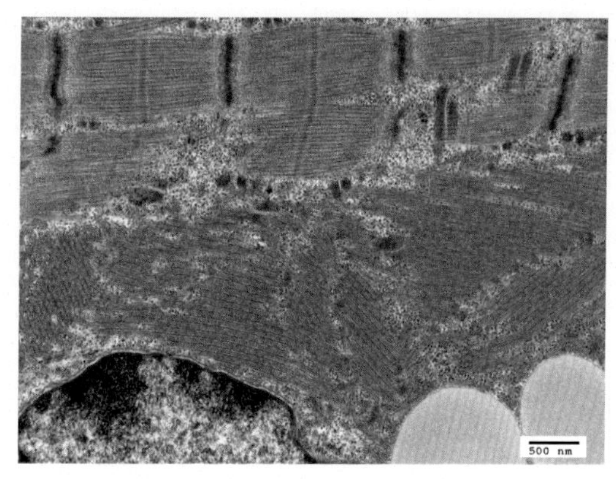

FIGURE 4 Electron micrograph of muscle biopsy showing prominent tubular aggregates and excess lipid in case 4 (MEGDEL).

apart from occasional fibers with pale staining for cytochrome oxidase. Electron microscopy, however, demonstrated highly unusual prominent tubular aggregates and excess lipid (see Figure 4). Complex III activity was normal in this second muscle sample, but there was a mild reduction of complex IV activity (0.013, reference range 0.014–0.034).

Subsequently, WES identified a homozygous known pathogenic mutation in *SERAC1* that was confirmed by Sanger sequencing [12]. Parents were both heterozygous. Mutations in *SERAC1* cause MEGDEL syndrome (3-methylglutaconic aciduria, deafness, and encephalopathy in the Leigh-like spectrum) [13], consistent with the clinical and biochemical features observed in this case. SERAC1 plays a role in remodeling of the mitochondrial inner membrane, and mutations result in ineffective lipid remodeling and consequent respiratory chain dysfunction [13]. Early hyperammonemia and hepatic dysfunction in a case with 3-methyl glutaconic aciduria who subsequently develops hearing loss and basal ganglia lesions should raise suspicion of underlying *SERAC1* mutations.

Case 5: X-Linked Pyruvate Dehydrogenase Complex Deficiency

A female infant was born at 42 weeks gestation to healthy unrelated parents. Three paternal-side half siblings were healthy. On day one of life, she had episodes of pallor and floppiness and was admitted to the neonatal intensive care unit with suspected seizures, and was treated with phenobarbital. She was found to have congenital lactic acidosis (plasma lactate 7.0–9.4 mmol/L), with a low-normal lactate/pyruvate ratio (ratio 11, normal < 25). CSF lactate was elevated at 15 mmol/L. Brain imaging demonstrated an absent corpus callosum and asymmetric ventriculomegaly. There was abnormal gyral patterning suggestive of polymicrogyria (Figure 5). Ophthalmic examination revealed bilateral optic nerve hypoplasia. She had progressive difficulties with seizures manifesting as apneas, with only partial response to escalating anticonvulsant therapy.

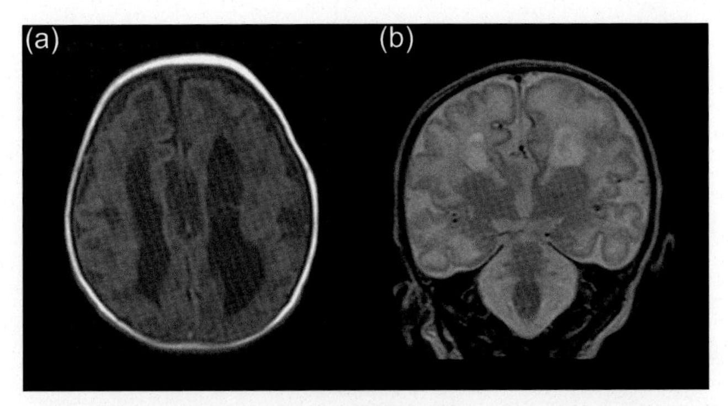

FIGURE 5 MRI brain scan from case 5 (pyruvate dehydrogenase deficiency). (a) Axial T1 weighted image showing ventriculomegaly and (b) coronal T2 weighted image showing absent corpus callosum and abnormal gyral patterning suggestive of polymicrogyria.

In view of the clinical, radiological, and biochemical features, molecular genetic testing was initially targeted at PDHc deficiency, and it confirmed a heterozygous mutation in *PDHA1*. This is an X-linked gene encoding the alpha subunit of the E1 enzyme of the PDH complex. In view of this finding, a ketogenic diet was initiated with good effect on seizure control. A muscle biopsy was not required in the diagnostic pathway for this child. A low or normal lactate/pyruvate ratio in a child with congenital lactic acidosis should alert suspicion of PDHc deficiency, and structural brain abnormalities are frequently seen in girls with this disorder [14]. Diagnosis can be either by enzyme assay in fibroblasts or direct molecular testing.

DISCUSSION

The presumed final common pathway of the diverse mitochondrial disorders is ATP depletion and cellular energy failure, although other aspects of mitochondrial function may be secondarily implicated particularly where the underlying etiology affects mitochondrial replication, membrane function, or turnover. The cellular level energy failure impacts on tissues with high energy demands, notably the central nervous system. Mitochondrial disorders caused by mutations of nuclear DNA encompass disorders of oxidative phosphorylation (OXPHOS) associated with defects in the ~80 nuclear-encoded structural subunits of the OXPHOS complexes, or in genes encoding proteins involved in OXPHOS complex assembly, as well as mtDNA replication and repair (Table 1). Leigh and Leigh-like syndrome can also be caused by mutations in genes responsible for coenzyme Q_{10} synthesis, associated energy-pathway enzymes, in particular PDHc, as well as genes such as *SERAC1* responsible for maintaining the integrity of the mitochondrial membranes (Table 2) [15].

The cases described illustrate the application of the diagnostic pipeline to pinpoint specific genetic causation for children with overlapping clinical phenotypes. The archetypal clinical feature is that of progressive central nervous system involvement, although the clinical signs and symptoms as well as evolution over time vary, as the cases illustrate. Certain clusters of clinical and biochemical features may point toward a specific diagnosis. Of particular help, the detection of isolated RCE complex deficiencies can narrow the search to potential causative genes, while the family history may help direct attention to either mitochondrial or nuclear genomes as the most likely candidates. Review by specialist multidisciplinary mitochondrial disorder services can be very helpful, by facilitating detailed clinical phenotyping and directing the laboratory investigations. Certain patterns of neuroimaging and biochemical features can also be nearly pathognomonic of particular disorders, as illustrated by case 4 (SERAC1 deficiency) and case 5 (PDHc deficiency), and direct molecular genetic testing in these cases can circumvent the need for more invasive investigations such as muscle biopsy.

Next-generation sequencing technologies have enabled the development of WES, which can play an important role in identifying candidate gene mutations, although verification and corroboration of findings from WES with the clinical and biochemical features are required in the interpretation. As case 4 illustrates, WES can help with the emergence of new disorders that have recognizable specific clinical phenotypes that can then be identified in future cases without the need for invasive investigations.

CLINICAL PEARLS

- Leigh syndrome is a progressive neurodegenerative disorder with onset usually during infancy and childhood and progression is often in a stepwise fashion characterized by periods of stability between acute or subacute progression.
- The most characteristic findings across genetic etiologies of Leigh syndrome are magnetic resonance imaging (MRI) features consistent with bilateral symmetrical necrotizing lesions in the basal ganglia, brainstem, and/or cerebellum. However, the syndrome is phenotypically and genetically very heterogeneous. Clinical and/or radiological evidence of basal ganglia and/or brainstem dysfunction, intellectual and motor delay, and elevated serum or CSF lactate indicating abnormal energy metabolism are suggestive of the diagnosis.
- There are multiple nuclear-encoded genes responsible for Leigh syndrome with mostly nonspecific genotype-phenotypic correlations. If no specific genotype is suggested from clinical features and initial investigations, broad genetic testing should be considered employing either a large gene panel using next generation sequencing or whole exome sequencing.
- The clinician should be aware of the possibility of multisystem involvement, and perform careful evaluation of multiple organ systems. Although specific treatments for mitochondrial disease are not available, symptomatic treatment increases life expectancy and more importantly quality of life.

ACKNOWLEDGMENT

Professor Rahman is supported by Great Ormond Street Hospital Children's Charity and receives research funding from the Lily Foundation, Muscular Dystrophy UK, and the Wellcome Trust.

REFERENCES

[1] Leigh D. Subacute necrotizing encephalomyelopathy in an infant. J Neurol Neurosurg Psychiatry August 1951;14(3):216–21.

[2] Rahman S, Blok RB, Dahl HH, Danks DM, Kirby DM, Chow CW, et al. Leigh syndrome: clinical features and biochemical and DNA abnormalities. Ann Neurol March 1996;39(3):343–51.

[3] Thorburn DR, Rahman S. Mitochondrial DNA-associated Leigh syndrome and NARP. 1993. In: Pagon RA, Adam MP, Ardinger HH, Wallace SE, Amemiya A, Bean LJH, Bird TD, Dolan CR, Fong CT, Smith RJH, Stephens K, editors. GeneReviews® (Internet). Seattle University of Washington: Seattle, WA; 1993-2015. 2003 October 30 [updated 2014 April 17].

[4] Alston CL, Davison JE, Meloni F, van der Westhuizen FH, He L, Hornig-Do HT, et al. Recessive germline SDHA and SDHB mutations causing leukodystrophy and isolated mitochondrial complex II deficiency. J Med Genet September 2012;49(9):569–77.

[5] Fassone E, Rahman S. Complex I deficiency: clinical features, biochemistry and molecular genetics. J Med Genet September 2012;49(9):578–90.

[6] Benit P, Chretien D, Kadhom N, De Lonlay-Debeney P, Cormier-Daire V, Cabral A, et al. Large-scale deletion and point mutations of the nuclear *NDUFV1* and *NDUFS1* genes in mitochondrial complex I deficiency. Am J Hum Genet June 2001;68(6):1344–52.

[7] Pagnamenta AT, Hargreaves IP, Duncan AJ, Taanman JW, Heales SJ, Land JM, et al. Phenotypic variability of mitochondrial disease caused by a nuclear mutation in complex II. Mol Genet Metab November 2006;89(3):214–21.

[8] Tiranti V, Jaksch M, Hofmann S, Galimberti C, Hoertnagel K, Lulli L, et al. Loss-of-function mutations of SURF-1 are specifically associated with Leigh syndrome with cytochrome c oxidase deficiency. Ann Neurol August 1999;46(2):161–6.

[9] Wedatilake Y, Brown R, McFarland R, Yaplito-Lee J, Morris AA, Champion M, et al. SURF1 deficiency: a multi-centre natural history study. Orphanet J Rare Dis July 5, 2013;8(1):96.

[10] Rahman S, Brown RM, Chong WK, Wilson CJ, Brown GK. A *SURF1* gene mutation presenting as isolated leukodystrophy. Ann Neurol June 2001;49(6):797–800.

[11] Atale A, Bonneau-Amati P, Rotig A, Fischer A, Perez-Martin S, de LP, et al. Tubulopathy and pancytopaenia with normal pancreatic function: a variant of Pearson syndrome. Eur J Med Genet January 2009;52(1):23–6.

[12] Wedatilake Y, Plagnol V, Anderson G, Paine S, Clayton P, Jacques TS, Rahman S. Tubular aggregates caused by serine active site containing 1 (*SERAC1*) mutations in a patient with a mitochondrial encephalopathy. Neuropathol Appl Neurobiol April 2015;41(3):399–402.

[13] Wortmann SB, Vaz FM, Gardeitchik T, Vissers LE, Renkema GH, Schuurs-Hoeijmakers JH, et al. Mutations in the phospholipid remodeling gene *SERAC1* impair mitochondrial function and intracellular cholesterol trafficking and cause dystonia and deafness. Nat Genet July 2012;44(7):797–802.

[14] Patel KP, O'Brien TW, Subramony SH, Shuster J, Stacpoole PW. The spectrum of pyruvate dehydrogenase complex deficiency: clinical, biochemical and genetic features in 371 patients. Mol Genet Metab July 2012;106(3):385–94.

[15] Rahman S, Thorburn D. Nuclear gene-encoded Leigh syndrome overview. In: Pagon RA, Adam MP, Ardinger HH, Wallace SE, Amemiya A, Bean LJH, Bird TD, Fong CT, Mefford HC, Smith RJH, Stephens K, editors. GeneReviews® [Internet]. Seattle (WA): University of Washington, Seattle; 1993–2015; October 1, 2015. Available from http://www.ncbi.nlm.nih.gov/books/NBK320989/ PubMed PMID: 26425749.

Chapter 15

Reversible Infantile Respiratory Chain Deficiency

Ulrike Schara[1], Adela Della Marina[1], Rita Horvath[2]
[1]Pediatric Neurology, University of Essen, Essen, Germany; [2]The John Walton Muscular Dystrophy Research Centre, MRC Centre for Neuromuscular Diseases, Institute of Genetic Medicine, Newcastle University, Newcastle upon Tyne, UK

CASE PRESENTATION

The now 24-year-old case is the third of four children of non-consanguineous German parents (Figure 1). Her two older brothers (siblings) died at the age of 4 days and 2 months, respectively, due to metabolic acidosis with high lactate levels (3.3 μmol/L, Norm < 2.0 μmol/L). Both brothers were preterm born (34th week of gestation). Muscle biopsy was performed and showed ragged red fibers with reduction in the activity of cytochrome *c* oxidase (COX) (data not shown).

The girl was born at term after an uneventful pregnancy and delivery. She exhibited mild muscular hypotonia since infancy with delayed motor milestones: sitting at 12 months, walking at 18 months of age with frequent falls; she could drive a two-wheel bicycle at 6 years of age. Syndromic features were documented, such as dysplasia of the midface and high arched palate. During childhood, her muscular endurance and strength were reduced compared to healthy children of the same age, and her CK levels were intermittently elevated (up to 294 U/L, normal range < 172 U/L). Her mental development was normal, and after completing her regular elementary schooling and secondary degree, she finished an education to become a kindergarten teacher.

At 1.5 years of age, she developed chronic diarrhea; however, her weight gradually increased, and at the age of 2 years and 9 months, her weight was 19.1 kg, which exceeded the 97th percentile, while her length (96 cm) was within the 50–75th percentiles. There was a marked worsening of her muscle strength after a carbohydrate-rich meal. Her diarrhea stopped under a reduction diet implying that it may have been caused by intolerance for milk, fruit, and sugar. In further support of a metabolic component for her symptoms, hypertriglyceridemia and insulin-resistance were diagnosed at 20 years of age. Currently, at the age of 24 years, she showed normal muscle strength and has a normal walking distance; she has no problems in her daily activity of life.

Mitochondrial Case Studies. http://dx.doi.org/10.1016/B978-0-12-800877-5.00015-2

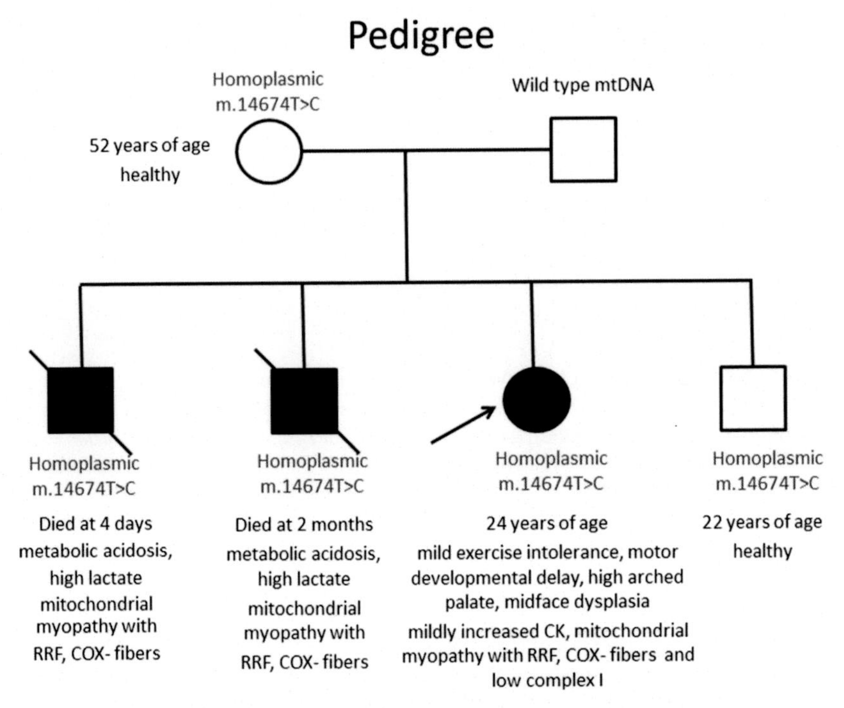

FIGURE 1 Pedigree and main clinical presentations.

On clinical examination, her metric data disclosed morbid obesity with length between the 10th and 25th centile (165 cm) and weight above the 97th centile (165 kg) at 24 years of age.

A muscle biopsy performed at 12 years of age showed signs of a mitochondrial myopathy with ragged red fibers and COX-negative fibers (Figure 2). Biochemical analysis of the respiratory chain (RC) enzymes showed a reduction of complex I with residual activity of 43% of normal. The activities of the other complexes II, III, and IV were normal. All substrate oxidation rates relative to citrate synthase (CS) activity were reduced. The production of ATP + CrP from pyruvate was reduced (17 mU CS, normal range 42–81 mU CS). The activity of complex I in fibroblasts was normal.

DIFFERENTIAL DIAGNOSIS

In this 24-year-old woman, a differential diagnosis depends on her leading clinical symptoms, which vary during the clinical course as follows:

Infancy:

- Congenital myopathy (hypotonia, motor developmental delay, facial hypotonia, high arched palate, slightly elevated CK, affected brothers who died early)

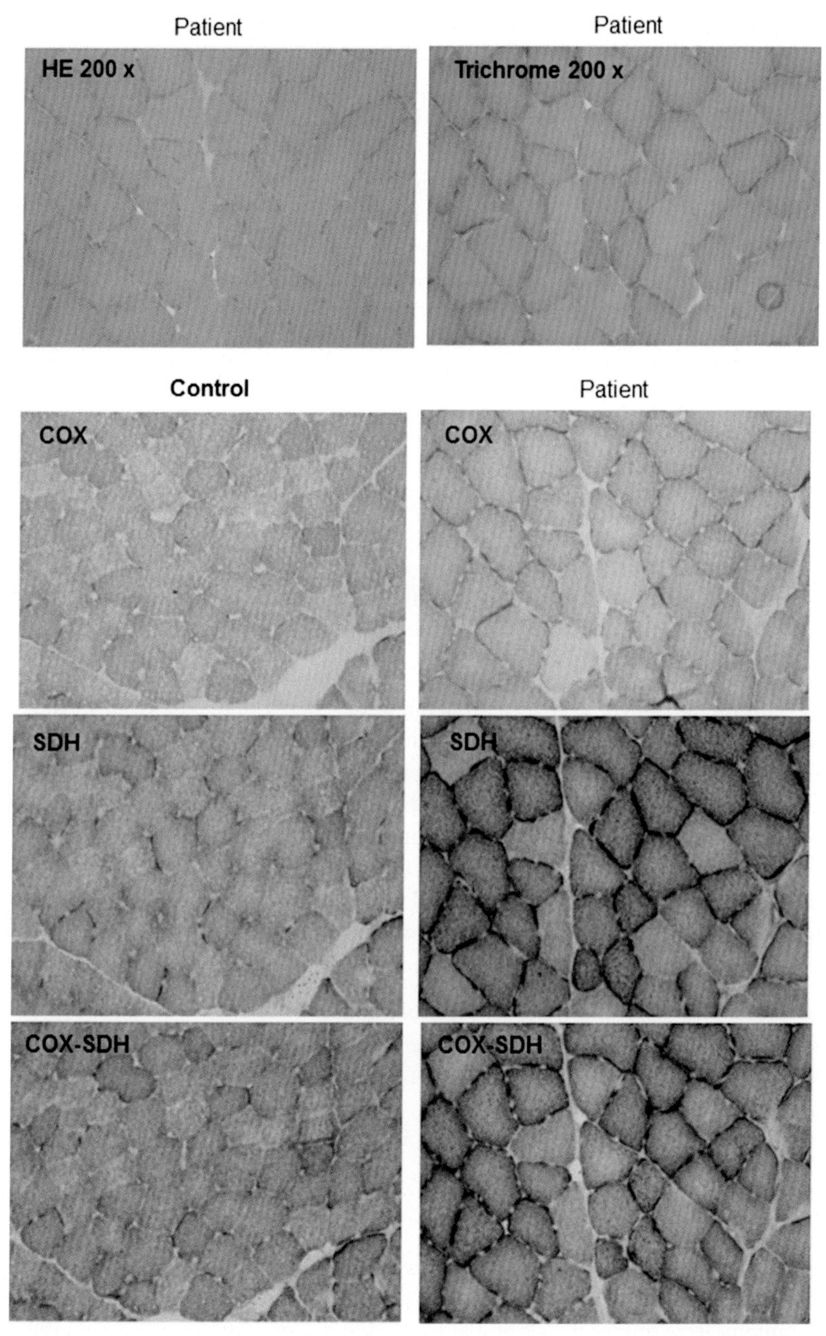

FIGURE 2 Muscle specimen of the case at 12 years of age: normal fiber variability (HE stain ×200), ragged red fibers (Trichrome stain, ×200), slightly reduced COX, and increased SDH reaction compared to controls, but clear COX-negative blue fibers in the combined COX/SDH reaction.

- Congenital muscular dystrophy (hypotonia, motor developmental delay, facial hypotonia, high arched palate, elevated CK, affected brothers who died early)
- Celiac disease (chronic diarrhea, hypotonia, developmental retardation)

Childhood:

- Allergic diathesis (recurrent pulmonary infections, dermatosis, intolerance of milk, fruits, sugars, chronic diarrhea)
- Immunodeficiency syndromes (recurrent infections)
- Metabolic disorder (exercise intolerance, developmental retardation, hypotonia, chronic diarrhea)
- Mitochondrial disorder (neurological symptoms, exercise intolerance, slightly elevated CK, recurrent infections and dermatosis, two affected brothers who died early)
- Syndromic diseases (developmental retardation, hypotonia, dysplasia of the midface)
- Hereditary neuropathy plus (dermatosis, slightly elevated CK, pes cavus)
- Phacomatosis (dermatosis and neurological symptoms)

Adolescence:

- Mitochondrial disorders (neurological symptoms, exercise intolerance, slightly elevated CK, recurrent infections, dermatosis, two affected brothers who died early)

During the clinical course, a lot of diagnoses were eliminated from consideration by different investigations. Intermittent, slight serum CK elevation was not typical for muscular dystrophy. Metabolic screening did not identify any other form of metabolic myopathy. Electrophysiological testing noted that the case has a myopathy, and it excluded neuropathy and neuromuscular transmission defect. Ragged red and COX-negative fibers and complex I defect in skeletal muscle biopsy confirmed mitochondrial myopathy. And because of reversibility of symptoms, a reversible infantile RC deficiency (RIRCD) was assumed.

The diagnosis could be made by the following:

- Reversible clinical symptoms described above
- Family history with affected brothers
- Investigations of other affected organs
- Muscle biopsy investigations
- Genetic analysis of the mt-tRNA^{Glu} (*MT-TE)* gene

DIAGNOSTIC APPROACH

In typical RIRCD presentations, the reversible myopathy would have been an important diagnostic clue. However, in our case, the severity of the early-onset symptoms was milder, masking the reversible disease course. The fatal course in the affected brothers was also misleading and implying rather a progressive mitochondrial disease. Similar milder clinical presentations were noted in family members of cases with severe RIRCD suggesting a wide range in the severity of the clinical manifestations, potentially influenced by genetic, epigenetic, or environmental factors. The two affected brothers who died of infantile mitochondrial myopathy, COX deficiency, and lactic acidosis at age 4 days and 3 months, respectively, were in support of a mitochondrial disease in our case. However, an autosomal recessive disease or a heteroplasmic mutation with different levels in siblings seemed more likely on the first sight rather than a homoplasmic mt-tRNA mutation.

Our case's weight gain is very atypical for this condition, and we are not aware of other cases with similar symptoms who carry the m.14674T>C mutation. Whether or not her gastrointestinal and nutritional problems are related to this mutation is currently unclear but may well be coincidental findings in our case.

Although the involvement of skeletal muscle was evident in the index case and showed improvement with age, a mitochondrial origin was only suggested by the lactic acidosis and mitochondrial changes in muscle biopsy of the affected brothers, who had a fatal and progressive disease. The discrepant clinical presentations in our index case and her brothers made it difficult to conclude that they have the same genetic disease, which was supported by the presence of mitochondrial changes in the muscle biopsy. The detection of a mitochondrial myopathy with RRF, COX-fibers, and combined RC deficiency in our case was the key and prompted us to perform genetic testing for mitochondrial DNA and specifically mt-tRNA mutations. The detection of the homoplasmic m.14674T>C mutation in the mitochondrial $tRNA^{Glu}$ gene (*MT-TE*) was unexpected in our case, but it confirmed the genetic diagnosis of RIRCD. The clinical presentation of fatal, infantile myopathy, mild myopathy with metabolic changes, and completely healthy state observed in homoplasmic mutations carriers of the m.14674T>C mutation highlight the difficulty of diagnosing these cases and suggest that so far unknown genetic, epigenetic, or environmental modifiers play a very important role in the clinical manifestation of the disease. This is well illustrated by the fact, that the case's mother, who never showed symptoms of a mitochondrial myopathy, also carries the homoplasmic m.14674T>C mutation.

TREATMENT STRATEGY

In contrast to fatal infantile neuromuscular diseases (such as spinal muscular atrophy, severe congenital myopathies, or muscular dystrophies) cases with RIRCD typically have an excellent prognosis and spontaneously improve and recover after the sixth month of life, therefore these cases should receive all intensive care and life-sustaining measures. Mechanical ventilation, tube feeding, monitoring, and treating other organ dysfunctions result in most RIRCD cases in regression of clinical symptoms. Because some cases presented with low levels of L-carnitine or mitochondrial cofactors (coenzyme Q10, thiamine, riboflavin), supplementing these factors may facilitate the clinical improvement of this condition. Low dietary cysteine intake has been suggested to compromise mitochondrial translation in infants carrying the homoplasmic m.14674T>C/G mutation, therefore supplementation with L-cysteine or N-acetyl-cysteine can be considered, especially in infants with clinical manifestations.

LONG-TERM OUTCOME

An excellent long-term outcome of cases with RIRCD is a hallmark of this condition. About half of the cases recover completely without any residual symptoms; however, mild muscle weakness, muscular hypotonia, and exercise intolerance can persist. Few cases have died in the severe, symptomatic phase; however, with adequate therapy the number of fatalities can be significantly reduced. An early diagnosis in this family may have prevented the death of the two brothers of the individual. The incomplete penetrance and tissue-specific manifestation of RIRCD raise the possibility of further genetic or epigenetic modifiers. However, even without identifying these modifiers, a careful genetic counseling and adequate perinatal care of children within families carrying the homoplasmic m.14674T>C/G mutation can prevent severe or fatal complications.

PATHOPHYSIOLOGY/NEUROBIOLOGY OF DISEASE

The first cases with this disease were reported by Salvatore DiMauro in 1981[1] followed by a few more case reports in the following years, until the identification of the m.14674T>C homoplasmic mt-tRNA mutation as the molecular cause of this disease [2]. The most common symptom is myopathy, which may be life-threatening in the first months of life resulting in respiratory failure and feeding difficulties, and motor developmental milestones are often delayed. A mild residual myopathy with ptosis, ophthalmoparesis, facial weakness, and limb weakness as remaining features may be present in later age. Muscle biopsies taken in the neonatal period detected numerous ragged red fibers and even more COX-negative fibers, as well as either isolated or combined deficiency of the mitochondrial RC enzymes. These changes significantly improved, but they usually did not completely disappear on later biopsies [2–4].

Another mitochondrial disease due to deficiency of the mitochondrial tRNA modifying enzyme 5-methylaminomethyl-2-thiouridylate methyltransferase (TRMU) causes severe liver failure in infancy, but similar to RIRCD, within the first years of life these infants may also recover completely [5,6]. Based on the potential functional link (thio-modification of mitochondrial tRNA^{Glu}) in reversible mitochondrial conditions, it has been suggested that the reversibility may be due to a compensatory improvement of mitochondrial translation [7]. In further support, partial recovery was noted in cases with mutations in other two recently identified mt-tRNA^{Glu} modifying genes (*EARS2* and *MTO1*) [8]. Understanding these mechanisms may provide the key to treatments of potential broader relevance in mitochondrial disease, where for the majority of the cases no effective treatment is currently available. If we understand what prevented the manifestation of the disease in the homoplasmic mother of the case, we may be able to develop better treatments for mitochondrial disease.

CLINICAL PEARLS

- Reversible mitochondrial myopathy (RIRCD) should be considered in infants presenting with muscle weakness (floppy baby) in the first days or months of life, and atypical or missing family history is not uncommon.
- Genetic testing for the m.14674T>C/G mutation is fast and, therefore, it should be performed prior to biopsy; however, in some cases, we cannot avoid a diagnostic muscle biopsy.
- Syndromic features and dysmorphologic signs are very rare in mitochondrial diseases; however, a myopathic facial appearance with an elongated face and chin, bilateral ptosis, narrowed up-slanting palpebral fissures, and an elongated, prominent nose are not uncommon in RIRCD.
- The possibility of a reversible mitochondrial disease should be always considered in infants with severe clinical symptoms such as muscular hypotonia and weakness (floppy baby), respiratory problems, and feeding difficulties. Genetic testing is simple and relatively cheap and, if positive, indicates intensive life-sustaining measures to keep babies alive who have a good prognosis.

ACKNOWLEDGMENT

RH is supported by the Medical Research Council (UK) (G1000848) and the European Research Council (309548).

REFERENCES

[1] DiMauro S, Nicholson JF, Hays AP, Eastwood AB, Koenigsberger R, DeVivo DC. Benign infantile mitochondrial myopathy due to reversible cytochrome c oxidase deficiency. Trans Am Neurol Assoc 1981;106:205–7.

[2] Horvath R, Kemp JP, Tuppen HA, et al. Molecular basis of infantile reversible cytochrome c oxidase deficiency myopathy. Brain 2009;132:3165–74.

[3] Mimaki M, Hatakeyama H, Komaki H, et al. Reversible infantile respiratory chain deficiency: a clinical and molecular study. Ann Neurol 2010;68:845–54.

[4] Uusimaa J, Jungbluth H, Fratter C, et al. Reversible infantile respiratory chain deficiency is a unique, genetically heterogenous mitochondrial disease. J Med Genet 2011;48:660–8.

[5] Zeharia A, Shaag A, Pappo O, et al. Acute infantile liver failure due to mutations in the TRMU gene. Am J Hum Genet 2009;85:401–7.

[6] Schara U, von Kleist-Retzow JC, Lainka E, et al. Acute liver failure with subsequent cirrhosis as the primary manifestation of TRMU mutations. J Inherit Metab Dis 2011;34:197–201.

[7] Boczonadi V, Smith PM, Pyle A, et al. Altered 2-thiouridylation impairs mitochondrial translation in reversible infantile respiratory chain deficiency. Hum Mol Genet 2013;22:4602–15.

[8] Boczonadi V, Bansagi B, Horvath R. Reversible infantile mitochondrial diseases. J Inherit Metab Dis November 19, 2014. [Epub ahead of print].

Chapter 16

Childhood Alpers-Huttenlocher Syndrome

Robert K. Naviaux[1,2]

[1]*The Mitochondrial and Metabolic Disease Center, Departments of Medicine, Pediatrics, and Pathology, University of California, San Diego School of Medicine, San Diego, CA, USA;* [2]*Veterans Affairs Center for Excellence in Stress and Mental Health (CESAMH), La Jolla, CA, USA*

CASE PRESENTATION

Birth and Delivery

BW (Figure 1) was the second child born to a 31-year-old, gravida 2, para 2 Caucasian woman and her 35-year-old husband. There was no consanguinity. Maternal ancestry was Irish Catholic. Paternal ancestry was Russian Jewish. The pregnancy was uncomplicated and full term. Labor, birth, and delivery were spontaneous and uncomplicated. Birth weight was 8.0 pounds. Physical and neurological examinations were normal, and the child was discharged to home without complications.

Developmental History

Growth and development were normal in the first 18 months. BW sat at 5 months, crawled at 7 months, walked at 14 months. He spoke his first words at 12 months. There was no history of failure to thrive.

Clinical Course

At 19 months of age, he experienced an acute febrile illness, with anorexia, vomiting, and diarrhea. This was complicated by acute truncal ataxia, lethargy, and hypertonicity. Hypoglycemia of 34 mg/dL was noted and treated. BW recovered from this episode, but with residual hypotonia, and mild ataxia. A similar episode occurred with an upper respiratory tract infection and encephalopathy at 22 months of age. Loss of expressive language milestones and worsened ataxia and hypotonia were noted. There were six more episodes of neurodegeneration over the next 2 years. These were typically associated with intercurrent coryzal

Mitochondrial Case Studies. http://dx.doi.org/10.1016/B978-0-12-800877-5.00016-4

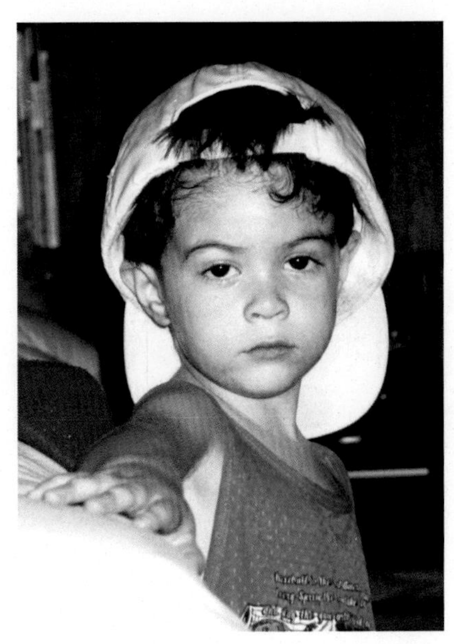

FIGURE 1 **BW was the first child with Alpers syndrome ever to be diagnosed at the protein [1] and DNA [2] levels.** Molecular studies in BW led to the discovery of the genetic cause of Alpers syndrome and brought to a close a medical mystery that began with the description of the first case in 1931 [3]. *Photo by parental permission.*

or diarrheal febrile illnesses. Liver transaminases were noted to be chronically elevated at 1.5- to 2-fold normal since 22 months of age. At 30 months, he developed a mixed-type seizure disorder with myoclonus. Nystagmus, worsened truncal ataxia with a 12-inch broad-based gait, and a reduction of expressive language to just four words were noted at 34 months of age. The boy used sign language to communicate at this time. By 38 months of age, he was admitted to the hospital with another acute episode of neurodegeneration. He could no longer walk, and cortical blindness occurred. At 41 months, he contracted a rotavirus gastroenteritis. At this time, he experienced a 2-h period of left-sided epilepsia partialis continua. The infection triggered a bout of acute fulminant liver failure with an AST of 1515, ALT of 749, GGT of 536 IU/L, ammonia of 183 μM, and total and direct bilirubin of 4.7 (normal ≤ 1.4) and 1.9 mg/dL (normal ≤ 0.3), respectively. Lactic acid peaked at 11.3 mM (normal ≤ 2.1) and fell back to 2.9–3.7 mM over the next week. Liver enzymes gradually fell to AST of 294 (normal 8–48), ALT of 96 (normal 7–55), and GGT of 69 (normal 7–19) over the next week. Chronic liver failure became apparent, and the individual died in hepatic coma, with worsening coagulopathy and jaundice at 42 months of age. The diagnosis of Alpers syndrome (also called Alpers-Huttenlocher syndrome; AHS) and the discovery of POLG enzymatic deficiency were made posthumously and reported first in an abstract [4], then later as a full clinical report [1]. The causal *POLG* DNA mutations were reported a few years later [2].

Metabolic Evaluation

The bout of encephalopathy and hypoglycemia at 22 months of age led to a comprehensive metabolic evaluation focusing on disorders of fatty acid oxidation. A monitored fast revealed hypoglycemia (33 mg/dL) after 15 h, with normal ketogenesis (fasting acetoacetate 0.3–1.0 mM; beta-hydroxybutyrate 0.4–1.2 mM). Non-specific elevations of C6–C10 urinary dicarboxylic acids, and *trans*-cinnamoyl glycine (TCG) occurred. TCG is a liver and microbiome metabolite of phenylalanine. Both medium chain and long chain fatty acid triglyceride loads both reproduced the increase in TCG. 3-hydroxy dicarboxylic acids also increased, indicating a relative block in the mitochondrial NADH-dehydrogenase step of fatty acid oxidation. Quantitative studies with ^{14}C-palmitate in skin fibroblasts were normal. Plasma free carnitine was mildly reduced at 13.6 μM. A blood lactate screen for mitochondrial disorders was normal at 2.1 mM (Normal $\leq$ 2.1 mM).

Neuroimaging

Brain magnetic resonance imagings (MRIs) at 30 months and 40 months of age were normal. Brain computed tomography (CT) at 42 months of age showed cerebral atrophy and a right thalamic hypodensity.

Cerebrospinal Analysis

Cerebrospinal fluid (CSF) protein was elevated at 157 mg/dL. CSF lactate and pyruvate were 3.22 and 0.046 mM, respectively. The calculated lactate to pyruvate ratio was 70 (normal ratio = 15–20).

Electrophysiology

Serial electroencephalograms (EEGs) between 30 and 39 months of age showed shifting epileptiform discharges that were first localized over the left hemisphere, then later migrated to the right posterior temporoparietal and other areas. The EEG at 39 months of age confirmed cortical blindness as the absence of an occipital driving response to photic stimulation. Continuous, high-amplitude (150–300 μV), slow-wave delta (1–3 Hz) background activity was present during wakefulness. This was intermixed with low-amplitude, sharply contoured (polyspike) beta (15–20 Hz) frequencies during the wake state.

Muscle Respiratory Chain Biochemistry

Respiratory chain enzymology showed 8–25% residual activity of complexes I, II/III, and IV, adjusted for citrate synthase activity. Complex IV (cytochrome c oxidase) was most severely affected with a residual activity of 8%. Isolated assay of complex II was normal. Citrate synthase activity showed 30% increase, consistent with mild mitochondrial proliferation.

Skeletal Muscle Mitochondrial POLG Enzymology

POLG catalytic activity [5] in mitochondria isolated from both muscle and liver was less than 5% of normal. Mixing experiments of case samples with controls confirmed the absence of any inhibitory activity.

mtDNA Mutation Analysis

Mitochondrial DNA analysis for six common point mutations (A3243G, T3271C, A8344G, T8356C, T8993G, and T8993C), and Southern analysis for mtDNA deletions, duplications, and rearrangements were normal.

mtDNA Content

Skeletal muscle mtDNA copy number at 42 months of age was 30% of age-matched controls. The liver contained 25% of normal mtDNA content.

Liver Histology and Electron Microscopy

Hematoxylin, trichrome, and oil red O stains of the liver at 42 months of age showed advanced micronodular cirrhosis, regenerative nodules, hepatocyte dropout, isolated necrotic hepatocytes, bile ductular proliferation, intrahepatic cholestasis, microvesicular steatosis, collapse of liver cell plates, and lobular disarray. Electron microscopy showed marked variation in mitochondrial numbers in adjoining hepatocytes. Some hepatocytes showed oncocytic change, with markedly proliferated, pleomorphic mitochondria, with tightly packed and occasionally concentric cristae. Neighboring cells had normal mitochondrial number and morphology.

Muscle Histology and Electron Microscopy

Skeletal muscle histology showed type II fiber atrophy, with increased intrafibrillar lipid predominantly affecting type I fibers. There were no ragged red fibers. However, modified Gomori-trichrome staining revealed diffusely distributed, eosinophilic filamentous bodies consistent with mitochondrial proliferation. The level of mitochondrial proliferation was insufficient to meet criteria for true ragged red fibers. Type I and type II fibers showed a decrease in the expected contrast observed by cytochrome *c* oxidase and succinate dehydrogenase staining. This was consistent with a reduction in fiber type specialization or differentiation compared to age-matched controls. Electron microscopy of skeletal muscle showed fiber disarray and mitochondrial proliferation with numerous pleomorphic forms between myofilaments. There were no subsarcolemmal accumulations of mitochondria or paracrystalline inclusions.

Neuropathology

Cerebellar atrophy was severe. Cerebral atrophy was milder. Microscopic analysis revealed marked cortical neuronal loss. The primary visual cortex in the region of the calcarine fissures showed diffuse, patchy neuronal dropout, principally affecting large pyramidal neurons in cortical lamina II, III, and V. This was associated with spongiform change in the same regions. Sections of the optic nerves showed prominent spongiform vacuolization in the chiasm and tracts. Spongy vacuolization was also present in the caudate, pons, and both anterior and posterior spinocerebellar tracts of the spinal cord. Cerebral subcortical white matter was normal. Global myelination was normal by Luxol fast blue staining. Reactive gliosis and Alzheimer type II glia were present. In the cerebellum, there was severe cortical atrophy, with near complete loss of cerebellar Purkinje cells. Focal sclerosis was present in the vermis. There was prominent Bergmann gliosis, with relative sparing of the granular layer. Prominent spongy vacuolization was noted in the interfolial white matter of the cerebellum.

Family History

The case had an older brother who died 7 years earlier with similar symptoms, at about 2.5 years of age (Figure 2). Development was normal until 15 months of age. Three days after a diarrheal illness, he became ataxic and presented to the hospital in status epilepticus. He developed cortical blindness and liver failure. Brain CT at 2 years of age showed severe cerebral and cerebellar atrophy with hydrocephalus ex vacuo. He died in hepatic coma with jaundice, ascites, and anasarca. Liver histology showed hepatocyte dropout, lobular disarray, micronodular cirrhosis, bile ductular proliferation, and microvesicular steatosis. Electron micrographs of the liver showed marked proliferation of pleomorphic mitochondria. Neuropathology showed reactive gliosis, neuronal and cerebellar Purkinje cell dropout. There was cortical pseudolaminar necrosis and spongiform change, principally affecting the large pyramidal neurons in layers III and V. The pseudolaminar spongiform change was most prominent in the occipital cortex.

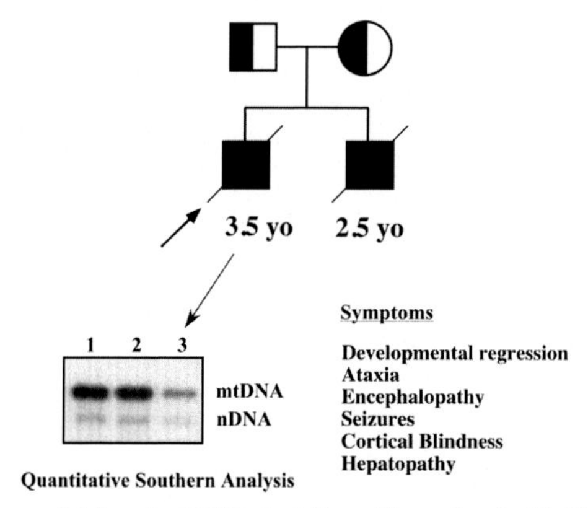

FIGURE 2 **Pedigree and delayed mtDNA depletion.** The proband with Alpers syndrome is indicated by the arrow. An older brother died at 2.5 years of age. Skeletal muscle Southern Analysis: Lane 1–Control. Lane 2–Proband at 1.5 years of age showing absence of mtDNA depletion. Lane 3–Proband at 3.5 years of age showing mtDNA depletion.

POLG DNA Sequencing

A total of 78 PCR primer pairs were designed and used to sequence the 4.5 kb cDNA expressed from the *POLG* gene. A total of 59,935 nucleotides of cDNA were sequenced. A total of 108 PCR primer pairs were designed and used to sequence the 21.1 kb contig of the *POLG* genomic DNA on chromosome 15q25. A total of 83,241 nucleotides of genomic DNA was sequenced. As this was the first large-scale sequencing of case DNA in Alpers syndrome, several incidental polymorphisms were found. One of these was a four nucleotide indel (GTAG) in intron 17, now known to have an allelic frequency of 53% in the population. The discovery of this indel and the redesign of PCR primers in this region confirmed that BW was heterozygous for a single nucleotide substitution in codon 873 in exon 17 that converted a glutamate (E) codon to a TAG stop codon. Other sequencing reactions identified a heterozygous substitution in *POLG* codon 467 in exon 7 that converted an alanine (A) to a threonine (T).

This overall DNA sequencing strategy led to the discovery that specific mutations in *POLG* were the cause of Alpers syndrome [2]. The term Alpers syndrome is used synonymously with AHS in this chapter. BW was found to carry two compound heterozygous *POLG* mutations. The genotype was A467T/E873X.

DIFFERENTIAL DIAGNOSIS

The age of onset and the symptoms at the time of presentation will direct the initial differential diagnosis of Alpers syndrome. The full symptom complex of AHS does not occur with the first presentation, and the clinical course is episodic, making early diagnosis difficult. The differential diagnosis includes the following: (1) specific mitochondrial respiratory chain defects, for example, Complex I deficiency or Complex IV deficiency; (2) mitochondrial respiratory chain assembly factor, tRNA modification, or translation defects; (3) non-Alpers mtDNA depletion syndromes such as those caused by mutations in deoxyguanosine kinase (*DGUOK*), *Twinkle* helicase (*PEO1*), and *MPV17*; (4) the mitochondrial phenylalanyl tRNA synthetase (*FARS2*); (5) SCN1A deficiency, with or without Dravet syndrome; (6) Pelizaeus-Merzbacher disease; (7) mitochondrial encephalomyopathy, with lactic acidemia and stroke-like episodes (MELAS); (8) Ceroid lipofuscinosis with *CLN3* mutations (Batten disease); (9) carbohydrate-deficient glycoprotein syndromes; and (10) Wilson syndrome. Occasionally, the appearance of fasting hypoglycemia will lead to an evaluation for congenital ACTH deficiency, adrenocortical insufficiency, or fatty acid oxidation disorder.

DIAGNOSTIC APPROACH

The only definitive diagnostic test for AHS is *POLG* DNA testing [6]. Quantitative mtDNA depletion studies are unreliable early in the course of disease

TABLE 1 Diagnostic Criteria for Alpers-Huttenlocher Syndrome

1. Clinical triad of refractory seizures (defined as requiring two or more anticonvulsants for control), psychomotor regression, and hepatopathy
2. If liver dysfunction is clinically apparent, then at least 2 of the 11 findings below must also be present. Frank liver failure is not required. If liver dysfunction is subclinical, then 3 of 11 findings and a compatible *POLG* genotype are required:
   a. Brain proton magnetic resonance spectroscopy indicating reduced N-acetyl aspartate (NAA), normal creatine, and elevated lactate
   b. Elevated cerebrospinal fluid protein (>100 mg/dL)
   c. Cerebral volume loss (central more than cortical, with ventriculomegaly) on repeated magnetic resonance imaging or computed tomography
   d. EEG with multifocal paroxysmal activity with high-amplitude delta (δ)-like slowing (150–1000 μV, 0.5–3 Hz), and asymmetric, low-amplitude polyspikes (10–100 μV, 12–25 Hz)
   e. Cortical blindness or optic atrophy
   f. Abnormal visual-evoked potentials and normal electroretinogram findings
   g. Quantitative mitochondrial DNA depletion in skeletal muscle or liver (≤35% of the mean)
   h. Deficiency in polymerase γ (POLG) enzymatic activity (≤10%) in skeletal muscle or liver
   i. Elevated blood or cerebrospinal fluid lactate (3 mM) on at least one occasion in the absence of acute liver failure
   j. Isolated Complex IV or a combination of Complex I, III, and IV electron transport defects (≤20% of normal) upon liver respiratory chain testing
   k. A sibling with confirmed Alpers-Huttenlocher syndrome

Modified from Ref. [7].

because organ-specific symptoms precede measureable depletion. *POLG* DNA testing confirms the diagnosis in 85% of the cases that are suspected on the basis of the compatible clinical triad of refractory seizures, dementia, and hepatopathy. Frank liver failure is not required for diagnosis of AHS. Some degree of liver dysfunction is universal, although this can be clinically mild or subclinical. The diagnosis of AHS can be established in the absence of liver findings when three of 11 additional signs and symptoms are present (Table 1). Most cases will have at least one of the following *POLG* substitutions: (1) A467T, (2) W748S, (3) R627Q, (4) G848S, (5) T251I-P587L, or (6) T914P. However, over 60 mutations in *POLG* are now known that cause AHS, so accurate diagnosis requires full sequencing of the *POLG* gene. Diagnosis of AHS can be confirmed without *POLG* gene testing by meeting the diagnostic triad plus 2 of 11 additional features (Table 1), or documenting 3 of 8 known Alpers-associated histological findings in the liver (Table 2). However, the current standard of care is to perform *POLG* DNA testing. About 15% of cases of AHS are phenocopies resulting from rare clinical presentations of mutations in other genes such as the *Twinkle* helicase or *FARS2*, better known for producing other syndromes.

TABLE 2 Specific Histologic Findings of the Liver in Alpers-Huttenlocher Syndrome [7]

At least 3 of 8 features must be present (Wilson disease has been excluded):

1. Bridging fibrosis or cirrhosis
2. Bile ductular proliferation
3. Collapse of liver cell plates
4. Hepatocyte dropout or focal necrosis, with or without portal inflammation
5. Microvesicular steatosis
6. Oncocytic change (mitochondrial proliferation associated with intensely eosinophilic cytoplasm in scattered hepatocytes)
7. Regenerative nodules
8. Parenchymal disease or disorganization of the normal lobular architecture

Respiratory Chain Biochemistry and mtDNA Copy Number

Although classified as a mitochondrial DNA (mtDNA) depletion disorder, mtDNA copy number in muscle and liver is normal in more than 70% of the cases with AHS at the first onset of symptoms, and it cannot be used to establish diagnosis. In contrast, the enzymatic activity of the mtDNA polymerase γ (POLG) is less than 10% of normal in every tissue measured throughout life. Later in the course of disease, a muscle or liver biopsy may reveal several abnormalities. Selective deficiency in mitochondrial cytochrome *c* oxidase (COX, Complex IV) activity is sometimes seen. Still later, mtDNA depletion (≤35% of normal mtDNA copy number) is seen. Multiple mtDNA deletions may be seen in older children and adults with AHS. Progressive mtDNA depletion leads, in turn, to global reductions in specific complexes of the mitochondrial respiratory chain. These are Complexes I, III, IV, and V. Complex II is preserved in Alpers syndrome.

Cerebrospinal Protein and Electrophysiology

Other tests may help support the diagnosis. For example, the CSF protein is typically elevated, often above 100 mg/dL. The EEGs of AHS cases frequently change over time. Often, there are certain features such as continuous, high-amplitude (150–1000 μV), symmetric or asymmetric slow-wave (0.5–3 Hz) activity during wakefulness. These are intermixed with sharply contoured, low-amplitude, high-frequency (12–25 Hz) polyspikes. Visual-evoked potentials are often asymmetric and abnormal, while the electroretinogram is normal. Peripheral neuropathy is rare in the childhood forms of *POLG* disease but common in adult forms.

Neuroimaging and Spectroscopy

The brain MRI is most often normal when done early in the disease course. Diffusion tensor imaging can sometimes reveal abnormalities after an acute clinical event that cannot be seen on routine T2/FLAIR imaging. Later, there is evidence of brain shrinkage with greater central volume loss than cortical loss characterized by enlarged ventricles. This can be severe and have the appearance of hydrocephalus *ex vacuo*. Cerebellar hypoplasia may be prominent and is associated with ataxia and Purkinje cell loss. The brain proton magnetic resonance spectroscopy typically shows reduced NAA and elevated lactate over affected areas (typically near the visual cortex or basal ganglia).

Variable Symptom Onset and Severity among Siblings

The extent of brain abnormalities and the severity of liver disease can differ radically between siblings [1,8]. The first symptom in one sibling might be status epilepticus, while acute fulminant liver failure can be the first presenting symptom in another sibling [9]. Over time, the symptom complexes converge on the fully developed AHS phenotype.

Liver Disease

Progressive liver disease can produce cirrhosis and biosynthetic failure. Hypoalbuminemia occurs late. Chronic, low-level elevations (1.5- to 2-fold increases) in liver transaminases may occur for months to years before other signs of liver dysfunction appear. Blood alpha-fetoprotein levels may be elevated as the result of hepatocyte dropout and reactive micronodular regeneration. Transient hyperammonemia may occur. Relative deficiency of liver-synthesized anticoagulant proteins like proteins C and S and anti-thrombin III can result in systemic episodes of deep venous thrombosis. This can complicate the maintenance of central intravenous lines in the critical care setting. Progressive liver failure ultimately leads to deficiency of clotting factors and difficulties with hemostasis, progressive hyperammonemia, and cirrhosis. This can progress to hepatic coma. Liver transplantation is contraindicated in AHS, as transplantation has no effect on brain function, and all cases to date have gone on to die of progressive AHS brain disease.

Gastrointestinal Abnormalities

Gastrointestinal dysmotility is common and can complicate feeding in the later stages of AHS, as in many other mitochondrial disorders. Selective atrophy of the outer longitudinal muscular layer with fibrous change may occur in the small and large intestines, leading to problems with intestinal pseudoobstruction and dysmotility.

Cardiomyopathy

Clinically apparent cardiomyopathy occurs as a late finding in 10% to 20% of AHS cases. This is typically a dilated cardiomyopathy characterized by fiber disarray, myocyte dropout, and fatty infiltration.

The Role of Intercurrent Infections

Twelve of the 15 children (12/15=80%; 95% CI=52–96%) evaluated in the first *POLG*-confirmed clinical series of AHS presented with their first symptoms within 3–10 days of an intercurrent infection [7]. The clinical course after presentation is frequently punctuated by episodic setbacks that occur after common childhood infections. The frequency of infections is not greater in children with AHS than in their siblings. However, the sequelae from the infections are often, although not always, more severe. Viral infections appear to cause more problems than bacterial infections. Mitochondria are now known to play an important role in innate immunity—the first steps in cellular defense and the antiviral response [10]. Viral infections do not cause AHS. However, the cellular response to infection causes a shift in mitochondrial and cellular metabolism that can unmask disease in a child with underlying *POLG* mutations that were previously well compensated. Much more will be learned about this intimate connection between mitochondria, innate immunity, and the cell danger response in years to come.

PATHOPHYSIOLOGY

Molecular Biology

The 140 kDa, catalytic (α) subunit of the mtDNA polymerase γ (POLG) exists as an α/β_2 heterotrimer with the 55 kDa accessory (β) subunit (POLG2). The holoenzyme has a mass of 250 kDa and associates along the inner mitochondrial membrane with several other proteins and mtDNA that constitute the mtDNA replisome. The POLG protein can be divided into three functional domains. The amino third of the protein contains 3′–5′ proofreading exonuclease. The middle third contains a linker domain that is involved in POLG2 binding and processivity. The carboxyl third of POLG contains the DNA polymerase domain required for both mtDNA replication and base excision repair [11]. The finding that Alpers symptoms appear before mtDNA depletion can be documented suggests that these early symptoms may be the result of other POLG functions such as base excision repair [11], or they may be the result of the patchy nature of early abnormalities. The common A467T substitution produces an unstable catalytic subunit of POLG that folds poorly, does not bind to POLG2, and results in an enzyme with just 4% of wild-type polymerase activity [12].

Genetics

Alpers syndrome is an autosomal recessive genetic disease with a frequency of about 1:100,000 live births. Many cases die before an accurate diagnosis is made, so the true frequency is still an estimate. The *POLG* gene is located on chromosome 15q25. The accessory subunit of the mitochondrial polymerase γ is called *POLG2*. *POLG2* is located on chromosome 17q24, and mutations described to date have not been involved in AHS. The *POLG* gene contains 23 exons, encodes a 4.5 kb mRNA, a 1239 amino acid protein, and is about

21,100 DNA base pairs long. Cases with homozygous A467T (A467T/A467T) or W748S (W748S/W748S) mutations have a milder clinical course than cases carrying the compound heterozygous (A467T/W748S) genotype [13]. Over 60 AHS-causing mutations have been documented. The NIEHS POLG Website is recommended as a valuable resource (http://tools.niehs.nih.gov/polg/).

A Note about A467T Homozygosity

The homozygous A467T/A467T genotype is highly protean [14]. It can present in at least four clinically distinct syndromes. It can present as a slightly older onset form of classical AHS in a 7–10 years old child. It can present in the teen or young adult years as an ataxia neuropathy spectrum disorder with or without myoclonus epilepsy, including sensory ataxia, neuropathy, dysarthria, and ophthalmoplegia (SANDO). It can present as an autosomal recessive progressive external ophthalmoplegia (arPEO) or SANDO in adults 25–40 years old. Finally, it can present as an MNGIE (mitochondrial neurogastrointestinal encephalomyopathy)-like disorder with significant gastrointestinal symptoms and ataxia, without leukoencephalopathy after 40 years of age. Overlap phenotypes are common. The presence of epilepsy is a strong determinant of disease severity. As with several mitochondrial disorders, the clinical symptoms are more correlated with the age of onset than with the underlying genotype. In the case of A467T, the sequence is as follows: (1) childhood brain and liver disease, (2) young adult brain disease and peripheral neuropathy, (3) adult PEO or SANDO, and (4) adult MNGIE-like disorder with ataxia. The allelic frequency of the A467T substitution in POLG ranges from less than 0.1% in Italian cohorts, 0.6% in Belgium, to 1% in Norway. The reason for this north–south gradient in gene frequencies is currently unknown, but it may relate to either founder effects or ecological selection.

Tissue Specificity

The basis for the selective involvement of the brain and liver in AHS is currently unknown. *POLG* mRNA and protein are expressed in all tissues during normal development. Enzymatic deficiency in POLG activity can be measured in every tissue in AHS, although frank mtDNA depletion is more selective. Late in the course of disease, mtDNA depletion is present in brain, liver, and skeletal muscle. However, mtDNA copy numbers are often preserved in heart, kidney, and skin fibroblasts, making these tissues unsuitable for confirming mtDNA depletion in Alpers syndrome.

CLINICAL PEARLS

Diagnosis

- The symptoms of AHS evolve over time. Periods of stasis and partial recovery are punctuated by episodes of neurodegeneration that often coincide with intercurrent infections.

- The pattern of symptoms, but not the sequence of their appearance, is diagnostic. The first symptoms to appear are typically nonspecific, making early diagnosis difficult or impossible without a high index of clinical suspicion.
- One of the common nonspecific symptoms is fasting hypoglycemia in the first 2 years of life. This is sometimes documented during the evaluation of a febrile seizure with encephalopathy that occurred during an illness. Fasting hypoglycemia in this context results from a combination of liver immaturity and secondary defects in mitochondrial fatty acid oxidation. Fasting hypoglycemia is observed in several different mitochondrial respiratory chain disorders in the first 2 years of life. When fasting hypoglycemia is the result of central or peripheral adrenal cortical insufficiency or primary fatty acid oxidation disorder, encephalopathy does not persist for more than a few hours after blood sugars are corrected.
- If encephalopathy or postictal neurological changes last more than 12–24 h after a presumed febrile seizure in a child under 2 years, despite correction of hypoglycemia, mitochondrial disease must be suspected, and *POLG* disease, in particular, must rank high in the differential diagnosis.

Classification as an mtDNA Depletion Disorder

- Neither quantitative mtDNA studies looking for mtDNA depletion nor mitochondrial respiratory chain biochemistry are consistently abnormal early in the course of disease and are, therefore, not useful for early diagnosis.
- Serious symptoms may be present in POLG cases despite documentation of normal mtDNA copy numbers and the absence of deletions in the affected tissues. While it is true that depletion and/or deletions often appear with disease progression, they are not universal.
- These molecular facts cast doubt on the quantitative role of mtDNA copy number or deletions in AHS pathogenesis and open the door for a more subtle, POLG-mediated mechanism. One such mechanism might be the progressive accumulation of base excision repair intermediates such as deoxyribose phosphate (dRP) residues in mtDNA, with downstream effects on transcription and translation. Future studies will be required to answer these questions.

Nomenclature

- The use of the terms Alpers disease or Alpers-like disease to describe the heterogeneous group of disorders with refractory seizures and dementia but without associated AHS symptoms (Table 1), hepatopathy (Table 2), or a *POLG* DNA diagnosis is discouraged and should be abandoned.
- Nonspecific usage of the Alpers diagnosis conveys a false sense of diagnostic security to the family, and it carries no useful genetic, prognostic, pathophysiologic, or medical guidance in case care.

- When the diagnostic criteria described in Tables 1 and 2 are followed, Alpers syndrome and AHS are synonymous and describe a single, classical neurogenetic disease that was first described in part over 80 years ago [3].

Genetic Counseling

- Recurrence risk of Alpers syndrome is 25% in each subsequent pregnancy.
- Prenatal diagnosis is available.
- First-degree relatives of parents of children with confirmed AHS and POLG mutations have a 50% risk of being silent carriers.
- The carrier frequency of Alpers-associated POLG mutation in the general population is about one in 150. Carrier frequency for one of the other 300+known, non-Alpers *POLG* mutations is about one in 50 (2%). See http://tools.niehs.nih.gov/polg for a catalog of pathogenic *POLG* mutations.

Valproic Acid (Depakote) Toxicity

- Children with Alpers syndrome are very sensitive to valproic acid (Depakote, Depakene, Valproate, Divalproate) toxicity, and they should not receive this drug.
- Valproic acid reproducibly leads to liver toxicity or failure within 6–16 weeks of starting therapy, and it produces other toxic effects.
- Toxicity is not prevented by giving carnitine.
- Symptoms of valproate toxicity may include visual hallucinations, behavioral and personality changes, unexplained temporary loss of consciousness (syncope), atonic seizures, or paradoxical worsening of preexisting seizures.

Treatment

- AHS must currently be considered a lethal genetic disorder. Survival typically ranges from a few months to 10–12 years after the onset of symptoms. Few children with onset under 5 years of age live beyond their teens.
- Cerebral folate deficiency may occur and exacerbate seizures. This is diagnosed by the presence of low 5′-methyl-tetrahydrofolate (mTHF) levels in the cerebrospinal fluid. Cerebral folate deficiency can be treated with oral folinic acid (leucovorin; 0.25–1 mg/kg PO BID) with some improvement in seizures and markers of cerebral inflammation [15].
- Selenium supplementation may help in rare cases of deficiency [16] or *POLG* mutations that produce a selenyl-cysteine tRNA suppressible UGA stop codon.
- Regular monitoring for known complications such as liver disease, gastrointestinal dysmotility, primary and secondary hypoventilation, and rare cardiomyopathy is indicated.
- Physical and occupational therapy, avoidance of group settings that promote the spread of common seasonal, childhood respiratory infections, avoidance of fasting to prevent hypoglycemia, and attention to good nutrition can help to

ease symptoms and reduce the frequency of neurodegenerative episodes, but they are not proven to improve the overall severe prognosis.

- The uses of mitochondrial cocktails and supplements such as CoQ10, B vitamins, carnitine, creatine, or lipoic acid have not been studied systematically. However, these supplements have not been found to alter the course of disease in several expert practices.
- Clinical trials of other drugs for mitochondrial disease are ongoing.
- Effective management is best performed by a collaborative team of specialists including neurology, biochemical genetics, gastroenterology, general pediatrics, nutrition, and physical, occupational, and speech therapy when appropriate [17].

REFERENCES

[1] Naviaux RK, et al. Mitochondrial DNA polymerase gamma deficiency and mtDNA depletion in a child with Alpers' syndrome. Ann Neurol 1999;45:54–8.
[2] Naviaux RK, Nguyen KV. POLG mutations associated with alpers' syndrome and mitochondrial DNA depletion. Ann Neurol 2004;55:706–12.
[3] Alpers BJ. Diffuse progressive degeneration of the gray matter of the cerebrum. Arch Neurol Psych 1931;25:469–505.
[4] Naviaux RK, et al. DNA polymerase gamma deficiency in mitochondrial disease (Abstract 39). Ann Neurol 1996;40:295–6.
[5] Naviaux RK, Markusic D, Barshop BA, Nyhan WL, Haas RH. Sensitive assay for mitochondrial DNA polymerase gamma. Clin Chem 1999;45:1725–33.
[6] Saneto RP, Cohen BH, Copeland WC, Naviaux RK. Alpers-Huttenlocher syndrome. Pediatr Neurol 2013;48:167–78.
[7] Nguyen KV, Sharief FS, Chan SS, Copeland WC, Naviaux RK. Molecular diagnosis of Alpers syndrome. J Hepatol 2006;45:108–16.
[8] Morse 2nd WI. Hereditary myoclonus epilepsy: two cases with pathological findings. Bull Johns Hopkins Hosp 1949;84:116–33.
[9] Nguyen KV, et al. POLG mutations in Alpers syndrome. Neurology 2005;65:1493–5.
[10] Seth RB, Sun L, Ea CK, Chen ZJ. Identification and characterization of MAVS, a mitochondrial antiviral signaling protein that activates NF-kappaB and IRF 3. Cell 2005;122:669–82.
[11] Copeland WC, Longley MJ. Mitochondrial genome maintenance in health and disease. DNA Repair 2014;19:190–8.
[12] Chan SS, Longley MJ, Copeland WC. The common A467T mutation in the human mitochondrial DNA polymerase (POLG) compromises catalytic efficiency and interaction with the accessory subunit. J Biol Chem 2005;280:31341–6.
[13] Tzoulis C, et al. The spectrum of clinical disease caused by the A467T and W748S POLG mutations: a study of 26 cases. Brain 2006;129:1685–92.
[14] Neeve VC, et al. What is influencing the phenotype of the common homozygous polymerase-gamma mutation p.Ala467Thr? Brain 2012;135:3614–26.
[15] Hasselmann O, et al. Cerebral folate deficiency and CNS inflammatory markers in Alpers disease. Molecular Genet Metab 2010;99:58–61.
[16] Ramaekers VT, Calomme M, Vanden Berghe D, Makropoulos W. Selenium deficiency triggering intractable seizures. Neuropediatrics 1994;25:217–23.
[17] Cohen BH, Naviaux RK. The clinical diagnosis of POLG disease and other mitochondrial DNA depletion disorders. Methods 2010;51:364–73.

Chapter 17

Juvenile Alpers-Huttenlocher Syndrome

Christopher Beatty, Russell P. Saneto
Department of Neurology/Division of Pediatric Neurology, Seattle Children's Hospital/University of Washington, Seattle, WA, USA

CASE PRESENTATION

Our case was the second-born child of unrelated parents of northern European descent. She was completely healthy until 17 years of age. During the fall of her junior year of high school, she began to experience progressively more frequent headaches and fatigue. The headaches were described as left temporal episodic pain that were thought to be migrainous and treated with aspirin. Two months later, she began to experience left visual field auras associated with kaleidoscopic shapes that were occurring once per week. Her symptoms progressed to daily photophobia, nausea, and episodes of emesis. Her headaches became severe enough to necessitate coming home from school early and avoiding activities that she enjoyed in order to rest.

One month later, she was roller-skating with her friends and began to experience left-sided arm and leg twitching associated with a particularly severe headache. She was observed at home for an hour and then taken to a local emergency department where she was found to be febrile, and her twitching was identified as focal seizure activity. She received lorazepam and fosphenytoin, which briefly controlled the seizures. In the next 24 h, her seizures became near continuous evolving into epilepsia partialis continua (EPC) and eventually progressed to secondary generalized seizure activity necessitating a medically induced coma with pentobarbital and midazolam. She was subsequently transferred to our institution.

After her braces were removed, a magnetic resonance image (MRI) of the brain demonstrated subtle T2 and fluid-attenuated inversion recovery (FLAIR) abnormalities in bilateral thalami (Figure 1). Magnetic resonance angiogram (MRA) was unremarkable. Proton magnetic resonance spectroscopy (MRS) was notable for presence of a lactate peak, which was likely cerebrospinal fluid (CSF) based on the large voxel encompassing the CSF space and surrounding brain parenchyma. Laboratory evaluation was extensive. Evaluation of the blood

Mitochondrial Case Studies. http://dx.doi.org/10.1016/B978-0-12-800877-5.00017-6

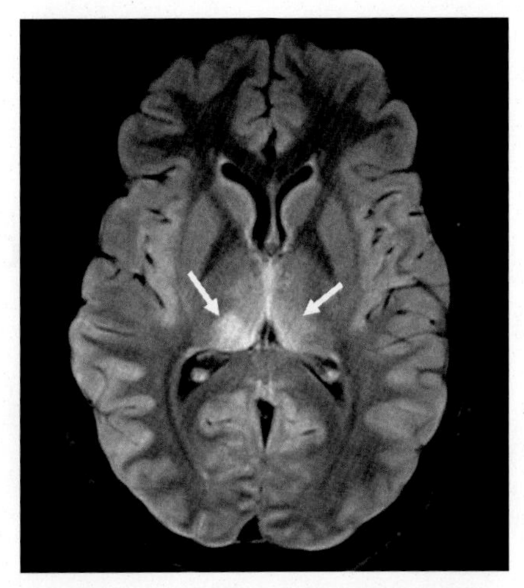

FIGURE 1 This is an axial MRI image (3Telsa) of the brain acquired on the day of seizure onset. The image is a T2 FLAIR sequence taken at the level of the basal ganglia. The increased signal is noted in the thalami, bilaterally, with the right medial thalamus more involved the than the left (arrows).

and urine demonstrated no evidence of infection. The CSF analysis showed elevated protein of 131 mg/dL (normal < 40 mg/dL). Autoimmune evaluation of N-methyl-D-aspartate receptor (NMDAR) and paraneoplastic antibodies were unremarkable. Metabolic evaluation yielded unremarkable plasma amino acids, urine organic acids, and acylcarnitine profile. Upon initial presentation her lactate was unremarkable; however within a week of seizure onset, her arterial lactate was 5 mM (<1.6 mM normal) and remained elevated for several days. Concerns of possible vasculitis prompted a brain biopsy, which was unrevealing.

Due to persistent electrographic seizures, the case remained in a medically induced coma with pentobarbital. Periodic reductions in the dosing of the pentobarbital induced electrographic seizures, and therefore suppression burst pattern continued for 6 weeks. Upon resolution of electroencephalogram (EEG) seizures, interictal EEG demonstrated rhythmic delta activity in the posterior quadrant at times with spike and wave components (Figure 2). The clinical features of EPC, status epilepticus, high CSF protein levels, high arterial lactate levels, preceded by a progressively worsening migraine with visual changes, prompted polymerase gamma 1 (*POLG*) gene sequencing. Sequencing demonstrated a likely homozygous mutation for p.A467T (c.1399G>A) mutation. The homozygosity was demonstrated as the mutation segregated with the mother and the father each having a single p.A467T mutation.

In addition to the visual field changes and kaleidoscopic shapes, in subsequent follow-ups she also complained of transitory vision loss. A repeat MRI scan demonstrated diffusion weighted imaging (DWI) changes in the right calcarine cortex, which corresponded to her transitory visual loss. Ocular coherence tomography was normal indicating an intact retina.

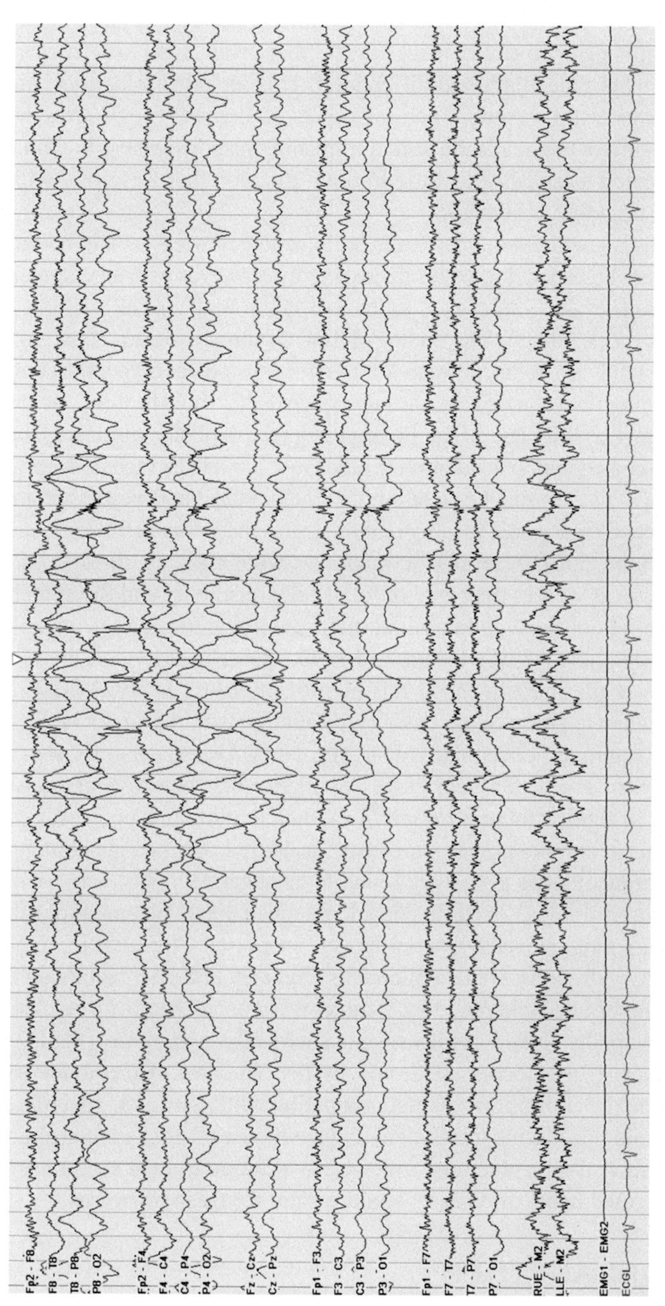

FIGURE 2 Electroencephalogram in an anterior–posterior bipolar montage demonstrating rhythmic delta activity in the right posterior quadrant with admixed spike components.

DIFFERENTIAL DIAGNOSIS

Our case of a 17-year-old girl with a prodrome of headaches and visual auras for 2–3 months prior to acute onset of EPC that progressed to refractory status epilepticus produced a broad differential. This included infectious, autoimmune, paraneoplastic, vascular, and metabolic etiologies. The progressive and explosive nature of the headache with diminishing responsiveness to pain medications indicated a serious alteration in brain function that was not compatible with typical migraine headaches.

There was a significant concern for an infectious process as she had a fever during her initial presentation with seizure activity; however with an unremarkable initial infectious evaluation, as well as the 2–3 month time course, a simple infectious process was less likely. The finding of EPC that progressed to refractory status epilepticus would also be uncommon for an infectious process. The most common etiology of a progressive headache and subsequent EPC would be tumor or vascular malformation. The MRI and MRA findings helped essentially eliminate these etiologies from consideration in addition to other vascular insults such as deep venous injury or vasculitis. The diagnosis of febrile illness-related epilepsy syndrome was considered, but the persistent focality of the seizures made this an unlikely etiology.

Autoimmune and paraneoplastic processes were strongly considered given the time course as well as the acute onset of seizure activity and the evidence of focal seizure semiology with her initial seizure activity. Rasmussen encephalitis was considered, however the location of EEG seizure activity and the bilateral involvement of the thalami made this significantly less likely. MRI findings during the subacute phase of our case's presentation would also not be the usual finding of Rasmussen encephalitis [1]. Anti-NMDAR encephalitis was a significant concern given the clear stages of illness that our case demonstrated that resulted in significant seizure activity. Other autoimmune processes such as postinfectious or paraneoplastic were also considered. A broad panel of autoimmune and paraneoplastic tests were found to be negative.

Metabolic etiologies were also considered given the involvement of bilateral thalami on imaging, elevated arterial lactate, increased CSF protein, explosive seizures, EPC, refractory status epilepticus, and elevated CSF lactate on MRS. The prodrome of headache and visual auras was concerning for mitochondrial encephalomyopathy, lactic acidosis, and stroke-like syndrome (MELAS) or Juvenile Alpers-Huttenlocher syndrome (AHS). The lack of metabolic strokes in a nonvascular distribution was not compatible with MELAS. Juvenile onset neuronal ceroid lipofuscinosis (Batten) and adult onset (Kufs) were also considered given the visual involvement followed by difficult to treat seizure activity; however the rapid onset of disease would be rare for these processes.

DIAGNOSTIC APPROACH

Juvenile AHS is a heptocerebral mitochondrial deoxyribonucleic acid (mtDNA) depletion disorder. The diagnosis is made by a combination of clinical findings, biochemical abnormalities, neuroimaging abnormalities, and gene sequencing of *POLG* DNA (Table 1). AHS, also known in the literature as Alpers syndrome, presents in children 2–4 years and in older adolescents to young adults, ages 15–25 [2–5]. First described by Harding et al., the latter group of cases have been named Juvenile AHS [6]. The older onset AHS cases have similar symptoms and diagnostic studies.

The important clues to the diagnosis of Juvenile AHS are the constellation of signs and symptoms that evolve as the disease progresses, similar to the childhood-onset type of AHS. Typically, early development is normal as in the childhood-onset form. There can be a variety of onset symptoms. Our case described a progressive migraine-like headache with unilateral visual changes. In case series of Juvenile AHS cases, the literature describes 6 of the 12 presented with migraine-like headache with visual changes [5–8], three presented with seizures, one had unilateral sensory loss and weakness, and another had ataxia [6–10]. The visual symptoms varied with descriptions of flashing lights, blurred vision, and scotoma in association with nausea and vomiting [6–9]. These symptoms typically preceded the onset of seizure activity by up

TABLE 1 Diagnosis of Alpers–Huttenlocher Syndrome

Naviaux Diagnostic Criteria for Alpers-Huttenlocher Syndrome Set Forth in 2006 [12,19]:

1. Clinical triad of refractory seizures, psychomotor regression, and hepatopathy
2. In the absence of hepatopathy or additional findings, the diagnosis can be confirmed by polymerase gamma gene sequencing, liver biopsy, or postmortem examination
3. Additional clinical findings (at least 2 of the 11 findings must be present):
   a. Cranial proton MRS indicating reduced N-acetyl aspartate, normal creatine, and elevated lactate
   b. Elevated cerebrospinal fluid protein (>100 mg/dL)
   c. Cerebral volume loss (central more than cortical, with ventriculomegaly) on repeated MRS imaging or computed tomography
   d. At least one electroencephalogram revealing multifocal paroxysmal activity with high amplitude delta slowing (200–1000 μV) and spikes/polyspikes (10–100 μV, 12–25 Hz)
   e. Cortical blindness or optic atrophy
   f. Abnormal visual-evoked potentials and normal electroretinogram findings
   g. Quantitate mitochondrial DNA depletion in skeletal muscle or liver (35% of the mean)
   h. Deficiency in polymerase gamma enzymatic activity (≤10%) in skeletal muscle or liver
   i. Elevated blood or cerebrospinal fluid lactate (3 mM) on at least one occasion in the absence of acute liver failure
   j. Isolated complex IV or a combination of complex I, III, IV electron transport defects (≤20% of normal) upon liver respiratory chain testing
   k. A sibling confirmed as manifesting AHS

to 11 months [7], and one case developed sensorineural hearing loss several years before seizure onset [10]. Five cases did not have complaints of headache during their reported disease course. Focal motor seizures are common at the onset of the epilepsy and can rapidly progress to EPC, but these cases can also develop generalized tonic-clonic seizures as well as status epilepticus [2,6–9]. All of the cases reported to date eventually developed intractable seizures with bouts of status epilepticus. Interictal EEG is similar to the childhood variant and demonstrates characteristic rhythmic unilateral or bilateral posterior delta activity with low amplitude spikes, or polyspikes as seen in our case [2–4,6–10].

The extent of liver involvement in Juvenile AHS is difficult to discern given the current literature. The majority of cases do not report liver function abnormalities at onset, and if present, are mild. As with the childhood form, liver failure is almost universal in cases exposed to valproic acid. This finding has led to the almost universal exclusion of valproic acid in all AHS cases [10]. Pathological findings are similar between onset age groups, suggesting a similar pathophysiological process in the different age groups [6]. Given the similarity with the childhood onset group, some degree of liver dysfunction is universal, but depending on timing of death, may not be clinically expressed. Few liver biopsies or autopsies have been reported, which may reduce how often hepatopathy has been reported in these cases. With so few Juvenile AHS cases reported and the lack of case descriptions without the use of valproic acid, we do not feel as though frank liver failure/hepatopathy is required for a diagnosis when a pathogenic *POLG* mutation is present. Our case is one of the only cases with Juvenile AHS that was not exposed to valproic acid. We are hopeful that liver involvement may not develop over time to the same extent that is seen in Childhood AHS.

Biochemical abnormalities have varied in reported cases. Elevations in blood lactate have been reported at onset of disease and were present in our case; however, more commonly it is within the normal range [6–8]. CSF lactate concentrations have been either normal or mildly elevated. In cases where proton magnetic resonance (MRS) was performed, as in our case, a lactate peak was seen [7–9].

Imaging of the brain with MRI has been fairly consistent across reported cases. T2/FLAIR hyperintensities are seen most commonly in unilateral or bilateral thalamic and occipital regions even early in the disease process. The early presence of basal ganglia and deep cerebellar nuclei lesions have also been reported in individual cases. DWI can demonstrate hyperintensity in the affected areas [2,3,5–9,11]. MRS in Juvenile AHS is not well documented in the literature, however, elevated lactate has been seen in the cerebrum, basal ganglia, and cerebellum [3,10].

Other testing can be valuable for diagnosis (Table 1). CSF analysis in Juvenile AHS may be valuable as most cases will likely have increased protein. In Childhood AHS, optic atrophy is often found with no retinal involvement and can serve as a discriminating finding unique to AHS when compared to other mitochondrial disease [12]. Our case had optic atrophy but with no retinal involvement.

Muscle biopsy in this population may or may not be helpful. Enzyme activity of the respiratory chain complexes is most likely abnormal only late in the disease. In one of our previously reported cases, a complex III dysfunction was noted [10]. However, the most likely complex to be affected is complex I, as there are seven subunits produced by the mitochondrial DNA [13]. The pathophysiology of *POLG* defects is from significant reduction in mtDNA, but early in the disease, the reduction may not be significant in all tissues. Depletion of mtDNA is most striking in the liver and muscle. In other *POLG* diseases, especially in older onset cases, mtDNA analysis demonstrated multiple deletions in non-AHS cases. Multiple deletions of mtDNA have not been described in AHS cases with *POLG* mutations.

With the discovery that AHS is caused by *POLG* mutations, genetic testing has become part of the diagnostic evaluation [14,15]. All the reported mutations giving rise to Juvenile AHS are autosomal recessive. The most common allelic mutations reported in Juvenile AHS are p. A467T homozygotes, with lesser numbers with autosomal recessive p. W748S and compound heterozygotes of the two loci [2,8,15,16]. Both of these mutations lie in the linker region of the protein and seem to give a milder clinical course than other mutations in the protein. This may be a partial reason for the later onset of Juvenile AHS. While the initial presentation of disease is similar in all the cohorts, those with homozygous p. A467T mutations appear to have decreased mortality with some cases surviving many years after onset. Many other *POLG* mutations have been identified in these cases [10], and thus gene sequencing should be performed as part of the evaluation as opposed to point mutation analysis.

Examining the Naviaux criteria, our case clearly had enough features to have a diagnosis of AHS even without frank clinical hepatopathy. She has AHS and given her age, Juvenile AHS.

PATHOPHYSIOLOGY

The holoenzyme of POLG exists as a heterotrimer composed of the catalytic subunit POLG with two subunits of polymerase gamma 2. This enzyme complex associates with the inner mitochondrial membrane and several other DNA replication proteins to form the mtDNA replisome. POLG represents the only mitochondrial specific replicase in mammals. The catalytic subunit of POLG has two functional domains; a 3′-5′ proofreading exonuclease and a DNA polymerase region. The latter is associated with both replication and base excision repair. These two domains are linked together by a linker region. This linker region also has a functional component of a binding site for polymerase gamma 2, which when bound to POLG enhances processing. The importance of the linker region can be seen with enzymatic activity studies demonstrating that with the common p. A467T mutation, activity is only 4% of wild type activity [17].

The reduced replication of mtDNA induces a loss of the gene products needed for proper mitochondrion function: 37 protein subunits of the respiratory chain and transfer ribonucleic acid (tRNA) and ribosomal ribonucleic acid (rRNA) needed for protein translation. Mutations in different regions of the *POLG* gene compromise mtDNA levels or function and eventually lead to adenine triphosphate (ATP) compromise. mtDNA depletion is found in younger cases with more severe disease, while less severe disease is expressed in older cases with mtDNA deletions. However, the exact mechanism is not clearly understood as the same mutations can lead to a varied severity of disease [17].

POLG mutations, including the ones leading to Juvenile AHS, produce multiple phenotypes. For example, the same homozygous p. A467T mutations can produce classic AHS, ataxia with neuropathy with or without myoclonic epilepsy, peripheral neuropathy, dysarthria, ophthalmoplegia, severe gastrointestinal dysmotility, Parkinsonism, ovarian failure, and leukodystrophy. Many of these symptoms have been found to overlap in individual cases. The specificity of organ involvement and age differential of onset is not clearly understood.

Some cases in Juvenile AHS are reported as having stroke-like lesions on neuroimaging and placed in an overlap syndrome with MELAS [18]. However, data is emerging from pathological studies that *POLG*-induced disease induces a distinct type of lesion that is actually necrosis and not apoptosis and is termed focal energy-dependent neuronal necrosis (FENN) [13]. So, although our case had DWI changes in the right calcarine cortex, this would be distinct to those lesions seen in MELAS, based on the pathophysiological etiology of the lesion in *POLG*-encoded AHS.

TREATMENT

Unfortunately, there is no universal treatment for the underlying mechanism of mitochondrial disease. Clinical treatment is based mostly on symptoms. Seizures are treated with antiseizure medications, with the strict withholding of valproic acid due to the risk of liver failure. Cerebral folate deficiency is often found and is treated with oral folinic acid (leucovorin). The addition of folinic acid can also be helpful in reducing seizures. Nutrition is very important for proper mitochondrial function. As the disease progresses, gastrointestinal dysmotility and delayed gastric emptying often prompts the need for gastrostomy placement. Migraine headaches are treated with over-the-counter pain medications and try to avoid prolonged use of acetaminophen, which can induce massive oxidation of glutathione in the liver.

Vitamins and cofactors are often used to try to limit the formation of excessive oxygen radicals and enhance respiratory chain activity. However, these supplements have not been found to alter the course of this disease in several expert practices, nor have they been studied systematically.

CLINICAL PEARLS

- The diagnosis of Juvenile AHS should be considered in a case with explosive seizures and especially if there is EPC or super/refractory status epilepticus.
- The EEG may be helpful in the differential diagnosis if the epileptiform discharges are predominantly in the posterior head region accompanied by slowing. These discharges and slowing are often shifting in hemisphere dominance.
- Valproic acid is an antiseizure medication that is commonly used in the setting of refractory seizures. However, in this case population, it has been linked to fulminant liver failure and should be avoided as it can drastically increase morbidity and alter the life span of affected individuals [10].
- Diagnosis of the disease cannot be based on routine metabolic screening labs as blood lactate, liver function tests, and CSF lactate can be normal. Additionally, muscle biopsy is not an effective tool in diagnosis given the limited data that show predominantly normal samples during the early part of the disorder. Diagnosis, therefore, must rest on a constellation of clinical and lab findings in addition to genetic testing. The multiple genetic mutations in the POLG gene makes sequencing the entire gene necessary.
- The varying phenotypes of specific allelic variants are important to grasp, given that the A467T homozygotes can have a very long life span if valproic acid is avoided. This understanding can dramatically alter goals of care and quality of life discussions.

REFERENCES

[1] Bien CG, Granata T, Antozzi C, et al. Pathogenesis, diagnosis and treatment of Rasmussen encephalitis: a European consensus statement. Brain: J Neurol 2005;128(Pt 3):454–71.

[2] Engelsen BA, Tzoulis C, Karlsen B, et al. POLG1 mutations cause a syndromic epilepsy with occipital lobe predilection. Brain: J Neurol 2008;131(Pt 3):818–28.

[3] Wolf NI, Rahman S, Schmitt B, et al. Status epilepticus in children with Alpers' disease caused by POLG1 mutations: EEG and MRI features. Epilepsia 2009;50(6):1596–607.

[4] SG B, Harden A, Egger J, Pampiglione G. Progressive neuronal degeneration of childhood with liver disease ("Alpers' disease"): characteristic neurophysiological features. Neuropediatrics 1986;17:75–80.

[5] Visser NA, Braun KP, Van den Bergh WM, et al. Juvenile-onset alpers syndrome: interpreting MRI findings. Neurology 2010;74:1231–3.

[6] Harding B, Alsanjari N, Smith S, et al. Progressive neuronal degeneration of childhood with liver disease (Alpers' disease) presenting in young adults. J Neurol Neurosurg Psychiatry 1995;58:320–5.

[7] Uusimaa J, Hinttala R, Rantala H, et al. Homozygous W748S mutation in the POLG1 gene in patients with juvenile-onset Alpers syndrome and status epilepticus. Epilepsia 2008;49(6):1038–45.

[8] Tzoulis C, Engelsen BA, Telstad W, et al. The spectrum of clinical disease caused by the A467T and W748S POLG mutations: a study of 26 cases. Brain: J Neurol 2006;129 (Pt 7):1685–92.

[9] Wiltshire EW, Davidzon G, DiMauro S, et al. Juvenile Alpers disease. Arch Neurol 2008;65(1):121–4.
[10] Saneto RP, Lee IC, Koenig MK, et al. POLG DNA testing as an emerging standard of care before instituting valproic acid therapy for pediatric seizure disorders. Seizure: J Brit Epilepsy Assoc 2010;19(3):140–6.
[11] Uusimaa J, Gowda V, McShane A, et al. Prospective study of POLG mutations presenting in children with intractable epilepsy-prevalence and clinical features. Epilepsia 2013;54(6):1002–11.
[12] Saneto RP, Cohen BH, Copeland WC, Naviaux RK. Alpers-huttenlocher syndrome. Pediat Neurol 2013;48(3):167–78.
[13] Tzoulis C, Tran GT, Coxhead J, et al. Molecular pathogenesis of polymerase gamma-related neurodegeneration. Ann Neurol 2014;76(1):66–81.
[14] Naviaux RK, Nyhan WL, Barshop BA, et al. Mitochondrial DNA polymerase y deficiency and mtDNA depletion in a child with Alpers' syndrome. Ann Neurol 1999;45(1):54–8.
[15] Naviaux RK, Nguyen KV. POLG mutations associated with Alpers' syndrome and mitochondrial DNA depletion. Ann Neurol 2004;55(5):706–12.
[16] Ferrari G, Lamantea E, Donati A, et al. Infantile hepatocerebral syndromes associated with mutations in the mitochondrial DNA polymerase-gammaA. Brain: J Neurol 2005;128 (Pt 4):723–31.
[17] Stumpf JD, Copeland WC. Mitochondrial DNA replication and disease: insights from DNA polymerase gamma mutations. Cellular and molecular life sciences. CMLS 2011;68(2):219–33.
[18] Deschauer M, Tennant S, Rokicka A, et al. MELAS associated with mutations in the POLG1 GENE. Neurology 2007;68:1741–2.
[19] Nguyen KV, Sharief FS, Chan SS, Copeland WC, Naviaux RK. Molecular diagnosis of Alpers syndrome. J Hepatol 2006;45(1):108–16.

SUGGESTED READING

[1] Harding B, Alsanjari N, Smith S, et al. Progressive neuronal degeneration of childhood with liver disease (Alpers' disease) presenting in young adults. J Neurol Neurosurg Psychiatry 1995;58:320–5.
[2] Saneto RP, Cohen BH, Copeland WC, Naviaux RK. Alpers-huttenlocher syndrome. Pediat Neurol 2013;48:167–78.
[3] Uusimaa J, Hinttala R, Rantala H, et al. Homozygous W748S mutation in the POLG1 gene in patients with juvenile-onset Alpers syndrome and status epilepticus. Epilepsia 2008;49:1038–45.
[4] Tzoulis C, Tran GT, Coxhead J, et al. Molecular pathogenesis of polymerase gamma-related neurodegeneration. Ann Neurol 2014;76:66–81.
[5] Stumpf JD, Copeland WC. Mitochondrial DNA replication and disease: insights from DNA polymerase gamma mutations. Cell Mol Life Sci 2011;68:219–33.
[6] Cohen BH, Chinnery PF, Copeland WC. POLG-Related Disorders. 2010 Mar 16 [updated 2014 Dec 18]. In: Pagon RA, Adam MP, Ardinger HH, Wallace SE, Amemiya A, Bean LJH, Bird TD, Dolan CR, Fong CT, Smith RJH, Stephens K, editors. GeneReviews® [Internet]. University of Washington, Seattle, WA, 1993–2015.

Chapter 18

Chronic Progressive External Ophthalmoplegia Secondary to Nuclear-Encoded Mitochondrial Genes

Patrick Yu-Wai-Man[1,2]

[1]Wellcome Trust Centre for Mitochondrial Research, Institute of Genetic Medicine, Newcastle University, UK; [2]Newcastle Eye Center, Royal Victoria Infirmary, Newcastle upon Tyne, UK

CASE PRESENTATION

A 48-year-old woman presented to her local ophthalmologist complaining of intermittent horizontal diplopia that was worse for near than for distance. Two years ago, prisms had been incorporated into her glasses by her optometrist when she first experienced these symptoms. The prisms proved effective, but over the past 6 months, the diplopia had recurred, and she decided to seek further medical advice. She had two older sisters, and both of them had been diagnosed with thyroid eye disease in their early 40s. Given the strong family history, she had an annual health check with her family physician, and 3 years ago, she was found to be hypothyroid, for which levothyroxine replacement was started. She was not on any other regular medication and there was no other past medical history of note. On examination, her visual acuities were documented at 20/20 bilaterally and visual fields were full. No eyelid retraction was documented. Instead, there was bilateral mild ptosis, which was thought to be involutional in nature secondary to aponeurotic dehiscence of the levator palpebrae superioris muscle. There was bilateral symmetrical limitation of eye movements that was worse on upgaze with an alternating exotropia. The patient reported horizontal diplopia in the primary position and on both dextroversion and levoversion. Dilated fundus examination was normal with no optic nerve or retinal abnormalities. An MRI scan of the brain and orbits with gadolinium contrast was performed and possible increased signals within the extraocular muscle bellies were noted bilaterally. Baseline blood tests, including inflammatory markers, thyroid function tests, and thyroid autoantibodies, were normal.

Mitochondrial Case Studies. http://dx.doi.org/10.1016/B978-0-12-800877-5.00018-8

A diagnosis of thyroid associated orbitopathy (TAO) was made and the patient was treated with a course of intravenous methylprednisolone followed by orbital radiotherapy. Six months later, she had bilateral strabismus surgery, which successfully controlled her diplopia with the continued use of corrective prisms in her glasses. One year later, the patient became aware of more persistent horizontal diplopia, and she was reviewed by the same ophthalmologist for further assessment. There was no clear diurnal pattern, and both single fiber electromyography (EMG) studies and anti-acetylcholinesterase antibodies were negative. Given the patient's persisting symptoms, she was referred to our institution for a second opinion in the neuro-ophthalmology clinic.

The patient attended with old photographs of herself, and it became evident that her mild ptosis was present in her late 20s, precluding age-related aponeurotic dehiscence of the levator palpebrae superioris muscle as the underlying cause. No fatiguability or Cogan lid twitch was elicited. There was slight weakness of the orbicularis oculi muscles, which did not extend to the other facial or bulbar muscles. Neurological examination was unremarkable with no proximal myopathy or ataxia. As noted in the patient's earlier assessment, there was bilateral, symmetrical limitation of eye movements, which was worse on upgaze (Figure 1). Ocular saccades were slow, but there was no cogwheel pursuits or dysmetria. The absence of lid retraction and normal thyroid autoantibody

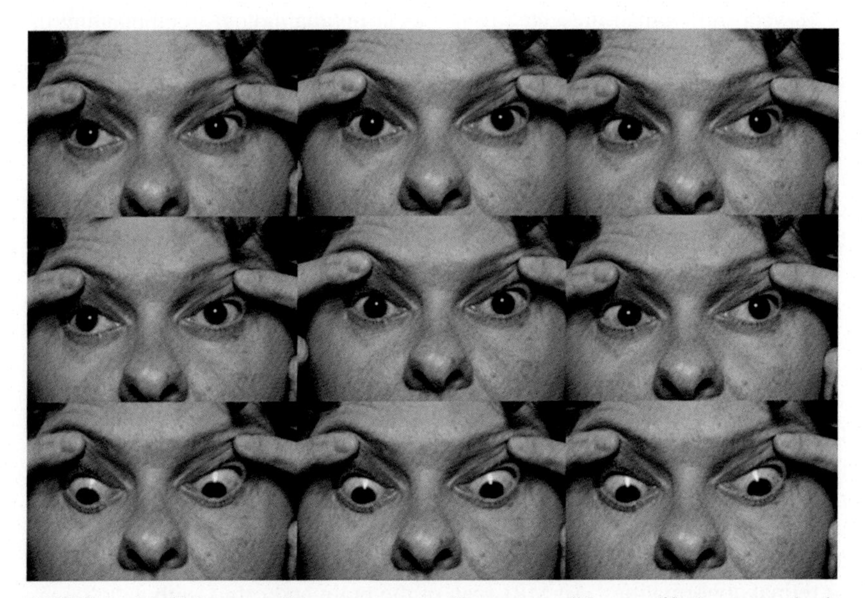

FIGURE 1 **Limited range of eye movements.** The upper eyelids were lifted to more clearly demonstrate the range of eye moments in the nine cardinal positions of gaze. There was significant generalized limitation of eye movements, which was worse on upgaze and mostly symmetrical. There is no manifest ocular deviation in the primary position as these pictures were taken after the patient had undergone bilateral strabismus surgery to correct an alternating exotropia.

profile were somewhat atypical for TAO, and a repeat MRI scan of the brain and orbit with gadolinium contrast was performed for review with the neuroradiology team. No abnormalities were detected within the brain parenchyma or in the orbit, and the extraocular muscles showed no high signal changes or altered morphology. Chronic progressive external ophthalmoplegia (CPEO) was considered in the differential diagnosis and a muscle biopsy was discussed with the patient. She phoned back 2 weeks later agreeing to the procedure, and in the meantime, additional blood tests, including serum lactate and creatine kinase levels, had come back as normal.

A fine needle biopsy of the anterior tibialis muscle was carried out under local anesthesia. The muscle biopsy specimen showed about 25% cytochrome *c* oxidase (COX)-negative muscle fibers—the pathological hallmark of an underlying mitochondrial cytopathy. Long-range PCR experiments did not identify a single mtDNA deletion, but there was clear evidence of multiple mtDNA deletions, prompting the search for a nuclear-encoded mitochondrial disease gene. An initial screen of common mutations within the polymerase gamma (*POLG*) gene identified a single heterozygous c.1399G>A (p.Ala467Thr) mutation. Sequencing of the entire *POLG* coding revealed a second heterozygous c.2209G>C (p.Gly737Arg) mutation, which confirmed a molecular diagnosis of autosomal recessive CPEO secondary to compound heterozygous *POLG* mutations. Genetic counselling was provided to the patient and to her two unaffected children. One of the patient's two sisters, who lived locally, was subsequently reviewed in the neuro-ophthalmology clinic, and she was also found to have CPEO secondary to the same compound heterozygous *POLG* mutations. The second sister with a diagnosis of thyroid eye disease was not contactable.

DIFFERENTIAL DIAGNOSIS

CPEO is a classical manifestation of mitochondrial disease, and it affects nearly half of all affected individuals harboring confirmed pathogenic mutations within nuclear-encoded mitochondrial genes. When assessing cases of bilateral external ophthalmoplegia, it is therefore important for clinicians to consider this diagnosis, especially when associated with myogenic ptosis and other clinical features pointing toward an underlying mitochondrial cytopathy. The majority of cases with CPEO will initially present complaining of progressive ptosis and/or diplopia, with a peak age of onset in the third and fourth decades of life depending on the specific genetic etiology. Both ptosis and diplopia are common symptoms that can be due to a broad spectrum of neuro-ophthalmological disorders. Furthermore, a diagnosis of CPEO is not always apparent in isolated cases without any obvious neurological or endocrinological features, and in the absence of a suggestive family history. Given these potential pitfalls, a rigorous approach is essential to avoid diagnostic delays, which are frequently compounded by unnecessary investigations and treatments.

CPEO is a slowly progressive disorder and few cases are aware of the limitation of eye movements unless symptoms of diplopia are present. Similarly, the ptosis develops insidiously, and the individual, not infrequently, only becomes aware of the reduced height of their upper eyelids when it is brought to their attention by a third party. A detailed clinical history will help to establish the chronology of the patient's symptoms, the presence of additional extra-ophthalmological features (if any), and other important clues pointing toward a primary mitochondrial etiology. The presence of obvious red flags such as pain, proptosis, periorbital swelling, and chemosis in the context of a rapidly progressive acute or subacute bilateral ophthalmoplegia, whether associated with ptosis or not, are not consistent with CPEO, and urgent investigations are warranted to exclude other acquired infectious, inflammatory, or neoplastic causes. Some specific disorders will be discussed below that can mimic CPEO, but there are usually distinguishing clinical and ancillary features that can help to differentiate them.

Myasthenia Gravis

As for CPEO, myasthenia gravis has a predilection for the extraocular muscles and associated weakness of the orbicularis oculi muscle is a common finding. More than 50% of cases with myasthenia gravis will present with pure ocular features characterized by variable ptosis, fluctuating patterns of ophthalmoplegia and diplopia, and frequently, a diurnal pattern with their symptoms being worse in the evening. An important caveat, however, is that involutional ptosis secondary to age-related aponeurotic dehiscence of the levator palpebrae superioris muscle also tends to become more prominent later in the day and with tiredness. In some reported case series, about 50% of cases with ocular myasthenia gravis develop generalized disease within 2 years of disease onset. Despite an extensive array of investigative approaches, a definite diagnosis can be elusive in the absence of a gold standard test. Anti-acetylcholinesterase antibodies are negative in 40% to 60% of cases. The ice pack test has gained increasingly popularity in the office setting for assessing ptosis reversibility. In specialist centers with skilled personnel, single fiber EMG of the orbicularis oculi muscle has a relatively high diagnostic sensitivity and specificity. A positive response to a trial course of acetylcholinesterase inhibitor and/or steroids also favors a diagnosis of myasthenia gravis. It is important to remember that the ophthalmoplegia in CPEO is generally bilateral, symmetrical, and fixed, with none of the variable pattern seen with ocular myasthenia gravis, which can mimic isolated cranial nerve palsies or more complex ocular motility limitations at different time points.

Thyroid-Associated Orbitopathy

TAO is usually a straightforward clinical diagnosis, particularly in the acute phase of the disease process when inflammation of the orbital and periorbital tissues result in well-recognized features such as proptosis, soft tissue swelling, conjunctival chemosis, and nonspecific ocular discomfort. The majority of cases (80%) are due to hyperthyroidism, predominantly in the context of Graves disease, but TAO can also be associated with the euthyroid (15%) and hypothyroid (5%) states. Other key distinguishing features are the presence of specific circulating thyroid autoantibodies and enlargement of the belly of the extraocular muscles that does not extend to the tendinous insertions. Lid retraction is virtually pathognomonic of Graves orbitopathy, although this can become less obvious as the thyrotoxicosis gets under medical control. Although uncommon, the fixed ophthalmoplegia pattern seen in longstanding inactive TAO can result in diagnostic uncertainties, and this possibility should be kept in mind when considering a diagnosis of CPEO. The clinical history, supplemented by autoantibody testing and imaging of the extraocular muscles when indicated, will usually suffice in excluding TAO.

Oculopharyngeal Muscular Dystrophy

Oculopharyngeal muscular dystrophy (OPMD) is an autosomal dominant disorder caused by pathological GCG trinucleotide repeat expansions within the *PABPN1* gene, which encodes for the polyadenylate-binding nuclear protein 1. Cases with OPMD usually present in the fifth decade of life with progressive ptosis, dysphagia, and proximal muscle weakness. About 60% of cases will develop significant ophthalmoplegia, but complete limitation of eye movements is rare and diplopia is uncommon. The diagnosis of OPMD is notoriously difficult to make, and clinicians should specifically probe for the presence of bulbar symptoms in patients who present with insidious extraocular features that could be consistent with CPEO.

Myotonic Muscular Dystrophy

Cases with type 1 Myotonic muscular dystrophy (MMD) carry pathological CTG trinucleotide repeat expansions in the *DMPK* gene. This dominantly inherited disorder not only affects skeletal and smooth muscle but is also associated with prominent ocular, cardiac, endocrine, and neurological sequelae. Ptosis and external ophthalmoplegia are well-recognized ocular features, and these invariably occur in the context of more generalized myofacial muscle weakness, resulting in a typical facies when accompanied with frontal balding. When MMD is being considered in the differential diagnosis, a dilated fundus examination can be very helpful, as a Christmas tree cataract, consisting of highly refractile, multicolored, spoke-like opacities, is present in nearly all affected individuals with this disorder. Cases with advanced MMD also have involvement of the oropharyngeal muscles that results in progressive dysphagia.

Congenital Cranial Dysinnervation Disorders

Congenital fibrosis of the extraocular muscles (CFEOM) is a genetically heterogeneous group of disorders characterized by nonprogressive ophthalmoplegia, which can be complicated with ptosis in some cases. Vertical gaze is usually more severely limited that horizontal gaze, and cases frequently adopt an abnormal head position to compensate for the incomitant strabismus. So far, eight nuclear genes causing CFEOM phenotypes have been identified, and a new classification, under the label of congenital cranial dysinnervation disorders (CCDDs), has been proposed to highlight our improved understanding of the primary pathological process. Although CFEOM is the end-stage phenomenon that accounts for the observed ophthalmoplegia, all known CCDDs genes regulate critical pathways involved in oculomotor neuron development at either the nuclear, brain stem, or peripheral nerve levels, clearly implicating a primary neurogenic etiology. The clinical presentation in early childhood and the nonprogressive nature of the ocular manifestations in CCDDs are usually sufficient pointers to allow a clear distinction to be made from classical CPEO.

DIAGNOSTIC APPROACH

Once a diagnosis of CPEO is suspected, the patient should have a comprehensive ophthalmological and systemic evaluation. A dilated fundus examination should be carried out to look for evidence of optic atrophy and/or pigmentary retinopathy (Figure 2). If the patient has subnormal vision, visual electrophysiology can be very useful as the electroretinogram (ERG) and visual evoked potentials (VEPs) can help pinpoint the anatomical location and extent of retinal dysfunction. Baseline assessment by a trained orthoptist is recommended, if possible, to document the range of eye movements, the presence of manifest or latent strabismus, and the need for corrective prisms for cases with symptomatic diplopia. The majority of cases with CPEO will have multisystemic

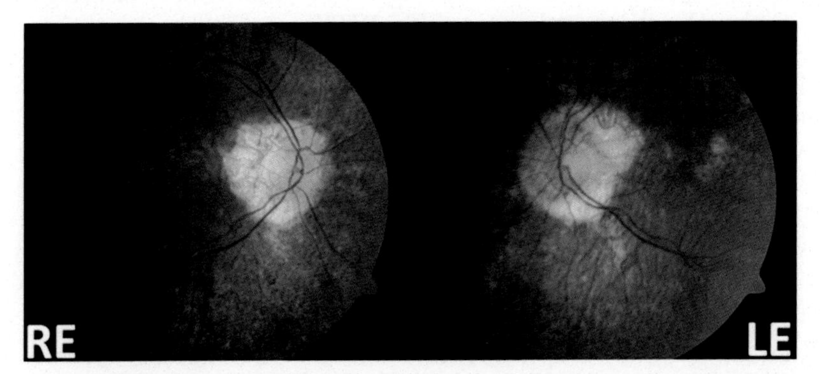

FIGURE 2 **Pigmentary retinopathy in mitochondrial cytopathies.** These fundus photographs illustrate the typical pigmentary retinopathy seen in cases with mitochondrial disease. In this particular example, the patient had CPEO as part of the Kearns-Sayre syndrome secondary to a single large-scale mtDNA deletion. There was generalized retinal pigment epithelial disturbance throughout the fundus, including the macula, with areas of hypo- and hyperpigmentation. A prominent ring of peripapillary atrophy was also present around both optic discs. LE, left eye; RE, right eye.

involvement, albeit to varying degrees of severity, and it is crucially important for a neurologist and/or internist to be involved to allow extraocular comorbidities to be detected and managed appropriately, to avoid secondary complications. An electrocardiogram is mandatory to identify possible cardiac conduction defects or evidence of ventricular hypertrophy.

Neuroimaging is frequently performed either as part of the patient's initial evaluation or after a diagnosis of mitochondrial disease has been considered. Cortical and cerebellar atrophy is commonly seen in cases with syndromic CPEO, in keeping with the higher burden of neurological disease in this specific patient population. Generalized atrophy of the extraocular muscles is also a characteristic feature and it is usually highly symmetrical between the two eyes of the same case, reflecting the pattern of ophthalmoplegia observed clinically (Figure 3).

A skeletal muscle biopsy remains the gold standard diagnostic procedure for cases with suspected CPEO. This minimally invasive procedure can be performed under local anaesthetic as a day case procedure, and complications are rare. Histological examination of muscle sections stained with specific markers is invaluable in detecting some of the pathological hallmarks of mitochondrial disease, namely a mosaic pattern of COX-negative muscle fibers and the presence of ragged red fibers that arise due to the proliferation and accumulation of mitochondria in the subsarcolemmal regions. Long-range PCR analysis performed on homogenate skeletal muscle DNA will clarify whether the case harbors a single mtDNA deletion, which remains the most common cause of CPEO, or multiple mtDNA deletions. If present, the latter strongly implicates a nuclear-encoded gene leading to secondary mtDNA instability, and depending on ease of access to molecular genetic testing, those genes known to cause CPEO should be screened (Table 1). CPEO can also occur in the context of mtDNA depletion syndromes, although it tends to be overshadowed by more severe multisystemic complications. The advent of next-generation exome

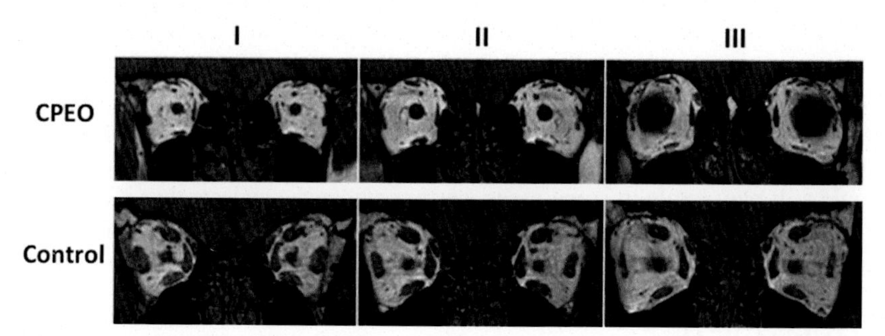

FIGURE 3 **Extraocular muscle atrophy in CPEO.** All four recti muscles in this particular patient with CPEO were atrophic compared with a healthy age-matched control. Representative cross-sections of the extraocular muscles have been provided at three different anatomical locations. Slice locations: (I) 4mm behind slice II toward the orbital apex; (II) central slice; (III) 4mm in front of slice II toward the extraocular muscle insertions onto the globe.

TABLE 1 Nuclear Genetic Causes of CPEO

	Autosomal Dominant	Autosomal Recessive
Multiple mtDNA deletions	*POLG* *C10orf2* *RRM2B* *OPA1* *SPG7* *ANT1*	*POLG* *POLG2* *C10orf2* *RRM2B* *SPG7* *TK2* *DGUOK* *MGME1*
MtDNA depletion		*POLG* *C10orf2* *RRM2B* *TYMP*

sequencing is starting to have a major impact on the investigation of cases with suspected mitochondrial disease, and it is hoped that a molecular diagnosis will soon be possible in the majority of cases of CPEO.

PATHOPHYSIOLOGY

The selective tissue-specific vulnerability of the extraocular muscles in mitochondrial disease has not yet been fully explained. Extraocular muscles are embryologically and genetically distinct from skeletal muscle with different anatomical and physiological properties. These intrinsic differences are thought to predispose extraocular muscle fibers to accumulate secondary mtDNA abnormalities, such as multiple mtDNA deletions, at a much faster rate compared with skeletal muscle fibers, and when combined with a lower mutational threshold, a biochemical defect of the respiratory chain is therefore more likely to become apparent with ensuing cellular dysfunction. Neuroimaging evidence of extraocular muscle atrophy supports a primary myopathic etiology for the progressive ophthalmoplegia that develops in CPEO (Figure 3). Although, degeneration of the oculomotor nuclei within the brainstem supranuclear pathways could still play a role in some cases, current evidence suggests that it is likely to be a relatively minor contributory component to the overall limitation of eye movements in this disorder.

CLINICAL PEARLS

Based on our own personal experience and previously published observations, a number of features strongly point toward a diagnosis of CPEO. These clues should not be viewed in isolation, but when faced with a patient with bilateral

TABLE 2 Extraocular Muscle Changes in CPEO Compared With Other Oculomotility Disorders

	CPEO	Ocular Myasthenia Gravis	Thyroid-Associated Ophthalmopathy	Congenital Fibrosis of the EOMs
EOM volume compared with controls	Decreased	No change	Increased	Decreased
MRI signal abnormalities in EOMs	None	None	Abnormal bright signals observed within the muscle belly on T1-weighted MRI slices	Abnormal bright signals observed within the muscle belly on T1-weighted MRI slices

CPEO, chronic progressive external ophthalmoplegia; EOM, extraocular muscle.

ophthalmoplegia and/or ptosis, they help build the case for further investigations, in particular the need for a diagnostic muscle biopsy.

- Patients with CPEO almost always exhibit an exotropia (divergent strabismus), and in one case series, 68% of all paired extraocular muscles had symmetry of movement within five degrees of each other. Unilateral or marked asymmetrical ophthalmoplegia are, therefore, rare features in classical CPEO.
- The presence of orbicularis weakness should be looked for specifically as it is a characteristic feature of CPEO.
- Formal eye movement recordings are only available in some specialist centers. Nevertheless, a careful examination will reveal slowed ocular saccades in most cases with CPEO.
- If present, pigmentary retinopathy is an important hallmark of mitochondrial disease, and in the context of CPEO, it constitutes part of the diagnostic triad for the Kearns-Sayre syndrome (Figure 2).
- A number of other oculomotility disorders can result in bilateral progressive ophthalmoplegia, and differentiating these from CPEO can sometimes be challenging. Imaging of the orbit can be a useful adjunct in the diagnostic evaluation of this group of patients as marked atrophy of the extraocular muscles is a striking morphological feature in CPEO (Table 2).

CASE MANAGEMENT

Genetic counselling is paramount, and the mode of inheritance of the causative genetic variant(s) identified in the proband needs to be carefully explained, sometimes in more than one clinic consultation. A confirmed molecular diagnosis of CPEO has obvious implications for other family members and specialist reproductive advice might be necessary in some circumstances. The management of the ocular symptoms in CPEO falls under three main categories, namely: (1) dry eye symptoms; (2) diplopia; and (3) ptosis.

Dry Eye Symptoms

Patients with CPEO often report ocular discomfort in the form of a gritty or dry eye sensation. There is sometimes objective evidence of lid margin disease (blepharitis) and insufficient ocular surface wetting with fluorescein staining. The management of dry eye symptoms is conservative with daily lid hygiene and the use of regular topical lubricating drops, which ideally should be preservative-free to reduce the risk of corneal epithelial toxicity with long-term treatment.

Diplopia

About a third of patients with CPEO will experience either intermittent or constant diplopia, which tends to be worse for near due to the eye's inability to converge properly. A full orthoptic assessment is essential to document the angles of deviation on both near and distance fixation, and the direction of gaze where symptoms of diplopia are maximal. Prisms are very effective in controlling symptoms of diplopia, and it can be incorporated into the patient's glasses once a stable prism strength has been achieved. Individuals will sometimes request surgical correction for a manifest strabismus to improve the cosmetic appearance, especially if a large angle deviation is present. Both the injection of botulinum toxin and maximal strabismus surgery on the horizontal recti muscles can significantly improve ocular alignment, as well as minimize symptoms of diplopia in CPEO. However, due to the progressive nature of the disease, the strabismus invariably recurs, and this fact needs to be made very clear to the patient prior to embarking on surgical correction for ocular misalignment.

Ptosis

Ptosis surgery can be highly effective in lifting the upper eyelid in CPEO, improving not only the patient's field of view, but also having a positive psychological impact on their body image. However, this procedure should only be carried out by an experienced oculoplastic surgeon who is aware of the dangers of overcorrection with CPEO, especially in the context of orbicularis weakness, dry eye symptoms, and a poor Bell's phenomenon, all of which increases the risks of corneal exposure and potentially blinding complications such as corneal ulcers or ocular perforation. The actual surgical intervention needs to be tailored based on the residual function of the levator palpebrae superioris muscle, the power of the frontalis muscle, and the patient's cosmetic preferences. The two most commonly performed ptosis procedures for CPEO are an anterior resection of the levator palpebrae superioris to maximize its upward muscle action, or if this is likely to be insufficient, a brow suspension can be performed that makes use of either a silicone sling or an autologous fascia lata to transfer the mechanical strength afforded by the upward movement of the frontalis muscle onto the upper eyelid.

FURTHER READING

[1] Richardson C, Smith T, Schaefer A, Turnbull D, Griffiths P. Ocular motility findings in chronic progressive external ophthalmoplegia. Eye 2005;19(3):258–63.

[2] Bau V, Zierz S. Update on chronic progressive external ophthalmoplegia. Strabismus 2005; 13(3):133–42.

[3] Schoser BG, Pongratz D. Extraocular mitochondrial myopathies and their differential diagnoses. Strabismus 2006;14(2):107–13.

[4] Yu Wai Man CY, Chinnery PF, Griffiths PG. Extraocular muscles have fundamentally distinct properties that make them selectively vulnerable to certain disorders. Neuromuscul Disord 2005;15(1):17–23.

[5] Fraser JA, Biousse V, Newman NJ. The neuro-ophthalmology of mitochondrial disease. Surv Ophthalmol 2010;55(4):299–334.

[6] Greaves LC, Yu-Wai-Man P, Blakely EL, et al. Mitochondrial DNA defects and selective extraocular muscle involvement in CPEO. Invest Ophthalmol Vis Sci 2010;51(7):3340–6.

[7] DiMauro S, Schon EA, Carelli V, Hirano M. The clinical maze of mitochondrial neurology. Nat Rev Neurol 2013;9(8):429–44.

[8] Stumpf JD, Saneto RP, Copeland WC. Clinical and molecular features of *POLG*-related mitochondrial disease. Cold Spring Harb Perspect Biol 2013;5(4):a011395.

Chapter 19

Infantile-Onset Spinocerebellar Ataxia (IOSCA)

Tuula Lönnqvist[1], Pirjo Isohanni[1,2], Anders Paetau[3]

[1]*Department of Child Neurology, Children's Hospital, Helsinki University and Helsinki University Hospital, Helsinki, Finland;* [2]*Research Programs Unit, Molecular Neurology, Biomedicum Helsinki, University of Helsinki, Helsinki, Finland;* [3]*Department of Pathology, HUSLAB, Helsinki University Hospital and University of Helsinki, Helsinki, Finland*

CASE PRESENTATION

Case 1

The case is a 2-year-old girl, who was sent to our hospital because of clumsiness at the age of 1.5 years. She was born at full-term after her mother's second uneventful pregnancy. Her early development was normal: she walked independently and spoke some words at the age of 1 year. After a common infection, when she was 1.5 years old, the child became clumsy: she walked with broad-based gait, fell easily, and had peculiar eye movements and a suspicion of squint from time to time. The development of expressive speech halted.

When examined, the child had ataxia, mild muscle hypotonia, and muscle stretch reflexes could not be elicited. Some oculomotor apraxia, but no permanent movement restriction, was noticed in her eyes. In spite of clumsiness, the child was joyful, cooperated well with her parents and the examiner, and her play and other behavior were age appropriate.

All routine laboratory and metabolic screening tests were normal, as well as electroneuromyography (ENMG), somatosensory-evoked potentials, and brain magnetic resonance imaging (MRI). The IOSCA (infantile-onset spinocerebellar ataxia) DNA test revealed that the case was homozygous for the Y508C mutation.

Case 2: A Historic Case: The Clinical Course of IOSCA

After normal early development of this second child of healthy parents, clumsiness was noticed at the age of 1.5 years. Ataxia, athetoid movements in hands and face, sensorineural hearing deficit, and loss of deep tendon reflexes was diagnosed at the age of 2.5 years. His older sister had similar symptoms. Squint was noticed at the age of 4 years, ophthalmoplegia by the age of 5.5 years, and

Mitochondrial Case Studies. http://dx.doi.org/10.1016/B978-0-12-800877-5.00019-X

optic atrophy during the second decade. He walked without support by the age of 8 years. Since then, he has used walking aids, and at the age of 15 was wheelchair bound. He had scoliosis, but he did not need operative treatment. Autonomic dysfunction, difficult urination, and constipation became evident with the progression of neuropathy.

He lost expressive speech. He communicated with sign language, but he understood written Finnish. He went to a special school for deaf people. His intellectual capacity was primarily normal, but decreased with the progression of the disease. After basic education, he lived at home and went daily to the local center for disabled.

He had severe migraine attacks since his teenage years, but no psychiatric symptoms. He had his first epileptic seizure at the age 28 years, and he died after prolonged status epilepticus at the age of 39 years.

His ENMG and sural nerve biopsy showed severe sensory axonal neuropathy at the age of 15 years. Muscle morphology, biochemical and histochemical enzyme analyses, and the structure and amount of mtDNA were normal. All routine laboratory and metabolic screening tests were normal including plasma and cerebrospinal fluid lactate until the start of epilepsy, when an occasional elevation of transaminase was discovered. Electroencephalograph (EEG) was normal or showed only mild background abnormality until the first epileptic seizure. The supratentorial brain remained normal until the appearance of epilepsy, but infratentorial cerebellar and brain stem atrophy was seen at the age 15 of years in MRI.

DIFFERENTIAL DIAGNOSIS

The differential diagnosis of subacute onset progressive ataxia is large and includes a variety of genetic and metabolic disorders. In the early phase of IOSCA disease, an erroneous diagnosis of postinfectious ataxia or acute encephalitis is common because the first clinical symptoms typically appear during an acute infection. However, the normality of CSF (cerebrospinal fluid) examination and continuation of clinical symptoms should raise a suspicion of a progressive disease. The loss of deep tendon reflexes and sensory axonal neuropathy in ENMG makes IOSCA one of the diseases in the group of hereditary sensory ataxias, of which the most common is Friedreich's ataxia (FRDA). Most FRDA cases have cardiomyopathy, not seen in IOSCA, and FRDA cases usually have neither hearing deficit nor ophthalmoplegia. Cases with *POLG (polymerase γ)* mutations may have sensory ataxia and ophthalmoplegia. Other early-onset ataxias, like ataxia telangiectasia (AT) and ataxia with oculomotor apraxia (AOA) should be included in the differential diagnosis, but increased serum concentration of α-fetoprotein, IgA deficiency, or hypoalbuminemia are often found in those conditions.

Dominant spinocerebellar ataxias (SCA) should be considered, though cerebellar atrophy is an early finding on MRI, and muscle stretch reflexes are usually brisk with positive Babinski sign. In contrast, in IOSCA, the muscle stretch reflexes are lost early, and the Babinski sign becomes positive only in a

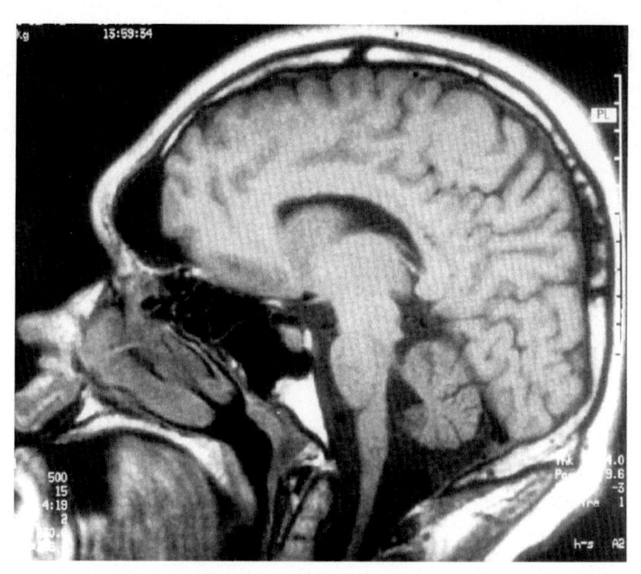

FIGURE 1 T1 MR image, sagittal view showing pontocerebellar atrophy in an 18-year-old male case.

later phase of the disease. Cerebellar atrophy becomes evident during the second decade (Figure 1). The family history of IOSCA cases supports a recessive mode of inheritance with healthy parents and no affected relatives except their own sisters or brothers. Select autosomal recessive SCA should also be considered.

Many metabolic disorders can also give rise to intermittent ataxia in early childhood, and ataxia may occur as a minor component of storage and other metabolic neurodegenerative disorders. These disorders can be excluded with normal metabolic screening tests.

Ataxia, ophthalmoplegia, and hearing deficits are common signs in other types of mitochondrial disorders as well. Elevation of lactate is usual in infantile-onset diseases with mitochondrial DNA (mtDNA) defects. However, childhood-onset mitochondrial encephalopathies are often nuclear encoded, and elevation of lactate is not a rule. Recessive mutations in *C10orf2* cause IOSCA-like clinical syndromes. Some of the described disorders or case reports are severe early-onset encephalopathies, while other resemble SCA. In Finland, we have described two cases with compound heterozygote mutations (A318T and Y508C). The clinical symptoms appear earlier in the compound heterozygote cases, and the disease progresses faster.

Epilepsia partialis continua with stroke-like episodes (SLE) is common in mitochondrial disorders, especially in MELAS (mitochondrial encephalopathy with lactic acidosis and stroke-like episodes) and *POLG*. In IOSCA, epilepsy is a late manifestation preceded by a progressive spinocerebellar degeneration (sensory ataxia with deafness and ophthalmoplegia and optic atrophy). The epilepsy in IOSCA is like in other mitochondrial disorders: epilepsia partialis continua with SLE and prolonged epileptic statuses (Figure 2(a)).

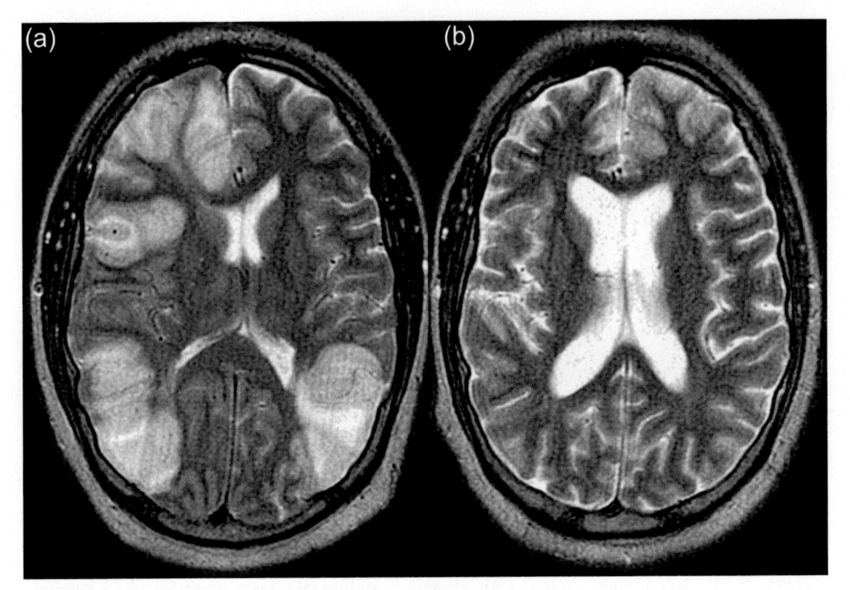

FIGURE 2 T2 MR axial image of 34-year-old male showing multiple oedematous lesions during status epilepticus (a) and atrophy 1 month later (b).

DIAGNOSTIC APPROACH

The clinical picture: presentation of ataxia with peculiar movements in the eyes and face, muscle hypotonia, and loss of muscle stretch reflexes around the age of 1.5 years in a previously healthy child is the leading diagnostic clue. As many childhood-onset metabolic disorders present with ataxia, metabolic screening tests have until now been the first-line diagnostic investigations. Brain MRI is mandatory in a child with ataxia to show possible infratentorial pathology. As special laboratory markers are lacking in IOSCA, ataxia panels and next-generation sequencing should be in the first line in diagnostic approach. Neurophysiological, ophthalmological, and audiological investigations remain more important in phenotyping the disease and its clinical course.

TREATMENT STRATEGY

There is no specific treatment for the progressive spinocerebellar degeneration or epileptic encephalopathy in IOSCA. Therapeutic support and physical and communicational aids are important to keep the cases ambulatory as long as possible. Because of the progressive hearing deficit, it is important to teach sign language from the very beginning of the clinical symptoms. Hearing aids are important especially in the early phase of the disease. Surgical correction does not cure the strabismus because of the progressive ophthalmoplegia. The optic disks become pale and visual evoked potentials become delayed with advancing age and progression of neuropathy, but visual accuracy remains normal until the very late phase of the disease. Orthopedic operations may be needed because of a progressive pes cavus deformity or neurogenic scoliosis. Autonomic dysfunction is part of the progressive neuropathy in IOSCA. Abnormal sweating, constipation, and difficulties to start urination and empty the bladder are common findings in IOSCA cases with advancing age. Medication for prevention of constipation and urinary infections are essential in adolescent cases.

Adolescent females need calcium and vitamin D substitution because of hypogonadism and low or lacking estrogen production. Migraine attacks are treated with common analgesics. Psychiatric symptoms include panic reactions and psychotic behavior, which require psychiatric medication, but in IOSCA, it is important to start with a careful dosage to prevent side effects.

The response for antiepileptic drugs is poor, especially among the prolonged epileptic statuses. Valproate is contraindicated because of liver toxicity, as is the case also in cases with *POLG* mutations. Benzodiazepines, specifically midazolam infusion, have occasionally been effective. Phenytoin or phosphenytoin are ineffective and may cause elevation of liver enzymes. Oxcarbazepine may have some effect on the seizures, but hyponatremia has been a troublesome side effect in all treated cases. Levetiracetam has been the drug of choice for long-term treatment, but unfortunately monotherapy is seldom enough to treat refractory seizures. Barbiturate anesthesia has been poor in treating epileptic statuses: the seizures do not stop, and cases develop pneumonia or other severe infections during anesthesia. However, in mitochondrial disorders, less medication is more useful as the cases with poorly working mitochondria are likely to have side effects.

LONG-TERM OUTCOME

A mild to moderate sensorineural hearing deficit and ophthalmoplegia with restriction of eye movements and inverted squint develop by the age of 5 years. Deafness, a hearing deficit of more than 80–100 dB, manifests within a few years. Severe sensory axonal neuropathy in ENMG and optic atrophy are found by the teenage years. Because of the progressive spinocerebellar degeneration, all IOSCA cases are severely physically handicapped and wheelchair bound by adolescence.

The cases' primary mental capacity is normal, but they have learning difficulties. With proper education and rehabilitation, they preserve their intelligence quite well, and only mild intellectual disability is discovered in adolescence. The start of epilepsy is a trigger for encephalopathy, supratentorial brain atrophy (Figure 2(a) and (b)), and loss of mental capacity. Thirteen of the 25 cases in Finland have died during or after the onset of severe epileptic encephalopathy. However, all IOSCA cases do not develop encephalopathy, and the age of the appearance of epileptic encephalopathy is variable. The oldest case with homozygous IOSCA mutations is 55 and has no epilepsy, but her older brother died at the age of 30 years during a prolonged status epilepticus. The compound heterozygote IOSCA cases died because of epileptic encephalopathy at the age 4 years.

PATHOPHYSIOLOGY/NEUROBIOLOGY OF DISEASE

The sensory system is severely affected in IOSCA. The light microscopic studies of sural nerves showed a severe loss of especially large myelinated axons with increasing age and progression of the disease. Postmortem studies revealed moderate atrophy of the brain stem and cerebellum, and severe atrophy in the dorsal roots, the posterior columns, and the posterior spinocerebellar tracts (Figures 3 and 4). There was a severe loss of myelinated axons in the vestibulocochlear (sensory) nerve, while in the facial (motor) nerve, the axons were well preserved. The supratentorial findings showed patchy laminar cortical necrosis (Figure 5), and the thalami, subthalamic nuclei, amygdala, and hippocampus were variably affected depending on the severity of the epileptic encephalopathy.

C10orf2 is a nuclear gene encoding the mtDNA helicase Twinkle, which is one of the proteins important for mtDNA maintenance. However, we do not yet know how the IOSCA *C10orf2* mutations (either homozygous Y508C or compound heterozygous Y508C and A318T) affect the helicase function. In postmortem studies we have been able to show tissue-specific mtDNA depletion in IOSCA: in the liver of compound heterozygous cases and in the brain and liver of homozygous IOSCA cases. In addition, in the brain, especially the large neurons showed respiratory chain complex I and IV (CI and CIV) deficiency. Yet, we do not know the pathophysiological mechanism behind the axonal damage and epileptic crisis.

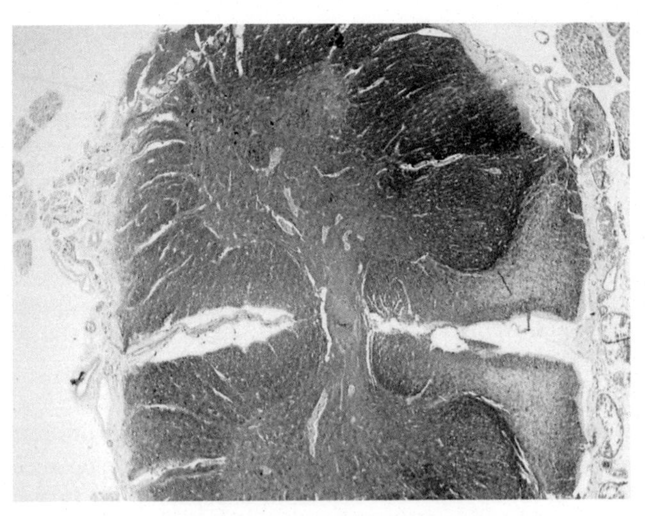

FIGURE 3 Lumbar spinal cord showing degeneration and fiber loss in posterior columns (right). SMI-311 neurofilament staining, original magnification x10.

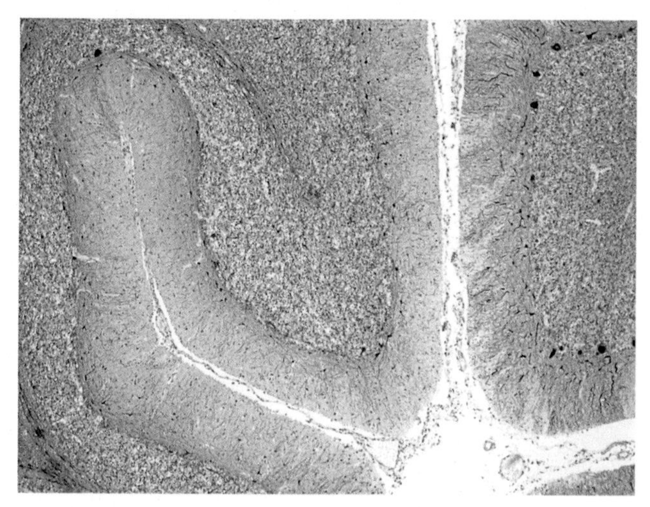

FIGURE 4 Cerebellar cortex with moderate drop-out of Purkinje cells. SMI-311 neurofilament staining, original magnification x40.

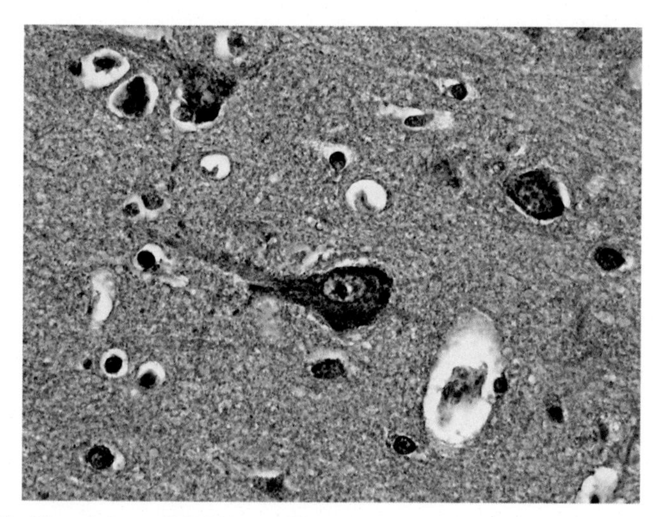

FIGURE 5 Frontal cortex, SMI-311 neurofilament staining shows severe depletion of pyramidal neurons in mid-laminae, original magnification x20.

CLINICAL PEARLS

- After an early normal development, appearance of ataxia, peculiar movements in hands and face, muscle hypotonia, and loss of deep tendon reflexes should raise a suspicion of IOSCA.
- Normality of metabolic screening tests supports the diagnosis.
- Development of sensory axonal neuropathy, hearing deficit, and ophthalmoplegia are pathognomonic for the disease.

- Secondary sex characteristics do not develop in females with IOSCA.
- Epileptic encephalopathy is a late manifestation of the disease.

FURTHER READING

[1] Koskinen T, Santavuori P, Sainio K, Lappi M, Kallio A-K, Pihko H. Infantile onset spinocerebellar ataxia with sensory neuropathy: a new inherited disease. J Neuro Sci 1994;121:50–6.

[2] Lönnqvist T, Paetau A, Nikali K, von Boguslawski K, Pihko H. Infantile onset spinocerebellar ataxia with sensory neuropathy (IOSCA): neuropathological features. J Neurol Sci 1998;161:57–65.

[3] Hakonen AH, Isohanni P, Paetau A, Herva R, Suomalainen A, Lönnqvist T. Recessive Twinkle mutations in early onset encephalopathy with mtDNA depletion. Brain 2007;130:3032–40.

[4] Sarzi E, Coffart S, Sertre V, Chretien D, Slama A, Munnish A, et al. Twinkle helicase (PEO1) gene mutation causes mitochondrial DNA depletion. Ann Neurol 2007;62:579–87.

[5] Hakonen AH, Goffart S, Marjavaara S, Paetau A, Cooper H, Mattila K, et al. Infantile-onset spinocerebellar ataxia and mitochondrial recessive ataxia syndrome are associated with neuronal complex I defect and mtDNA depletion. Hum Mol Genet 2008;17:3822–35.

[6] Lönnqvist T, Paetau A, Valanne L, Pihko H. Recessive twinkle mutations cause severe epileptic encephalopathy. Brain 2009;132:1553–62.

[7] Hartley JN, Booth FA, Del Bigio MR, Mhanni AA. Novel autosomal recessive *c10orf2* mutations causing infantile-onset spinocerebellar ataxia. Case Rep Pediatr 2012:1–4.

[8] Park M-H, Woo H-M, Hong JB, park JH, Yoon BR, Park J-M, et al. Recessive *C10orf2* mutations in a family with infantile-onset spinocerebellar ataxia, sensorimotor polyneuropathy, and myopathy. Neurogenetics 2014;15:171–82.

Chapter 20

MPV17-Related Hepatocerebral Mitochondrial DNA (mtDNA) Depletion Syndrome

Amy Goldstein
Neurogenetics & Metabolism, Division of Child Neurology, Children's Hospital of Pittsburgh, Pittsburgh, PA, USA

CASE PRESENTATION

Our case is a 4-month-old child with elevated liver enzymes alanine aminotransferase (ALT) and aspartate aminotransferase (AST) in the 200–300 IU/L range (normal values ALT 14–54 IU/L; AST 22–63 IU/L) and failure to gain weight. After an uncomplicated pregnancy, labor, and delivery, she developed jaundice in the first week of life. Laboratory studies revealed a direct hyperbilirubinemia. An intraoperative cholangiogram was performed to rule out biliary atresia. She was placed on Actigall but developed failure to thrive and poor weight gain. She had recurrent episodes of hypoglycemia despite continuous feeds with Enfaport formula (specialized formula with higher medium chain fats for easier absorption). Her liver failure persisted, and she developed a coagulopathy requiring cryoprecipitate and anemia (hemoglobin of 7 g/dL) requiring a transfusion. Infectious evaluation was negative.

Her neurological examination revealed hypotonia, but she was not overtly encephalopathic, and her mental status was appropriate for age. However, the EEG showed diffuse slowing indicative of a mild encephalopathy. Her brain magnetic resonance imaging (MRI) showed abnormal restricted diffusion in the splenium of the corpus callosum, globus pallidus, fornix, and cerebral peduncles. The white matter appeared healthy (Figure 1).

Liver ultrasound showed an echogenic liver consistent with fatty infiltration as well as abdominal ascites, with no abnormality on Doppler.

Metabolic studies revealed a persistent lactic acidosis between 9.7 and 13.8 mMol/L (normal range 0.5–2.2 mMol/L), an elevated alanine on plasma amino acids, and urine amino acids analysis indicated possible liver failure. Urine organic acids showed lactic aciduria, ketonuria, a massive elevation of tyrosine metabolites (*p*-hydroxyphenyllactate, 4-hydroxyphenylpyruvate)

Mitochondrial Case Studies. http://dx.doi.org/10.1016/B978-0-12-800877-5.00020-6

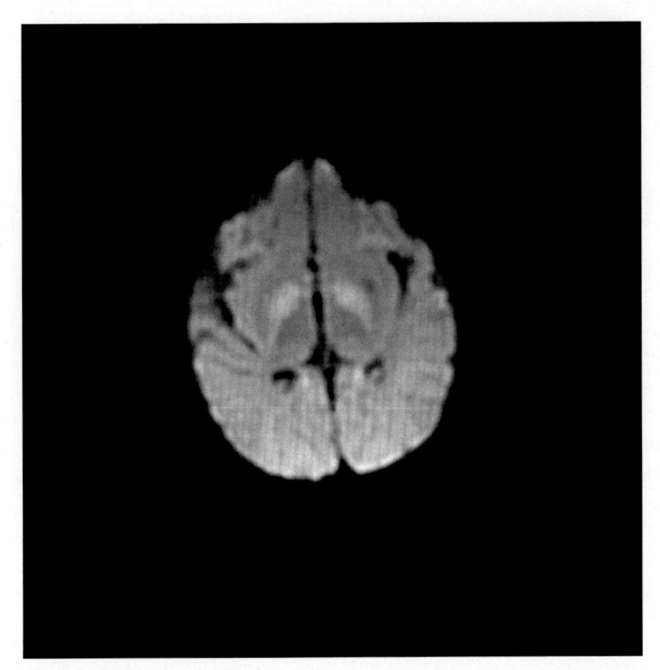

FIGURE 1 MRI brain: axial diffusion weighted imaging at level of globus pallidi.

without an elevation of succinylacetone (therefore, removing tyrosinemia from consideration), and elevations in glutaric acid, 3-methylglutaric acid, and 3-methylglutaconic acid. The acylcarnitine profile displayed elevations of multiple long chain species including hydroxylated species of C14–C18 (but not as high as seen in LCHAD) and a low total carnitine level (28 μM/L, normal range 38–68). Creatine phosphokinase (CPK) and ammonia levels were normal. Testing via isoelectric focusing of glycosylated transferrin was negative for N-glycosylated forms of congenital disorders of glycosylation.

Family history indicates that the parents are consanguineous. Her SNP-array showed multiple segmental regions of homozygosity (>8%) without large deletions or duplications (Agilent 180 K ISCA CGH + SNP on Agilent; whole genome, 200 kb in average).

Her liver biopsy revealed moderate hepatocellular and canalicular cholestasis, ballooning degeneration, and minimal steatosis with mild/early cholangiolar proliferation. Electron microscopy of the liver showed autolysis, distorted and enlarged and abnormally shapened mitochondria without more specific abnormalities; cristae showed a normal pattern.

Frozen liver tissue from this biopsy was sent for mitochondrial DNA copy number analysis (via qualitative PCR) showing mitochondrial depletion at 8% of the control sample.

Medical Genetics and Neurology services were consulted. Single gene testing of *POLG* did not reveal any mutations, and it was tested in an expedited manner in case liver transplant became necessary. Targeted whole exome sequencing revealed a homozygous c.284G>A (p.G95D) variant of unknown clinical significance (VUS) in the *MPV17* gene. This was confirmed by Sanger sequencing. Sanger sequencing also showed that both parents are mutation carriers, or heterozygous, for this VUS. Given her symptoms, liver biopsy results, and the predicted in silico pathogenicity of this variant of unknown significance, *MPV17* was determined to be her diagnosis.

DIFFERENTIAL DIAGNOSIS

Neonatal/childhood hepatopathies may present as acute liver failure, steatosis, hepatitis, cholestasis, or chronic liver failure with cirrhosis. If liver disease presents in the neonatal period, a primary mitochondrial disorder may not be suspected since liver dysfunction can be seen in a number of conditions in a sick newborn, such as sepsis, infection, and hypoxic-ischemic injury [1]. Abnormal lab findings consistent with a mitochondrial etiology may include lactic acidosis, hypoglycemia, elevated liver enzymes, and elevated bilirubin. It needs to be discussed that frank liver failure can present with similar biochemical labs from a number of etiologies. However, the additional clues to a mitochondrial etiology include intrauterine growth retardation, poor feeding, neurological features such as encephalopathy or hypotonia, and renal tubulopathy (typically Fanconi type) [2].

Primary mitochondrial diseases causing a hepatopathy include mtDNA deletion (Pearson marrow syndrome) or mitochondrial DNA depletion syndromes. Mitochondrial DNA (mtDNA) depletion syndromes (MDS) are typically inherited in an autosomal recessive fashion and lead to a significant reduction in the quantity of mtDNA in affected tissues. Specific nuclear genes involved in MDS are involved in mtDNA maintenance and include the following: *POLG, POLG2, DGUOK, MPV17*, Twinkle (*PEO1*), *TK2, SCO1, BCS1L, TYMP, SUCLA2, SUCLG1, OPA1*, and *RRM2B*. These nuclear genes are either involved directly in mtDNA maintenance or replication or involved in the nucleoside pool turnover [3].

There are four major categories of MDS disorders, characterized by variable phenotypes as myopathic, encephalomyopathic, hepatocerebral, or neurogastrointestinal. The hepatocerebral phenotype presents with early-onset liver involvement followed by neurological symptoms and can be caused by mutations in *DGUOK, MPV17, POLG,* and *C10orf2/PEO1* (Twinkle), and also by mutations in *BCS1L* and *SCO1* [3].

Nuclear genes involved in mtDNA translation such as *TRMU, GFM1, EFG1*, and *EFTu* may also cause a mitochondrial hepatopathy. Mutations in *TRMU* are known to cause a reversible infantile hepatopathy but without mtDNA depletion seen. Genes involved in the electron transport chain that can also result in hepatopathy include *BCS1L* (Complex III), SCO1 (Complex IV), and *ACAD9* (Complex I assembly factor) [3].

When testing liver for depletion, it can sometimes be difficult to separate effects of mutations in *DGUOK* and *MPV17*. Cases with *DGUOK* mutations typically have between 3% and 7% of the control, and those with *MPV17* have between 2% and 18%. Cases with MPV17 can have an isolated hepatopathy, but cases with DGUOK typically have neurological involvement with hypotonia and developmental delay [3,4].

DIAGNOSTIC APPROACH

Children with unexplained liver dysfunction may have a liver biopsy as part of their evaluation; histological findings will be nonspecific and may include findings such as steatosis, cholestasis, disrupted architecture, and cytoplasmic crowding by atypical mitochondria with swollen cristae [5,6].

Snap frozen tissue must be saved and sent for mtDNA depletion (mtDNA copy number analysis). If mtDNA depletion is found, nuclear genes need to be examined for mutations and/or deletions/duplications [5,6].

To date, only 31 children have been reported to have *MPV17* mutations, and this includes those with Navajo neurohepatopathy (NNH). Because there are now many nuclear genes involved in hepatopathy, a next-generation sequencing panel or whole exome sequencing is more efficient and economical than sequencing of individual genes [7].

Arriving at the correct molecular diagnosis has important implications on future treatment options. For example, since mutations in *TRMU* can be reversible, these children would not need a liver transplant if they can be supported though the period of liver failure. Children with *MPV17* may do fine after a liver transplant if necessary, but children with *DGUOK* have a poor prognosis overall. Children with *POLG* mutations have been noted historically to have rapid neurological progression after liver transplant [8]. This topic deserves further study.

TREATMENT STRATEGY

Currently, there is no proven effective treatment for mitochondrial diseases. Current care management is aimed at treating the associated symptoms such as failure to thrive, developmental delay, and epilepsy. A multidisciplinary team may be involved, including metabolic geneticist, metabolic dietician, hepatologist, neurologist, and developmental/rehabilitation specialists.

The mortality rate for *MPV17* disease is about 50%. The role for liver transplantation in children with mitochondrial hepatopathies due to *MPV17* mutations has been explored and has not been favorable, but there have been individual successes. When the symptoms include hepatopathy and severe encephalopathy or other significant multisystemic disease, transplantation needs to be more carefully considered weighing the benefits with overall quality of life. In one review, liver transplantation was performed in 10 children with mutations in MPV17 (one-third of the total cases) and five survived. Three of the five were homozygous for the p.Arg50Gln mutation in MPV17

(Navajo neurohepatopathy). This mutation may confer a more favorable prognosis [9]. Those who died during the post-transplantation period were in multiorgan failure and/or had severe sepsis. Those who did not receive a transplant died from liver failure. There have been several cases with NNH who survived into their teens, to date.

Dietary management of MPV17-related hepatopathy has been considered by Kaji et al., who placed a second sibling on a high fat diet, along with succinate and ubiquinone [10]. The other brother had died post-liver transplant after recurrent metabolic crises during intercurrent viral infections and use of a high carbohydrate diet. The sibling on the high fat diet seemed to be doing well on this combination therapy. Avoidance of fasting is necessary to prevent hypoglycemia. Adequate caloric and nutritional intake should be closely monitored.

As some cases with MPV17 mutations have developed hepatocellular carcinoma, surveillance and early treatment is indicated.

LONG-TERM OUTCOME

The overall mortality for MPV17 disease is 50%. The liver disease is typically progressive, and the cause of death within the first year of life is from liver failure. Some children do survive beyond this and usually develop neurological symptoms and later die in the first to second decade; thus the manifestations are seen in two stages [11].

PATHOPHYSIOLOGY/NEUROBIOLOGY OF DISEASE

The *MPV17* gene codes the MPV17 protein, which is located on the inner mitochondrial membrane protein, but its function and role in mtDNA depletion syndromes remains unknown. Mutations in MPV17 lead to mtDNA depletion, disruption of ATP production by the respiratory chain, and end-organ dysfunction [11].

CLINICAL PEARLS

- Suspect a mitochondrial etiology in a case with early-onset liver dysfunction (typically progressive), especially when accompanied by metabolic crisis (lactic acidosis and recurrent hypoglycemia), along with neurological symptoms (developmental delay, hypotonia, myopathy, and seizures), failure to thrive, and poor feeding. MPV17 should be suspected as the cause of early-onset liver failure and mild neurological symptoms at the onset.
- Other symptoms may include the following: peripheral neuropathy, ataxia, dystonia, nystagmus, corneal synesthesia and scaring, acral ulceration and osteomyelitis leading to autoamputation, leukoencephalopathy, recurrent metabolic acidosis with intercurrent infections (Reye-like presentation), renal tubulopathy, hypoparathyroidism, and gastrointestinal dysmotility.

- *MPV17* mutations are also the genetic etiology of NNH, typically from the homozygous p.R50Q mutation.
- Liver transplantation has been completed successfully in select cases with mitochondrial disorders, specifically those with *MPV17*-related NNH. Liver transplantation may be considered on an individual basis and should be discussed between the liver transplantation team and a clinician with experience in mitochondrial disorders to help take into consideration the case's long-term prognosis.

REFERENCES

[1] Fellman V, Kotarsky H. Mitochondrial hepatopathies in the newborn period. Semin Fetal Neonatal Med August 2011;16(4):222–8. [Epub June 15, 2011].

[2] Al-Hussaini A, Faqeih E, El-Hattab AW, Alfadhel M, Asery A, Alsaleem B, et al. Clinical and molecular characteristics of mitochondrial DNA depletion syndrome associated with neonatal cholestasis and liver failure. J Pediatr March 2014;164(3):553–9. e1-2. doi: 10.1016/j.jpeds.2013.10.082. [Epub December 8, 2013].

[3] El-Hattab AW, Scaglia F. Mitochondrial DNA depletion syndromes: review and updates of genetic basis, manifestations, and therapeutic options. Neurotherapeutics April 2013;10(2):186–98. http://dx.doi.org/10.1007/s13311-013-0177-6.

[4] Dimmock DP, Zhang Q, Dionisi-Vici C, Carrozzo R, Shieh J, Tang LY, et al. Clinical and molecular features of mitochondrial DNA depletion due to mutations in deoxyguanosine kinase. Hum Mutat 2008;29:330–1. [PubMed: 18205204].

[5] Molleston JP, Sokol RJ, Karnsakul W, Miethke A, Horslen S, Magee JC, et al. Childhood liver disease research education network (Children).Evaluation of the child with suspected mitochondrial liver disease. J Pediatr Gastroenterol Nutr September 2013;57(3):269–76. http://dx.doi.org/10.1097/MPG.0b013e31829ef67a. PMID: 23783016.

[6] Hazard FK, Ficicioglu CH, Ganesh J, Ruchelli ED. Liver pathology in infantile mitochondrial DNA depletion syndrome. Pediatr Dev Pathol November–December 2013;16(6):415–24. http://dx.doi.org/10.2350/12-07-1229-OA.1. [Epub September 19, 2013. PMID: 24050659].

[7] Spinazzola A, Santer R, Akman OH, Tsiakas K, Schaefer H, Ding X, et al. Hepatocerebral form of mitochondrial DNA depletion syndrome: novel MPV17 mutations. Arch Neurol August 2008;65(8):1108–13. http://dx.doi.org/10.1001/archneur.65.8.1108.

[8] Gordon N. Alpers syndrome: progressive neuronal degeneration of children with liver disease. Dev Med Child Neurol 2006;48:1001–3.

[9] El-Hattab AW, Li FY, Schmitt E, Zhang S, Craigen WJ, Wong LJ. *MPV17*-associated hepatocerebral mitochondrial DNA depletion syndrome: new patients and novel mutations. Mol Genet Metab March 2010;99(3):300–8. http://dx.doi.org/10.1016/j.ymgme.2009.10.003. [Epub October 13, 2009].

[10] Kaji S, Murayama K, Nagata I, Nagasaka H, Takayanagi M, Ohtake A, et al. Fluctuating liver functions in siblings with MPV17 mutations and possible improvement associated with dietary and pharmaceutical treatments targeting respiratory chain complex II. Mol Genet Metab August 2009;97(4):292–6. http://dx.doi.org/10.1016/j.ymgme.2009.04.014. [Epub May 12, 2009].

[11] Uusimaa J, Evans J, Smith C, Butterworth A, Craig K, Ashley N, et al. Clinical, biochemical, cellular and molecular characterization of mitochondrial DNA depletion syndrome due to novel mutations in the MPV17 gene. Eur J Hum Genet 2014;22:184–91. http://dx.doi.org/10.1038/ejhg.2013.112. [published online May 29, 2013].

FURTHER READING

[1] Lee WS, Sokol RJ. Mitochondrial hepatopathies: advances in genetics and pathogenesis. Hepatology June 2007;45(6):1555–65.

[2] Goldstein AC, Bhatia P, Vento JM. Mitochondrial disease in childhood: nuclear encoded. Neurotherapeutics April 2013;10(2):212–26. http://dx.doi.org/10.1007/s13311-013-0185-6. [Review. PMID].

[3] Lee WS, Sokol RJ. Liver disease in mitochondrial disorders. Semin Liver Dis August 2007; 27(3):259–73.

[4] Spinazzola A, Viscomi C, Fernandez-Vizarra E, Carrara F, D'Adamo P, Calvo S, et al. MPV17 encodes an inner mitochondrial membrane protein and is mutated in infantile hepatic mitochondrial DNA depletion. Nat Genet 2006;38:570–5. [PubMed: 16582910, related citations] [Full Text: Nature Publishing Group].

[5] Viscomi C, Spinazzola A, Maggioni M, Fernandez-Vizarra E, Massa V, Pagano C, et al. Early-onset liver mtDNA depletion and late-onset proteinuric nephropathy in Mpv17 knockout mice. Hum Molec Genet 2009;18:12–26. [PubMed: 18818194, images, related citations].

Chapter 21

Mitochondrial DNA Depletion Syndromes Presenting in Childhood

Shana E. McCormack[1,2], Xiaowu Gai[3,4], Emily Place[4,5], Marni J. Falk[2,5]
[1]Division of Endocrinology and Diabetes, Department of Pediatrics, The Children's Hospital of Philadelphia, Philadelphia, PA, USA; [2]Department of Pediatrics, University of Pennsylvania Perelman School of Medicine, Philadelphia, PA, USA; [3]Center for Biomedical Informatics, The Children's Hospital of Philadelphia, Philadelphia, PA, USA; [4]Ocular Genomics Institute, Ophthalmology, Massachusetts Eye and Ear Infirmary, Harvard Medical School, Boston, MA, USA; [5]Division of Human Genetics, The Children's Hospital of Philadelphia, Philadelphia, PA, USA

CASE PRESENTATION

A 6-week-old Caucasian girl presented for diagnostic evaluation of bilateral severe to profound sensorineural hearing loss (SNHL).

The case was the 3.45-kg, 55.8-cm product of a 35-week uncomplicated pregnancy to a 36-year-old mother who had previously had two early miscarriages and three preterm deliveries. With respect to growth, her length at presentation was at the 10th percentile and weight was below the fifth percentile (around the fifth percentile, when adjusted for gestational age, at the 50th percentile for a full-term newborn). Bilateral SNHL had been detected on routine screening in the newborn period. She otherwise had appropriate development through the first 6 weeks of life when corrected for gestational age. Her physical examination was notable for the absence of cardiac murmur, organomegaly, or focal neurologic deficit. She had minimal head control consistent with corrected age, with fair to moderately decreased central tone in ventral suspension but no slip-through when upright. She moved all extremities equally with good strength. She had brisk (3+) patellar deep tendon reflexes, without ankle clonus. She had normal Moro and stepping reflexes, consistent with corrected age. A comprehensive chemistry profile performed at this time was normal.

Her family history was significant for her mother, father, and two older sisters all in good health. A 4-year-old maternal female cousin had developmental delay, speech and behavior problems, and hypotonia. An 8-year-old

Mitochondrial Case Studies. http://dx.doi.org/10.1016/B978-0-12-800877-5.00021-8

paternal female cousin had autism-spectrum disorder, bipolar disorder, and wore glasses. Most notably, however, was the proband's older brother, who was born at 35 weeks, gestation and found on newborn screening to have congenital bilateral severe hearing loss (55 dB in left ear, 65 dB in right ear). Detailed review of his medical history revealed that a follow-up audiology evaluation at age 2 months was significant for absent oto-acoustic emissions, normal electrocardiogram, and normal CT scan of the temporal bones. The ophthalmology evaluation, including dilated examination, detected no retinal changes but was significant for hyperopia, unilateral nasolacrimal duct obstruction, and exotropia. He was otherwise reportedly in good health with normal early development, including social smiling and cooing. However, he presented at 3 months of age to his pediatrician with a 2-day history of head lag, lethargy, poor feeding, and poor sucking. He was admitted to the hospital for concern for possible sepsis, at which time he was found to have elevated blood lactate (6.3 mmol/L, normal < 2 mmol/L), elevated CSF protein (131 mg/dL, normal < 45 mg/dL), and suspected demyelination on brain MRI. He was transferred to a tertiary care hospital for persistent irritability and recurrent bilious emesis. Abnormal labs during his hospitalization included plasma amino acid analysis showing mildly elevated alanine (557.3 mmol/mL, normal < 523), as well as elevations of 3-hydroxybutyrate (2.4 nmol/L, normal < 0.4 nmol/L), blood lactate ranging on multiple occasions from 15 to 20 mmol/L (normal 0.5–2.2), blood pyruvate (0.18 mmol/L, normal 0.03–0.08 mmol/L), and urine lactate. His blood lactate decreased to the 1–2.5 mmol/L range following three doses of dichloroacetate. While a congenital ventricular septal defect had closed spontaneously at birth, he reportedly developed a new murmur during this time as well as progressive hearing loss, asymmetric and slowly dilating pupils, and bilateral ptosis. He died from respiratory failure at 4 months of age. No autopsy was performed. However, several days before his death, a skeletal muscle biopsy was obtained. Muscle histology revealed marked accumulation of intracellular lipid in all fibers, atrophic fibers, disrupted sarcoplasmic vacuolation, increased sarcoplasmic basophilia, weak cytochrome oxidase activity staining, and poor distinction of muscle fiber type on ATPase stain. There was also glycogen accumulation in the muscle; with respect to location, glycogen was primarily sub-sarcolemmal and non-membrane bound. Electron transport chain enzyme activity analysis on frozen muscle was notable for reduced complex IV (cytochrome *c* oxidase) activity, 0.24 mmol/min/g (normal control activity mean 2.80 mmol/min/g and range 1.76–3.84 mmol/min/g), which was 8.6% of the control mean. Muscle mitochondrial DNA content analysis showed he had 99% depletion relative to age-matched controls mean. Genetic sequencing of *TK2* in muscle did not reveal any pathogenic mutations. Mitochondrial DNA common point mutations and deletion testing in blood by Sanger sequencing, *TK2* sequencing, and *SURF1* sequencing were also unrevealing of a specific genetic etiology for his disease.

The proband developed progressive medical problems beginning at age 3 months of life, involving worsening hypotonia, feeding difficulty, and respiratory distress. Her CSF studies revealed elevated lactate (39.5 mg/L, normal 4.5–25) and ratio of lactate to pyruvate (29, normal < 20), alanine (56.8 μmol/L, 16.2–40.6), and the branched-chain amino acids isoleucine (11.8 μmol/L, 4.2–8.9) and leucine 22.1 μmol/L (7.3–19.1). Skeletal muscle biopsy performed at that time showed mitochondrial proliferation (elevated citrate synthase, CS, at 227% of control mean), as well as markedly decreased complex I and IV activities (10% and 5% of control values, respectively, when corrected for CS). She also developed progressive lactic acidosis up to 11.1 mmol/L. A diagnostic genetic test was performed, but she died at 4 months of age before results were known. Ultimately, a posthumous molecular diagnosis was made.

DIFFERENTIAL DIAGNOSIS

In this case, the treating clinicians and case's family were initially concerned that her SNHL might be caused by the same genetic condition that ultimately led to her brother's early demise. Therefore, attention was directed toward identifying a unifying genetic etiology for her brother's dramatic clinical presentation. There exist broad differential diagnoses both for infantile-onset bilateral SNHL [1] and for congenital lactic acidosis [2]. However, the identification of profound mitochondrial DNA depletion in her brother's muscle focused the diagnostic investigation. Mitochondrial DNA depletion indicates there are too few copies of the mitochondrial DNA genome relative to nuclear DNA. Although estimates vary, each mammalian cell contains dozens to thousands of mitochondrial DNA genome copies, where the exact ratio of mtDNA copies per diploid nuclear genome varies significantly by tissue [3]. Identification of near-total mtDNA depletion in the muscle of the proband's brother strongly suggested that he had an autosomal recessive disorder caused by biparentally inherited mutations within both alleles of one of many candidate nuclear genes that are critical for the regulation of mtDNA replication and integrity.

Indeed, a growing number of nuclear genes have been recognized to serve essential roles in maintaining mtDNA content and integrity [4]. As our understanding of the molecular machinery required for mtDNA maintenance improves, it is likely that additional genetic conditions causing mtDNA depletion will be recognized. Phenotypically, the age at symptom onset (e.g., infancy in this case) and the specific constellation of tissues affected (e.g., brain, ear, heart, and skeletal muscle in this case) may enable the list of likely gene candidates to be narrowed. Overall, the genetic causes of mtDNA depletion syndromes can be divided into two overarching categories. First are genes encoding proteins that directly regulate mtDNA genome replication and repair; second are genes whose products determine the availability of the nucleotide pools necessary for mtDNA synthesis. In addition, a growing number of additional genes critical for diverse mitochondrial functions appear to be associated with mtDNA depletion when mutated.

The prototypic example of a nuclear gene in which mutations cause an mtDNA depletion syndrome with multi-organ system involvement is *POLG*. *POLG* encodes the catalytic subunit of polymerase gamma (POLG), the major mitochondrial DNA polymerase necessary for mtDNA replication. Affected individuals may present with *POLG*-related disease manifestations at any age, from infancy through late adulthood, due to mutations that follow either autosomal recessive or autosomal dominant patterns of inheritance. Epilepsy and hepatopathy are prominent features typical of *POLG*-related autosomal recessive disease, where cases present in early childhood with clinical manifestations of Alpers-Hüttonlocher syndrome, a disorder characterized by a clinical triad of psychomotor regression with Leigh-like brain disease, seizures, and liver disease [5]. In contrast, affected adults may develop manifestations caused by either autosomal recessive or autosomal dominant *POLG* mutations, including mitochondrial recessive ataxia syndrome, Parkinsonism, or progressive external ophthalmoplegia [6]. More than 20 additional nuclear genes involved in the regulation of mtDNA replication, including maintenance of the nucleotide pool, have also been implicated in mtDNA depletion syndromes (Figure 1), as discussed at length in Section "Pathophysiology" below [7,8]. One relevant example that was considered in the present case is *TK2*, which encodes a mitochondrial thymidine kinase that phosphorylates deoxynucleosides that are ultimately incorporated into mtDNA. Pathogenic mutations in *TK2* lead to an autosomal recessive mitochondrial depletion syndrome involving profound muscle weakness in infancy, progressive encephalomyopathy, elevated creatine kinase activity values, and respiratory failure within the first 3 years of life [9]. Sensorineural hearing loss and low-to-absent cytochrome *c* oxidase activity, as were seen in the case's brother, may occur in *TK2* disorders as well [10]. As detailed in the case report, *TK2* gene sequencing was normal in the proband's brother. Figure 1 lists other genes that regulate the nucleotide pools available for mtDNA synthesis; mutations in these genes have also been implicated in mtDNA depletion syndromes, as discussed in additional detail below. While age at onset and tissue-specific manifestations of mtDNA depletion syndromes may provide clues as to the particular genetic etiology, clinical presentations can be similarly heterogeneous as is typical of other classes of mitochondrial disease.

Mitochondrial complex IV (cytochrome *c* oxidase) deficiency at 10% of the control mean in skeletal muscle was also detected in the case's affected brother, a biochemical finding that may guide the differential diagnosis of mitochondrial disease. The last enzyme in the electron transport system, complex IV is a 13-subunit enzyme that functions to pass electrons to molecular oxygen as the final electron acceptor, and it also contributes to the proton gradient that generates a charge differential across the inner mitochondrial membrane. This membrane potential provides a proton motive force that is drives dissociated to energy production through complex V in the chemical form of ATP. Three of the 13 polypeptide subunits of complex IV are encoded by mitochondrial DNA [11], and the rest are nuclear-encoded polypeptides. Complex IV deficiency has

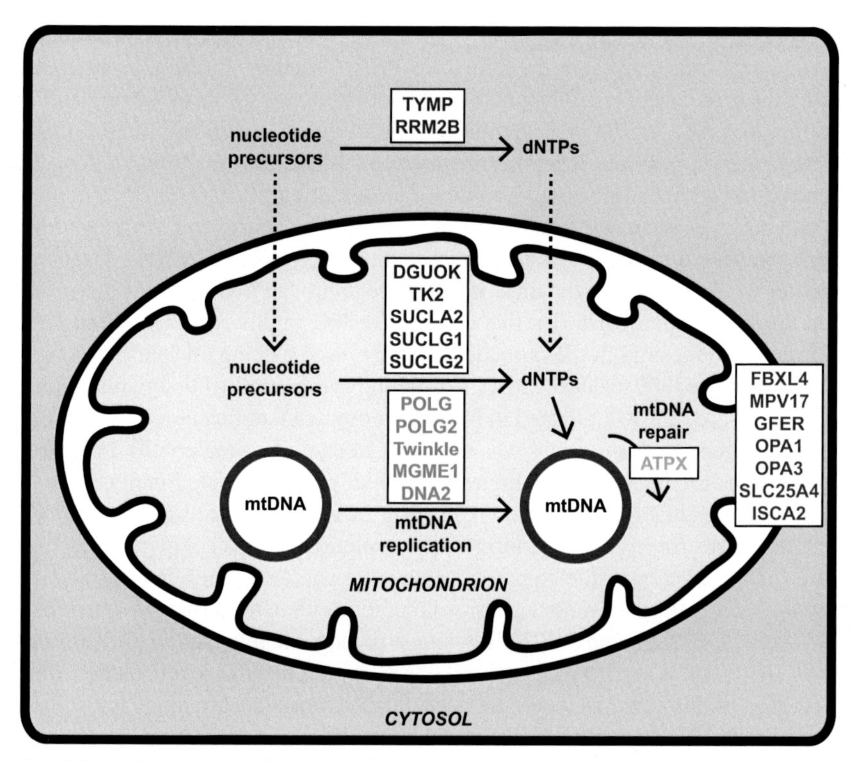

FIGURE 1 Genes and associated metabolic pathways implicated in childhood-onset mitochondrial DNA (mtDNA) depletion syndromes. Both cytosolic- and mitochondria-localized proteins involved in deoxyribonucleotide triphosphates (dNTP) synthesis or salvage that are encoded by nuclear genes are critical for synthesis of new mtDNA genomes (blue font). Nuclear genes involved in the mtDNA replication machinery are also frequently implicated causes of mtDNA depletion syndromes (green font). In addition, there is ongoing recognition of the role different proteins involved in inner mitochondrial membrane integrity play in regulating mtDNA maintenance (purple font). *Adapted from Ref. [4].*

been shown to result indirectly from mitochondrial DNA depletion, in which case there are insufficient mtDNA genomes to produce the 13 polypeptides that are all subunits of the respiratory chain within complex I, III, IV, or V. Mutations within a nuclear gene that encodes a complex IV assembly factor, such as the one that is most commonly implicated in complex IV deficiency, *SURF1*, or rarely, mutations in one of the three mtDNA-encoded subunits of complex IV, can also lead to complex IV deficiency [12–15]. As detailed in the case history, both mtDNA genome sequencing and *SURF1* sequencing were normal in the proband's brother.

DIAGNOSTIC APPROACH

The diagnosis of mitochondrial DNA depletion is made based on quantifying mtDNA relative to a specific nuclear reference gene within a specific target tissue. Diagnostic testing is typically done using Southern blot analysis or quantitative real-time polymerase chain (qRT-PCR) reaction. In the current case, given the presence of profound mtDNA depletion in the muscle of the proband's brother and the familial presentation of this condition in two children of unaffected parents, genetic testing in the proband's blood for mutations in nuclear genes that had been known at the time of presentation to cause mtDNA depletion syndromes in an autosomal recessive fashion was prioritized. Although initiating genetic diagnostic testing from remaining DNA from the case's deceased brother would have been the most informative approach to then test if the sister had the same genetic disorder, this was not feasible in this case. Therefore, initial diagnostic testing in the proband focused on sequencing nuclear genes that were known in 2009 to cause mtDNA depletion syndromes. Although mutations in mtDNA are rarely implicated in isolated complex IV deficiency, whole mitochondrial genome sequencing was also sent to exclude a maternally inherited mtDNA disorder that would change the familial recurrence risk. Finally, since it was possible, although much less likely, that the case's brother had two distinct genetic causes for his mitochondrial DNA depletion and his congenital hearing loss, for completeness, the common sequencing panel of genes causing SNHL was also sent. Ultimately, no deleterious mutations were found upon sequencing of *SURF1, DGUOK,* and *SUCLA2*. A single pathogenic mutation was identified in *SCO1* (c.16C>G; p.L6V), but no second mutation or exon-level deletion was identified in this gene, as would be necessary to cause disease in an autosomal recessive fashion typical of this disorder. In addition, no deleterious mutations were detected in the proband's blood upon mtDNA whole genome sequencing by PCR with Sanger sequencing.

However, two known pathogenic mutations in *RRM2B* that had previously been associated with mitochondrial disease were ultimately identified posthumously in the proband's blood DNA. *RRM2B* is a ribonucleotide reductase that catalyzes the conversion of deoxynucleoside diphosphates into the corresponding deoxynucleoside diphosphates necessary for DNA synthesis. Thus, its function directly affects the availability of the nucleotide pool [16,17]. *RRM2B* is controlled by p53, a tumor suppressor gene that is frequently inactivated in human cancers. Metabolic stressors, including, for example, radiation-induced DNA damage, can induce p53, and in turn, RRM2B, to facilitate DNA repair. In this case, the proband was found to be heterozygous for two mutations in *RRM2B*: c.122G>A (p.R41Q) was identified by Sanger sequencing in a clinical diagnostic laboratory, whereas a second mutation c.671T>C (p.I224S) was first identified on research-based exome next-generation sequencing analysis and subsequently verified by Sanger sequencing on a clinical diagnostic basis. The first *RRM2B* variant was previously reported in a case with Kearns-Sayre Syndrome and progressive external ophthalmoplegia [18]. Homozygosity of the c.617T>C (p.I224S) change in *RRM2B* was reported in a case with mtDNA depletion [19]. Biparental inheritance of these two heterozygous mutations in both the proband and in cellular DNA from her similarly affected brother was confirmed. Autosomal recessive inheritance of compound heterozygous mutations in *RRM2B*, as in the present case, is associated with severe, early-onset, multisystemic neurologic disease [20]. In contrast, adults with *RRM2B*-related mitochondrial disease may be more likely to have dominantly inherited single heterozygous mutations, with a corresponding later age of symptom onset. Adults with *RRM2B* mutations may also primarily experience myopathy, hearing loss, and/or gastrointestinal disturbances [20]. The proband's parents were asymptomatic.

PATHOPHYSIOLOGY

Mutations in nuclear-encoded proteins can affect the mtDNA replication and repair processes either by direct physical interaction with the molecular machinery or by regulating the availability of nucleotide pools. Figure 1 illustrates the intersection of these major processes. In addition to *POLG*, as was discussed above, other nuclear genes directly influence the molecular machinery of mtDNA replication. For example, Twinkle is a mitochondrial protein encoded by *C10ORF2* that is necessary for mtDNA replication. The Twinkle helicase co-localizes with mtDNA to unzip the two mtDNA strands in anticipation of replication. Dominant mutations in *C10ORF2* lead to progressive external ophthalmoplegia, whereas, as often occurs in children, autosomal recessive *C10ORF2* mutations lead to encephalopathy or hepatoencephalopathy, resembling the presentation of *POLG* disease. Second, mutations in *POLG2,* a 55-kilodalton accessory subunit of POLG, can stall the replication machinery, leading to an accumulation of mtDNA deletions in the context of clinical symptoms such as progressive external ophthalmoplegia. *MGME1*, or mitochondrial genome maintenance exonuclease 1, encodes a mitochondrial RecB-type exonuclease that participates in mtDNA maintenance by cleaving single-stranded DNA and processing DNA flap substrates. In vitro, mutations in *MGME1* leave cells unable to recover from chemically induced mtDNA depletion; in vivo, *MGME1* disorders lead to respiratory failure, failure to thrive, and progressive external ophthalmoplegia [21]. *DNA2* is a DNA nuclease/helicase thought to be important for mtDNA replication, in which mutations cause a combination of typically adult-onset, progressive myopathy along with mtDNA depletion [22]. Along with mtDNA replication, ongoing mtDNA repair is also key to maintaining adequate mtDNA genome copy number. The DNA-strand break repair protein aprataxin is encoded by the gene *APTX*; gene knock-down in vitro leads to reduced mtDNA copy number, and *APTX* gene disorders cause ataxia with oculomotor apraxia [23].

Maintenance of nucleotide pools (i.e., deoxyribonucleoside 5′-triphosphates, dNTPs) for use in the synthesis of mtDNA is also critical to maintain normal mtDNA content. Mutations in genes encoding important enzymes in this process can also lead to mtDNA depletion syndromes that manifest in childhood. As shown in Figure 1, similar to *TK2* and *RRM2B* as discussed previously, deoxyguanosine kinase *(DGUOK)* and thymidine phosphorylase *(TYMP)* also facilitate important steps in the synthesis of either purine or pyrimidine nucleotides, respectively, that are necessary for mtDNA genome synthesis. Mitochondrial DGUOK phosphorylates purine deoxyribonucleosides. Autosomal recessive mutations in *DGUOK,* particularly when manifest in the newborn period, can cause liver failure and liver-specific mtDNA depletion in the context of hepatocerebral syndrome [24]. TYMP is a cytosolic enzyme that participates in the salvage of nucleotides by catalyzing the phosphorylation of thymidine to thymine or of deoxyuracil to uracil [4]. Autosomal recessive mutations in thymidine phosphorylase lead to the mitochondrial neurogastrointestinal encephalomyopathy (MNGIE) syndrome, a disorder that presents with severe gastrointestinal dysfunction, pseudo-obstruction, and cachexia [25]. In addition, succinate-CoA ligases *(SUCLA2, SUCLG1)* are enzymes that participate in the citric acid cycle and may also interact with, and stabilize, some of the enzymes that regulate mtDNA nucleotide pools. *SUCLA2* recessive mutations can lead to Leigh-like encephalomyopathy and deafness [26]. Elevated lactate, mild methylmalonic aciduria, and abnormal carnitine esters are biochemical findings that may be identified on laboratory evaluations of *SUCLA2* cases. A common pathogenesis in all of the above conditions is the absence of adequate dNTP availability, such that mtDNA synthesis cannot proceed, and mtDNA genome depletion results. Recognition of this common pathogenesis may inform targeted therapies, as supplementation with deoxynucleoside monophosphates has been shown to rescue mitochondrial DNA depletion in some circumstances both in vitro [27] and in vivo in the case of TK2 deficiency [28].

Mutations in nuclear genes encoding mitochondrial inner membrane proteins that contribute to overall mitochondrial bioenergetics and network function have also demonstrated associations with mtDNA genome depletion. For example, *SCL25A4* (encoding the adenine nucleotide translocase, ANT1) encodes subunits of a dimeric transporter that brings ADP into mitochondria from the cytoplasm, and it allows ATP generated within the mitochondrial matrix to get out to the rest of the cell. Mutations in *SLC25A4* have been associated with a mtDNA depletion syndrome, whose clinical features include autosomal dominant progressive external ophthalmoplegia, myopathy, as well as cardiomyopathy when two pathogenic mutations are inherited in an autosomal recessive fashion [29]. Another gene, *GFER*, encodes a member of the disulfide relay system that, once transported into the mitochondrial intermembrane space, is thought to transfer its electrons to molecular oxygen via interactions with cytochrome *c* and cytochrome *c* oxidase. Clinically, mutations in *GFER* cause childhood-onset progressive myopathy, SNHL, congenital cataracts, and intellectual disability; biochemical deficiencies in the activities of complexes I, III, and IV activities along with mtDNA depletion have been noted [30]. An A-type mitochondrial iron-sulfur cluster protein that functions within the electron transport chain is encoded by *ISCA2*, which when mutated causes an infantile neurodegenerative mitochondrial disorder involving mtDNA genome depletion and decreased complex I activity [8]. *OPA1* encodes a dynamin GTPase localized to the inner mitochondrial membrane that is critical for a number of processes, including mitochondrial network dynamics (i.e., mitochondrial fusion). *OPA1* mutations initially present with vision loss due to dominant optic atrophy followed by development of progressive external ophthalmoplegia, ataxia, and deafness, along with COX-deficient muscle fibers and mtDNA genome depletion [31]. Mutations in *OPA3* can also cause optic atrophy, although considerable clinical heterogeneity has been described [32].

Other genes implicated in mtDNA depletion disorders encode mitochondrial membrane proteins of unknown function. Mutations in *MPV17* have been identified in neonates presenting with hepatic failure and liver-specific mitochondrial DNA depletion [33,34]. Mechanistic studies of the yeast homolog, *Sym1*, suggest that it may function to maintain the integrity of the inner mitochondrial membrane, such that mutations may indirectly lead to vulnerability to stresses that produce mtDNA depletion [35]. Mutations in *FBXL4*, a mitochondrial inter membrane space protein of uncertain function, have also recently been recognized to cause mtDNA depletion and a severe neonatal-onset mitochondrial encephalomyopathy with lactic acidemia and multisystem involvement [36,37].

Regardless of etiology, mtDNA genome depletion commonly results in the failure to properly synthesize the 13 mtDNA-encoded protein subunits of the mitochondrial respiratory chain complexes I, III, IV, and V, ultimately diminishing mitochondrial energy production and leading to tissue-specific energy deficiency states. Metabolically active tissues including brain, skeletal and cardiac muscle, liver, and the renal tubules are the ones most severely affected by mtDNA depletion disorders. The tissue specificity of mtDNA depletion, along with the degree and rapidity with which it occurs, appear to drive the ultimate clinical phenotype. Infants with widespread disease can present with a range of findings typically involving at least lactic acidosis, myopathy, hypotonia, and failure to thrive. Older children or adults with genetic disorders that cause mtDNA depletion are more likely to have epilepsy, headaches, intellectual disability with cognitive impairment, gastrointestinal symptoms, and/or ophthalmoplegia.

CLINICAL PEARLS

This case presentation of two siblings similarly affected with congenital, severe to profound SNHL and early demise within the first 6 months of life from developmental regression, progressive hypotonia, and respiratory failure, clearly demonstrates the idiosyncratic and multi-systemic nature of autosomal recessive mtDNA depletion syndromes. Several important clinical pearls can be gleaned from the case presentation, as outlined below.

- Identification of mtDNA depletion on tissue testing generates a specific differential diagnosis that, in 2015, invokes more than 20 distinct nuclear gene disorders.

- Each disorder can have variable manifestations of multiorgan involvement but typically involves progressive hepatopathy, encephalopathy, and/or myopathy with variable age of onset.
- mtDNA content in blood does not typically correlate with tissue mtDNA content, and therefore it cannot typically be used in isolation to identify mtDNA deletion disorders. Genetic diagnostic testing for mitochondrial depletion syndromes now relies heavily on next-generation sequencing analyses of multiple gene panels, or increasingly, whole exomes [38–40].
- Heterozygous (single) mutations within these nuclear genes can lead to adult-onset disease manifestations, whereas autosomal recessive mutations (where both alleles of a given gene are affected by compound heterozygous or homozygous mutations) typically lead to early-onset infantile or childhood diseases.
- Although no specific therapies exist for all mtDNA syndromes, early consideration would appropriately prompt careful multisystem evaluation and guide supportive care, including nutrition. For example, dietary strategies such as continuous glucose infusion or regular feedings have been shown to improve the clinical course of children affected with *MPV17* mutations [41].
- In addition, medications with known mitochondrial toxicities can be avoided or used with extreme caution. For example, onset of acute liver failure requiring transplantation has been described after treatment with valproate in the setting of *POLG* mutations [42], a drug that should clearly be avoided in the setting of mtDNA depletion.
- Finally, recent evidence is suggestive that nucleotide bypass therapies may offer effective treatments for mitochondrial depletion syndromes caused by mutations in nuclear genes that adversely affect the nucleotide pool.

ACKNOWLEDGMENT

This work was funded in part by the National Institutes of Health (R03-DK082446 to M.J.F., K12-DK094723-01 to S.E.M.), the Foerderer Award for Excellence from the Children's Hospital of Philadelphia Research Institute (M.J.F. and X.G.), and the Pediatric Endocrine Society (S.E.M.). The content is solely the responsibility of the authors and does not necessarily represent the official views of the National Institutes of Health.

REFERENCES

[1] Cohen M, Phillips 3rd JA. Genetic approach to evaluation of hearing loss. Otolaryngol Clin North Am 2012;45(1):25–39.

[2] Dimauro S, Garone C. Metabolic disorders of fetal life: glycogenoses and mitochondrial defects of the mitochondrial respiratory chain. Semin Fetal Neonatal Med 2011;16(4):181–9.

[3] Miller FJ, et al. Precise determination of mitochondrial DNA copy number in human skeletal and cardiac muscle by a PCR-based assay: lack of change of copy number with age. Nucleic Acids Res 2003;31(11):e61.

[4] Suomalainen A, Isohanni P. Mitochondrial DNA depletion syndromes–many genes, common mechanisms. Neuromuscul Disord 2010;20(7):429–37.
[5] Saneto RP, et al. Alpers-huttenlocher syndrome. Pediatr Neurol 2013;48(3):167–78.
[6] Isohanni P, et al. POLG1 manifestations in childhood. Neurology 2011;76(9):811–5.
[7] El-Hattab AW, Scaglia F. Mitochondrial DNA depletion syndromes: review and updates of genetic basis, manifestations, and therapeutic options. Neurotherapeutics 2013;10(2): 186–98.
[8] Al-Hassnan ZN, et al. ISCA2 mutation causes infantile neurodegenerative mitochondrial disorder. J Med Genet 2014;52(3):186–94.
[9] Gotz A, et al. Thymidine kinase 2 defects can cause multi-tissue mtDNA depletion syndrome. Brain 2008;131(Pt 11):2841–50.
[10] Chanprasert S, et al. TK2-Related mitochondrial DNA depletion syndrome, myopathic form. In: Pagon RA, et al., editors. GeneReviews(R). 1993. Seattle, WA.
[11] Shoubridge EA. Cytochrome c oxidase deficiency. Am J Med Genet 2001;106(1):46–52.
[12] Baertling F, et al. Mutations in COA6 cause cytochrome c oxidase deficiency and neonatal hypertrophic cardiomyopathy. Hum Mutat 2015;36(1):34–8.
[13] Olahova M, et al. A truncating PET100 variant causing fatal infantile lactic acidosis and isolated cytochrome c oxidase deficiency. Eur J Hum Genet 2014;23(7):935–9.
[14] Wedatilake Y, et al. SURF1 deficiency: a multi-centre natural history study. Orphanet J Rare Dis 2013;8:96.
[15] DiMauro S, Tanji K, Schon EA. The many clinical faces of cytochrome c oxidase deficiency. Adv Exp Med Biol 2012;748:341–57.
[16] Bourdon A, et al. Mutation of RRM2B, encoding p53-controlled ribonucleotide reductase (p53R2), causes severe mitochondrial DNA depletion. Nat Genet 2007;39(6):776–80.
[17] Gorman GS, Taylor RW. RRM2B-Related mitochondrial disease. In: Pagon RA, et al., editors. GeneReviews(R). 1993. Seattle, WA.
[18] Pitceathly RD, et al. Kearns-Sayre syndrome caused by defective R1/p53R2 assembly. J Med Genet 2011;48(9):610–7.
[19] Bornstein B, et al. Mitochondrial DNA depletion syndrome due to mutations in the RRM2B gene. Neuromuscul Disord 2008;18(6):453–9.
[20] Pitceathly RD, et al. Adults with RRM2B-related mitochondrial disease have distinct clinical and molecular characteristics. Brain 2012;135(Pt 11):3392–403.
[21] Kornblum C, et al. Loss-of-function mutations in MGME1 impair mtDNA replication and cause multisystemic mitochondrial disease. Nat Genet 2013;45(2):214–9.
[22] Ronchi D, et al. Mutations in DNA2 link progressive myopathy to mitochondrial DNA instability. Am J Hum Genet 2013;92(2):293–300.
[23] Sykora P, et al. Aprataxin localizes to mitochondria and preserves mitochondrial function. Proc Natl Acad Sci USA 2011;108(18):7437–42.
[24] Nobre S, et al. Neonatal liver failure due to deoxyguanosine kinase deficiency. BMJ Case Rep 2012;2012.
[25] Hirano M, Garone C, Quinzii CM. CoQ(10) deficiencies and MNGIE: two treatable mitochondrial disorders. Biochim Biophys Acta 2012;1820(5):625–31.
[26] Carrozzo R, et al. SUCLA2 mutations are associated with mild methylmalonic aciduria, Leigh-like encephalomyopathy, dystonia and deafness. Brain 2007;130(Pt 3):862–74.
[27] Bulst S, et al. In vitro supplementation with deoxynucleoside monophosphates rescues mitochondrial DNA depletion. Mol Genet Metab 2012;107(1–2):95–103.
[28] Garone C, et al. Deoxypyrimidine monophosphate bypass therapy for thymidine kinase 2 deficiency. EMBO Mol Med 2014;6(8):1016–27.

[29] Kaukonen J, et al. Role of adenine nucleotide translocator 1 in mtDNA maintenance. Science 2000;289(5480):782–5.

[30] Di Fonzo A, et al. The mitochondrial disulfide relay system protein GFER is mutated in autosomal-recessive myopathy with cataract and combined respiratory-chain deficiency. Am J Hum Genet 2009;84(5):594–604.

[31] Hudson G, et al. Mutation of OPA1 causes dominant optic atrophy with external ophthalmoplegia, ataxia, deafness and multiple mitochondrial DNA deletions: a novel disorder of mtDNA maintenance. Brain 2008;131(Pt 2):329–37.

[32] Sergouniotis PI, et al. Clinical and molecular genetic findings in autosomal dominant OPA3-related optic neuropathy. Neurogenetics 2015;16(1):69–75.

[33] Spinazzola A, et al. MPV17 encodes an inner mitochondrial membrane protein and is mutated in infantile hepatic mitochondrial DNA depletion. Nat Genet 2006;38(5):570–5.

[34] Wong LJ, et al. Mutations in the MPV17 gene are responsible for rapidly progressive liver failure in infancy. Hepatology 2007;46(4):1218–27.

[35] Dallabona C, et al. Sym1, the yeast ortholog of the MPV17 human disease protein, is a stress-induced bioenergetic and morphogenetic mitochondrial modulator. Hum Mol Genet 2010;19(6):1098–107.

[36] Bonnen PE, et al. Mutations in FBXL4 cause mitochondrial encephalopathy and a disorder of mitochondrial DNA maintenance. Am J Hum Genet 2013;93(3):471–81.

[37] Gai X, et al. Mutations in FBXL4, encoding a mitochondrial protein, cause early-onset mitochondrial encephalomyopathy. Am J Hum Genet 2013;93(3):482–95.

[38] Parikh S, et al. Diagnosis and management of mitochondrial disease: a consensus statement from the Mitochondrial Medicine Society. Genet Med 2014. http://dx.doi.org/10.1038/gim.2014.177, (Epub ahead of print).

[39] McCormick E, Place E, Falk MJ. Molecular genetic testing for mitochondrial disease: from one generation to the next. Neurotherapeutics 2013;10(2):251–61.

[40] Taylor RW, et al. Use of whole-exome sequencing to determine the genetic basis of multiple mitochondrial respiratory chain complex deficiencies. JAMA 2014;312(1):68–77.

[41] Parini R, et al. Glucose metabolism and diet-based prevention of liver dysfunction in MPV17 mutant patients. J Hepatol 2009;50(1):215–21.

[42] Hynynen J, et al. Acute liver failure after valproate exposure in patients with POLG1 mutations and the prognosis after liver transplantation. Liver Transpl 2014;20(11):1402–12.

Chapter 22

Mitochondrial Neurogastrointestinal Encephalomyopathy (MNGIE)

Michio Hirano, Beatriz García Díaz
Columbia University Medical Center, Department of Neurology, New York, NY, USA

CASE PRESENTATION

A 21-year-old Mexican-American man was the product of a normal spontaneous vaginal delivery. Early development was unremarkable. At age 3, he developed intermittent abdominal pain and vomiting with subsequent poor weight gain. He was hospitalized numerous times for recurrent emesis treated with antibiotics for presumed infections. He avoided spicy and greasy foods, which exacerbated his abdominal pain. He attended school through eighth grade but dropped out due to his medical problems.

At age 18, he reached his maximum weight of 96 pounds. At age 19, he lost about 10 pounds unintentionally over 1 month with increased lower abdominal pain and 1 week of increased vomiting. He was transferred to a tertiary medical center, where he was noted to be cachectic; height was 159.5 cm, weight was 35.4 kg, and body mass index was 13.9. General medical examination was otherwise unremarkable. Neurological examination revealed a normal mental status, bilateral ptosis, left greater than right ophthalmoparesis with restriction in all directions. Fundus was normal. Facial weakness was evident with inability to bury eyelashes. Other cranial nerve functions were intact. Other than facial weakness, muscle strength was normal. He had stocking sensory loss to vibration, but other sensory modalities were intact. Muscle stretch reflexes were absent. Coordination was normal. Romberg and Babinski signs were absent.

Family history was unremarkable; parents were non-consanguineous, and a 5-year-old brother was healthy.

The following blood tests were normal or negative: complete blood count, basic metabolic panel, hepatic function panel, venous lactate, creatine kinase,

Mitochondrial Case Studies. http://dx.doi.org/10.1016/B978-0-12-800877-5.00022-X

erythrocyte sedimentation rate (ESR), antinuclear antibody, C-reactive protein, and vitamin B12.

Upper gastrointestinal series with small bowel follow-through revealed esophageal dysmotility, a massively dilated stomach, and markedly dilated descending duodenum, but the remainder of the duodenum was not visible due to delayed duodenal emptying indicating duodenal pseudo-obstruction or obstruction. Five hours after the start of the study, contrast was seen in decompressed loops of distal small bowel. A gastric emptying study revealed markedly delayed emptying. A brain MRI showed diffuse increased T2 and fluid attenuation inversion recover (FLAIR) hyperintensity in white matter indicating a leukoencephalopathy. Nerve conduction studies revealed evidence of a diffuse demyelinating sensorimotor peripheral neuropathy with chronic axonal loss and denervation.

On a follow-up examination at age 21, he was noted to have weakness of hip and ankle flexors suggesting myopathy and motor neuropathy in addition to the previously noted ptosis, ophthalmoparesis, facial weakness, and signs of sensory neuropathy (stocking vibratory loss and areflexia).

In summary, this 21-year-old gentleman had early childhood-onset gastrointestinal dysmotility with severe cachexia and, at age 19, he was noted to have ptosis, ophthalmoparesis, sensorimotor peripheral neuropathy, and leukoencephalopathy with later development of proximal limb weakness indicating a myopathy.

DIFFERENTIAL DIAGNOSIS

The differential diagnosis of this complex multisystemic disease includes mitochondrial neurogastrointestinal encephalomyopathy (MNGIE) [1–3] and MNGIE-like disorders [4–11]. MNGIE is clinically identifiable by the combination of gastrointestinal dysmotility, cachexia, ptosis, ophthalmoparesis, diffuse leukoencephalopathy, with evidence of mitochondrial pathology (elevated blood or cerebrospinal fluid lactate, muscle biopsy showing ragged red fibers, cytochrome *c* oxidase [COX] deficient fibers, decreased activities of multiple respiratory chain enzymes [variable combinations of decreased complexes I, III, and IV], and mitochondrial DNA depletion, multiple deletions, or both). Because this case had a normal venous lactate level and did not undergo a muscle biopsy, his mitochondrial disease was not readily apparent. MNGIE-like cases may manifest all of the clinical features of MNGIE, but they lack diffuse leukoencephalopathy. Because MNGIE cases may show patchy white matter changes prior to diffuse leukoencephalopathy, it may be difficult to clinically distinguish MNGIE-like cases from individuals who will eventually develop MNGIE. In contrast to MNGIE, which appears to be a monogenic disorder, MNGIE-like cases have had pathogenic mutations in mtDNA (most notably the m.3243A>G mutation) as well as nuclear genes required for mtDNA maintenance (*POLG, RRM2B,* and *MGME1*) [4–11].

Other causes of extra-ocular weakness may be considered such as chronic progressive external ophthalmoplegia due to mtDNA single deletions, point mutations, or autosomal gene defects causing multiple mtDNA deletions. Oculopharyngeal muscular dystrophy (OPMD) causes ptosis and oropharyngeal dysphagia, which are seen in MNGIE, but onset of OPMD is typically in middle age, much later than in typical MNGIE [12].

The striking leukoencephalopathy in this case raises the possibility of numerous genetic causes of leukodystrophies [13,14]; however, the absence of cognitive manifestations and corticospinal tract signs would be atypical for most leukodystrophies with the exception of MNGIE and merosin deficiency. The latter typically presents as a congenital myopathy, which is strikingly different from our case's complex multisystemic disease.

DIAGNOSTIC APPROACH

Diagnostic evaluation of cases suspected of having MNGIE should begin with standard clinical testing to define the full clinical extent of the disease. Thus, testing should include the following: lactate measurement in blood, cerebral spinal fluid (CSF), or both; brain MRI to assess for leukoencephalopathy; nerve conduction (NCV)/electromyogaphy (EMG) to confirm the peripheral neuropathy, which is predominantly demyelinating in MNGIE; audiometry for hearing loss; evaluation for gastrointestinal dysmotility (including swallow, gastric emptying, and intestinal motility assessments); and hepatic assessment by blood liver function tests and abdominal ultrasound to screen for hepatic steatosis and cirrhosis.

Muscle biopsies had been performed routinely to screen for mitochondrial pathology (described above) in cases with MNGIE; however, after the identification of *TYMP* (previously *ECGF1*) mutations as the cause of this disease [15], molecular genetic and biochemical tests on blood or urine have become the gold standard for diagnosis. Because of the wide availability of genetic testing, most clinicians prefer to send blood for *TYMP* sequencing, which is highly accurate when known pathogenic mutations are identified; however, when variants of uncertain significance are detected, functional assessment of thymidine phosphorylase (TP) activity and/or measurement of plasma or urine thymidine and deoxyuridine levels can be provide the definitive diagnosis. Buffy coat TP activity in typical MNGIE cases is <10% of normal control means (mean 10 ± standard deviation 15 nmol thymine formed/h/mg-protein), while cases with a late-onset milder form of the disease have ~10–20% of normal buffy coat TP activity (85 ± 17) [16–18]. Defects of TP activity have also been detected in fibroblasts of MNGIE cases, but values for controls and cases have been less well characterized than in buffy coat samples [18,19].

In MNGIE cases, markedly elevations of plasma thymidine (>3 μmol/L, normal < 0.05) and deoxyuridine (>5 μmol/L, normal < 0.05) are diagnostic of classical MNGIE due to TP deficiency, while cases with the late-onset form of the disease have milder elevations of thymidine (≥0.4 μmol/L) and deoxyuridine (≥1 μmol/L). Urine levels of thymidine and deoxyuridine are also undetectable in normal individuals and elevated in MNGIE cases [20,21].

TREATMENT STRATEGY

Treatment of MNGIE is primarily symptomatic. Cases often benefit from nutritional supplementation early in the disease. Gastroparesis causing early satiety is a common manifestation that limits food intake, which can be enhanced by eating frequent small portions through the day. Jejunal feeding tubes may be placed to bypass the stomach; however, this therapy is often ineffective when intestinal motility is impaired. Bacterial overgrowth in the intestine can be detected by a CO_2 breath test and can be treated with antibiotics. Parenteral nutrition can be used to supplement enteral intake; however, this may cause severe hypertriglyceridemia and often becomes ineffective as the disease progresses.

Dialysis has been used to eliminate the toxic metabolites thymidine and deoxyuridine. Hemodialysis was ineffective in one case, while two other cases reportedly had mild clinical improvement with peritoneal dialysis [22–24]. However, peritoneal dialysis poses risks of infection in MNGIE cases, who are at risk of peritonitis from ruptured diverticula [1]. Erythrocyte-encapsulated TP has shown preliminary hints of efficacy in one case, but it must be infused multiple times yearly and is currently available at only one academic institution [25].

Hematopoietic stem cell transplant (HSCT including bone marrow transplant) has been performed on over 20 cases with MNGIE. When successful, HSCT restores circulating TP activity and dramatically reduces plasma thymidine and deoxyuridine [26]; however, morbidity and mortality have been unacceptably high with HSCT performed under a variety of transplant regimen [22,26–30]. An international consensus HSCT protocol has been proposed and is currently being tested in a phase I safety study (clinicaltrials.gov identifier NCT02363881).

Gene therapy, using an adeno-associated virus vectors to deliver human *TYMP* to hemapoietic stem cells and to liver, has shown promise in a mouse model of MNGIE [31,32].

LONG-TERM OUTCOME

MNGIE is a progressive disease that is ultimately fatal. Mean age-at-onset has been reported to be 17.9 years (range from infancy to age 59) with a mean age-at-death of 35 years (range 15–54 years) with rare cases surviving into their 60s [1]. The course of the disease is not uniform as cases often experience months to years of relative clinical stability with intermittent periods of rapid deterioration that may be triggered by stress such as infections or surgery.

PATHOPHYSIOLOGY/NEUROBIOLOGY OF DISEASE

The pathophysiology of MNGIE has been studied extensively studied, starting with the identification of causative mutations in *TYMP* [15], which encodes the cytosolic enzyme TP. Loss of TP activity leads to toxic accumulations of the nucleosides thymidine and deoxyuridine in plasma and tissues of cases [19,33–35]. The excess nucleosides are incorporated by the mitochondrial pyrimidine salvage pathway leading to deoxynucleotide pool imbalances that lead to instability of mtDNA as demonstrated in cell and mouse models of MNGIE [33,36–40].

The mtDNA instability manifests as depletion, multiple deletions, and site-specific point mutations predominantly in post-mitotic tissues [2,41–43]. The external muscle layer of the intestine appears to be particularly vulnerable to thymidine and deoxyuridine toxicity and enteric myopathy is the apparent cause of the intestinal dysmotility [44–46].

CLINICAL PEARLS

- Mitochondrial neurogastrointestinal encephalomyopathy (MNGIE) is a clinically recognizable autosomal recessive disorder characterized by ptosis, ophthalmoparesis, gastrointestinal dysmotility, cachexia, peripheral neuropathy, and leukoencephalopathy.
- MNGIE-like cases lack the diffuse leukoencephalopathy that is a hallmark of MNGIE.
- MNGIE can be diagnosed by genetic testing (detection of *TYMP* mutations) or by biochemical tests of blood (detection of TP deficiency in buffy coat, elevation of plasma thymidine and deoxyuridine, or both).
- HSCT has restored TP activity and produced clinical benefits, but safety of HSCT must be improved.
- TP gene therapy and erythrocyte-encapsulated TP have shown promise, but require further investigation.

REFERENCES

[1] Garone C, Tadesse S, Hirano M. Clinical and genetic spectrum of mitochondrial neurogastrointestinal encephalomyopathy. Brain 2011;134(11):326–32.

[2] Hirano M, Silvestri G, Blake DM, et al. Mitochondrial neurogastrointestinal encephalomyopathy (MNGIE): clinical, biochemical, and genetic features of an autosomal recessive mitochondrial disorder. Neurology 1994;44(4):721–7.

[3] Millar WS, Lignelli A, Hirano M. MRI of five patients with mitochondrial neurogastrointestinal encephalomyopathy. Am J Roentgenol 2004;182(6):1537–41.

[4] Chang TM, Chi CS, Tsai CR, Lee HF, Li MC. Paralytic ileus in MELAS with phenotypic features of MNGIE. Pediatr Neurol 2004;31(5):374–7.

[5] Gamez J, Lara MC, Mearin F, et al. A novel thymidine phosphorylase mutation in a Spanish MNGIE patient. J Neurol Sci 2005;228(1):35–9.

[6] Horvath R, Bender A, Abicht A, et al. Heteroplasmic mutation in the anticodon-stem of mitochondrial tRNA(Val) causing MNGIE-like gastrointestinal dysmotility and cachexia. J Neurol 2009;256(5):810–5.

[7] Kornblum C, Nicholls TJ, Haack TB, et al. Loss-of-function mutations in MGME1 impair mtDNA replication and cause multisystemic mitochondrial disease. Nat Genet 2013;45(2):214–9.

[8] Prasun P, Koeberl DD. Mitochondrial neurogastrointestinal encephalomyopathy (MNGIE)-like phenotype in a patient with a novel heterozygous POLG mutation. J Neurol 2014;261(9):1818–9.

[9] Shaibani A, Shchelochkov OA, Zhang S, et al. Mitochondrial neurogastrointestinal encephalopathy due to mutations in RRM2B. Arch Neurol 2009;66(8):1028–32.

[10] Tang S, Dimberg EL, Milone M, Wong LJ. Mitochondrial neurogastrointestinal encephalomyopathy (MNGIE)-like phenotype: an expanded clinical spectrum of POLG1 mutations. J Neurol 2012;259(5):862–8.
[11] Van Goethem G, Lofgren A, Dermaut B, Ceuterick C, Martin JJ, Van Broeckhoven C. Digenic progressive external ophthalmoplegia in a sporadic patient: recessive mutations in POLG and C10orf2/Twinkle. Hum Mutat 2003;22(2):175–6.
[12] Brais B. Oculopharyngeal muscular dystrophy. Handb Clin Neurol 2011;101:181–92.
[13] Parikh S, Bernard G, Leventer RJ, et al. A clinical approach to the diagnosis of patients with leukodystrophies and genetic leukoencephelopathies. Mol Genet Metab 2014.
[14] Schiffmann R, van der Knaap MS. The latest on leukodystrophies. Curr Opin Neurol 2004;17(2):187–92.
[15] Nishino I, Spinazzola A, Hirano M. Thymidine phosphorylase gene mutations in MNGIE, a human mitochondrial disorder. Science 1999;283(5402):689–92.
[16] Marti R, Spinazzola A, Tadesse S, Nishino I, Nishigaki Y, Hirano M. Definitive diagnosis of mitochondrial neurogastrointestinal encephalomyopathy by biochemical assays. Clin Chem 2004;50(1):120–4.
[17] Martí R, Verschuuren JJ, Buchman A, et al. Late-onset MNGIE due to partial loss of thymidine phosphorylase activity. Ann Neurol 2005;58(4):649–52.
[18] Massa R, Tessa A, Margollicci M, et al. Late-onset MNGIE without peripheral neuropathy due to incomplete loss of thymidine phosphorylase activity. Neuromuscul Disord 2009;19(12):837–40.
[19] Spinazzola A, Marti R, Nishino I, et al. Altered thymidine metabolism due to defects of thymidine phosphorylase. J Biol Chem 2002;277(6):4128–33.
[20] Schüpbach WM, Vadday KM, Schaller A, et al. Mitochondrial neurogastrointestinal encephalomyopathy in three siblings: clinical, genetic and neuroradiological features. J Neurol 2007;254(2):146–53.
[21] Lara MC, Weiss B, Illa I, et al. Infusion of platelets transiently reduces nucleoside overload in MNGIE. Neurology 2006;67(8):1461–3.
[22] Ariaudo C, Daidola G, Ferrero B, et al. Mitochondrial neurogastrointestinal encephalomyopathy treated with peritoneal dialysis and bone marrow transplantation. J Nephrol 2014.
[23] la Marca G, Malvagia S, Casetta B, et al. Pre- and post-dialysis quantitative dosage of thymidine in urine and plasma of a MNGIE patient by using HPLC-ESI-MS/MS. J Mass Spectrom 2006;41(5):586–92.
[24] Yavuz H, Ozel A, Christensen M, et al. Treatment of mitochondrial neurogastrointestinal encephalomyopathy with dialysis. Arch Neurol 2007;64(3):435–8.
[25] Moran NF, Bain MD, Muqit MM, Bax BE. Carrier erythrocyte entrapped thymidine phosphorylase therapy for MNGIE. Neurology 2008;71(9):686–8.
[26] Halter JP, Schüpbach MW, Mandel H, et al. Allogeneic haematopoetic stem cell transplantation for mitochondrial neurogastrointestinal encephalomyopathy. Brain 2015; Epub ahead of print.
[27] Filosto M, Scarpelli M, Tonin P, et al. Course and management of allogeneic stem cell transplantation in patients with mitochondrial neurogastrointestinal encephalomyopathy. J Neurol 2012;259(12):2699–706.
[28] Hirano M, Marti R, Casali C, et al. Allogeneic stem cell transplantation corrects biochemical derangements in MNGIE. Neurology 2006;67(8):1458–60.
[29] Peedikayil MC, Kagevi EI, Abufarhaneh E, Alsayed MD, Alzahrani HA. Mitochondrial neurogastrointestinal encephalomyopathy treated with stem cell transplantation: a case report and review of literature. Hematol Oncol Stem Cell Ther 2015.

[30] Halter J, Schupbach WM, Casali C, et al. Allogeneic hematopoietic SCT as treatment option for patients with mitochondrial neurogastrointestinal encephalomyopathy (MNGIE): a consensus conference proposal for a standardized approach. Bone Marrow Transplant 2011;46(3):330–7.

[31] Torres-Torronteras J, Gomez A, Eixarch H, et al. Hematopoietic gene therapy restores thymidine phosphorylase activity in a cell culture and a murine model of MNGIE. Gene Ther 2011;18(8):795–806.

[32] Torres-Torronteras J, Viscomi C, Cabrera-Perez R, et al. Gene therapy using a liver-targeted AAV vector restores nucleoside and nucleotide homeostasis in a murine model of MNGIE. Mol Ther 2014;22(5):901–7.

[33] López LC, Akman HO, Garcia-Cazorla A, et al. Unbalanced deoxynucleotide pools cause mitochondrial DNA instability in thymidine phosphorylase deficient mice. Hum Mol Genet 2009;18(4):714–22.

[34] Martí R, Nishigaki Y, Hirano M. Elevated plasma deoxyuridine in patients with thymidine phosphorylase deficiency. Biochem Biophys Res Commun 2003;303(1):14–8.

[35] Valentino ML, Marti R, Tadesse S, et al. Thymidine and deoxyuridine accumulate in tissues of patients with mitochondrial neurogastrointestinal encephalomyopathy (MNGIE). FEBS Lett 2007;581(18):3410–4.

[36] Ferraro P, Pontarin G, Crocco L, Fabris S, Reichard P, Bianchi V. Mitochondrial deoxynucleotide pools in quiescent fibroblasts: a possible model for mitochondrial neurogastrointestinal encephalomyopathy (MNGIE). J Biol Chem 2005;280(26):24472–80.

[37] Pontarin G, Ferraro P, Valentino ML, Hirano M, Reichard P, Bianchi V. Mitochondrial DNA depletion and thymidine phosphate pool dynamics in a cellular model of mitochondrial neurogastrointestinal encephalomyopathy (MNGIE). J Biol Chem 2006;281:22720–8.

[38] Garcia-Diaz B, Garone C, Barca E, et al. Deoxynucleoside stress exacerbates the phenotype of a mouse model of mitochondrial neurogastrointestinal encephalopathy. Brain 2014;137(Pt 5): 1337–49.

[39] Rampazzo C, Ferraro P, Pontarin G, Fabris S, Reichard P, Bianchi V. Mitochondrial deoxyribonucleotides, pool sizes, synthesis, and regulation. J Biol Chem 2004;279(17): 17019–26.

[40] Song S, Wheeler LJ, Mathews CK. Deoxyribonucleotide pool imbalance stimulates deletions in HeLa cell mitochondrial DNA. J Biol Chem 2003;278(45):43893–6.

[41] Nishigaki Y, Marti R, Copeland WC, Hirano M. Site-specific somatic mitochondrial DNA point mutations in patients with thymidine phosphorylase deficiency. J Clin Invest 2003;111(12):1913–21.

[42] Nishigaki Y, Marti R, Hirano M. ND5 is a hot-spot for multiple atypical mitochondrial DNA deletions in mitochondrial neurogastrointestinal encephalomyopathy. Hum Mol Genet 2004;13(1):91–101.

[43] Papadimitriou A, Comi GP, Hadjigeorgiou GM, et al. Partial depletion and multiple deletions of muscle mtDNA in familial MNGIE syndrome. Neurology 1998;51(4):1086–92.

[44] Giordano C, d'Amati G. Evaluation of gastrointestinal mtDNA depletion in mitochondrial neurogastrointestinal encephalomyopathy (MNGIE). Methods Mol Biol 2011;755:223–32.

[45] Giordano C, Sebastiani M, De Giorgio R, et al. Gastrointestinal dysmotility in mitochondrial neurogastrointestinal encephalomyopathy is caused by mitochondrial DNA depletion. Am J Pathol 2008;173(4):1120–8.

[46] Giordano C, Sebastiani M, Plazzi G, et al. Mitochondrial neurogastrointestinal encephalomyopathy: evidence of mitochondrial DNA depletion in the small intestine. Gastroenterology 2006;130(3):893–901.

FURTHER READING

[1] Garone C, Tadesse S, Hirano M. Clinical and genetic spectrum of mitochondrial neurogastro-intestinal encephalomyopathy. Brain 2011;134(11):326–32.

[2] Halter J, Schupbach WM, Casali C, et al. Allogeneic hematopoietic SCT as treatment option for patients with mitochondrial neurogastrointestinal encephalomyopathy (MNGIE): a consensus conference proposal for a standardized approach. Bone Marrow Transplant 2011;46(3):330–7.

[3] Hirano M, Marti R, Casali C, et al. Allogeneic stem cell transplantation corrects biochemical derangements in MNGIE. Neurology 2006;67(8):1458–60.

[4] Moran NF, Bain MD, Muqit MM, Bax BE. Carrier erythrocyte entrapped thymidine phosphorylase therapy for MNGIE. Neurology 2008;71(9):686–8.

Chapter 23

TK2-Related Mitochondrial DNA Depletion Syndrome, Myopathic Form

Sirisak Chanprasert, Fernando Scaglia
Department of Molecular and Human Genetics, Baylor College of Medicine, Houston, TX, USA; Texas Children's Hospital, Houston, TX, USA

CASE PRESENTATION

A 5-year-old boy, born to a Caucasian, non-consanguineous couple, presents to the Genetics Clinic with progressive muscle weakness. Pregnancy history was uncomplicated. He was delivered via C-section due to fetal distress. Apgar scores were 7 and 9 at 1 and 5 min, respectively. Gastroesophageal reflux was diagnosed at around 3 weeks of age. Treatment with a proton pump inhibitor was prescribed resulting in significant improvement of symptoms.

He was developmentally normal until 17 months of age when he exhibited unsteady gait and frequent falls. He continued to ambulate with difficulty. He was not able to get up from his bed several months after the onset of symptoms. Initial evaluation at the age of 2 years revealed hypotonia of proximal limb muscles with retained muscle stretch reflexes (DTRs). Blood chemistry revealed mild elevation of aspartate aminotransferase (AST) at 105 IU/L (normal value 10–40 IU/L) and alanine aminotransferase (ALT) at 72 IU/L (normal value < 35 IU/L). The serum creatine phosphokinase (CK) level was markedly elevated at 2400 U/L (normal range 38–120 U/L). Magnetic resonance imaging of the brain was within normal limits. Electromyography (EMG) and nerve conduction velocities demonstrated reduced motor unit potentials (MUPs) amplitude and duration that were consistent with a myopathic process. Various forms of muscular dystrophies, including Duchenne and early-onset limb-girdle muscular dystrophies were suspected. Muscle biopsy was recommended, but parents declined.

By 3 years of age, the case's muscle disease progressed to include bulbar and respiratory muscles, and he became respirator- and wheelchair-dependent. Laboratory investigation demonstrated modestly elevated AST and ALT values at 240 IU/L (normal value 10–40 IU/L) and 174 IU/L (normal value < 35 IU/L),

Mitochondrial Case Studies. http://dx.doi.org/10.1016/B978-0-12-800877-5.00023-1

respectively. Serum CK was 3000U/L (normal range 38–120U/L). Serum lactate was within normal limits. Because of persistent elevation of aminotransferases and progressive muscle weakness, the case underwent liver and muscle biopsies. The liver histopathology revealed a minimal infiltrate of nonspecific inflammatory cells in some portal areas. The muscle biopsy specimen showed prominent variance in fiber size. Histochemistry revealed cytochrome *c* oxidase (COX)-deficient fibers and ragged red fibers (RRF). Mitochondrial respiratory chain enzyme analyses on skeletal muscle was performed. The results revealed reduced activities of complex I, I+III, II+III, and IV of 25%, 15.2%, 11.7%, and 16% when compared to the mean of control values, respectively. The mtDNA content in the muscle was only 10% of age- and tissue-matched controls. The mtDNA copy number analysis in liver biopsy was not performed.

A mitochondrial depletion syndrome (MDS) was strongly suspected. Given that muscle disease is the prominent clinical feature, molecular genetic testing of *TK2* gene was performed. Two heterozygous mutations, c129_132delAGAA (p.K44Nfsx9) and c.389G>A (p.R130Q), were detected. Both mutations have been previously reported in MDS cases with *TK2* deficiency. These two mutations were confirmed to be in trans configuration by parental testing. A diagnosis of *TK2*-related mtDNA depletion syndrome was made.

The case continued to be respirator- and wheelchair-dependent. His muscles were severely wasted despite extensive physical therapy. He had multiple respiratory tract infections, requiring frequent hospitalizations. A gastrostomy tube was inserted by 4years of age because he was not able to eat or drink without aspiration and became severely malnourished. He died at 5years of age due to respiratory failure from pneumonia after a prolonged course in the intensive care unit.

DIFFERENTIAL DIAGNOSIS

Mitochondrial DNA (mtDNA) depletion syndromes (MDS) are a genetically and clinically heterogeneous group of disorders characterized by a significant reduction of mtDNA copy number in various tissues. They are transmitted in an autosomal recessive fashion. The clinical spectrum is very broad. However, the disease tends to affect high energy-requiring organ systems such as the brain, the skeletal muscle, and the liver. MDS can be categorized into myopathic, hepatocerebral, encephalomyopathic, and multisystemic forms. The molecular defects of nuclear genes responsible for mtDNA biogenesis and maintenance of deoxynucleotide pools are the cause of MDS [1]. To date, mutations in the nuclear genes, including *POLG*, *DGUOK*, *TK2*, *SUCLA2*, *SUCLG1*, *MPV17*, *TYMP*, *RRM2B*, *C10orf2*, *MGME1*, *FBXL4*, *ISCA2*, and *SERAC1* have been associated with MDS [2–5].

In the majority of cases with *TK2* deficiency, the age of onset of disease typically occurs within the first 2years of life [6]. The most common manifestations include progressive proximal muscle weakness and hypotonia [6].

DTRs are typically preserved. Some cases may develop epilepsy, encephalopathy, or sensorineural hearing loss [7,8]. The disease tends to be progressive with variable rates of progression. Affected individuals will eventually succumb to complications of severe muscle weakness including poor feeding, wasting, and respiratory failure. The majority of cases die within a few years after diagnosis. However, adult-onset disease with protracted clinical course and prolonged survival has been well described [9,10]. In addition, *TK2* mutations have been documented in cases who presented with progressive external ophthalmoplegia with mtDNA deletions [11]. Interestingly, these cases had normal mtDNA content [11]. *TK2*-related MDS typically afflicts skeletal muscle, but recent study has shown that cardiomyopathy and conduction defects presenting with prolonged QT interval and ventricular fibrillation may be associated with this condition [2].

The differential diagnosis of *TK2*-related MDS includes disorders that mainly cause hypotonia and progressive proximal muscle weakness. The following conditions should be considered.

1. Prader-Willi syndrome (PWS). PWS is characterized by severe hypotonia and feeding difficulties early in infancy followed by improvement of muscle tone and excessive eating and obesity in early childhood and adolescent. All cases with PWS have some degree of cognitive impairment, whereas in the majority of cases with *TK2*-related MDS, the cognitive function is spared, and the muscle disease is almost always progressive. In addition, although not apparent at birth, individuals with PWS will slowly develop characteristic facial features (narrow bifrontal diameter, almond-shaped palpebral fissures, narrow nasal bridge, thin upper lip, and downturned corners of the mouth) over time.
2. Dystrophinopathies. The clinical features of dystrophinopathies include a spectrum of muscle disease that ranges from asymptomatic increases in serum CK with muscle cramps and myoglobinuria to progressive proximal muscle weakness with cardiomyopathy that are classified as Duchenne/Becker muscular dystrophy. Individuals with Duchenne muscular dystrophy (DMD) typically present in early childhood with delayed motor milestones including delays in sitting and standing independently. These features are difficult to distinguish from *TK2*-related MDS. However, unlike cases with DMD, *TK2*-related MDS cases do not have calf hypertrophy. Serum CK levels are markedly elevated in DMD, and muscle biopsy demonstrates a complete or almost complete absence of dystrophin by immunohistochemistry.
3. Limb-girdle muscular dystrophy (LGMD). LGMD is a genetically heterogeneous group of neuromuscular disorders characterized by progressive proximal muscle weakness similar to *TK2*-related MDS. These disorders are inherited in an autosomal dominant and autosomal recessive fashion and are caused by mutations in the genes encoding sarcoglycans and related proteins in the muscle cell membranes. LGMD has variable age of onset but tends to affect individuals in late childhood and early adult

life. Molecular testing of the genes encoding proteins in the transmembrane sarcoglycan complex and other related proteins will distinguish LGMD from *TK2*-related MDS.

4. Spinal muscular atrophy (SMA). SMA is characterized by progressive muscle weakness resulting from the degeneration of anterior horn cells in the spinal cord. SMA is clinically classified into several types according to the age of onset and maximum function attained. The clinical features including poor muscle tone and muscle disease are similar to *TK2*-related MDS. However, in SMA, DTRs are absent or reduced, and EMG demonstrates reduced compound motor action potentials, widespread broad and polyphasic MUPs, and a denervation pattern. The diagnosis of SMA is based on molecular genetic testing of the two genes, *SMN1* and *SMN2*.

Although there is no definitive treatment of this condition, recent oral treatment with deoxycytidine and deoxythymidine monophosphates (dCMP+dTMP) in the H126N *Tk2* knock-in mice resulted in increasing dTTP concentrations and mtDNA content, significantly prolonging their lifespan [12]. The effect of this compound on *TK2*-related MDS cases remains to be explored.

DIAGNOSTIC APPROACH

The diagnosis of *TK2*-related MDS should be suspected in any cases who present with progressive muscle weakness and hypotonia regardless of age of onset. The diagnosis can be supported by series of laboratory investigations. Serum CK is modestly elevated, typically 5–10 times of the upper limit of normal. Elevation of AST and ALT can be seen and lead to unnecessary investigations of liver disease given that the elevation of liver aminotransferases are likely to be due to muscle destruction rather than true hepatic injury as demonstrated in this case presentation. However, three siblings from non-consanguineous parents were found to have true hepatic disease characterized by hepatomegaly, markedly elevated liver aminotransferase, and patchy periportal inflammation on liver biopsy [13].

Lactic acidemia, one of the hallmarks of mitochondrial disease, is variably present. Therefore, the absence of hyperlactatemia should not preclude a diagnosis of this condition. Serum CK is typically elevated, but in cases with severe muscle wasting, serum CK can be found within the normal range. When muscle biopsy is performed, RRF and COX-deficient fibers are considered to be frequently observed histological features [2]. Electron transport chain (ETC) activity in skeletal muscle shows decreased activities of multiple complexes with complex I, I+III, and IV being the most affected [6]. MtDNA content in muscle tissue is severely reduced, typically in the range from 5% to 30% of tissue- and age-matched controls [14]. However, according to the recent study that compiled data from known reported cases to date, there are two cases with normal mtDNA content in the muscle [2].

To confirm a diagnosis, molecular genetic testing needs to be performed. The gene *TK2* is located on chromosome 16q22-q23.1. It comprises 10 coding exons.

At least 36 mutations have been associated with this condition. Taken together, the most common mutations are missense mutations that account for approximately 67% of mutations [2]. The remaining mutations are frame-shift, splice site, a large deletion, or nonsense mutations. Full gene sequencing is the recommended first test of choice. However, if only one mutation is detected in cases with clinical and laboratory findings consistent with *TK2*-related MDS, gene dosage analysis should be performed to detect a large deletion involving the coding exons of this gene.

TREATMENT STRATEGY

Treatment of *TK2*-related MDS is largely supportive and requires a multidisciplinary approach. Feeding difficulties should be managed aggressively. A nasogastric tube or gastrostomy tube are frequently needed to avoid aspiration. Physical therapy can help maintain muscle function and prevent joint contractures. Pulmonary infection should be treated immediately to prevent deterioration in pulmonary function capacity.

TESTING STRATEGY

One strategy for making a diagnosis of *TK2*-related MDS in individuals with progressive muscle weakness and elevated serum CK concentration is the following:

1. Perform a skeletal muscle biopsy to measure ETC activities and mtDNA content analysis in the skeletal muscle. The diagnosis is strongly suspected when mtDNA content is severely reduced (typically less than 20% of age- and tissue-matched healthy controls).
2. Perform sequence analysis of the entire coding and exon/intron junction regions of *TK2* gene.
   a. If compound heterozygous or homozygous deleterious mutations are identified, the diagnosis is confirmed.
   b. If sequence analysis does not identify two deleterious mutations, gene dosage analysis should be considered.

PATHOPHYSIOLOGY

The *TK2* gene encodes thymidine kinase 2, the enzyme that catalyzes the first and rate-limiting step in the phosphorylation of deoxypyrimidine nucleosides in the salvage pathway of deoxynucleotide synthesis in the mitochondria [15]. Deficiency of *TK2* can result in an inability to synthesize mtDNA leading to mtDNA depletion. Mutations in *TK2* are associated with the myopathic form of MDS. The underlying pathophysiology as to why mainly skeletal muscle is affected is not known. The possible mechanisms include increased mtDNA content in muscle relative to other tissues reflecting the high rate of mtDNA synthesis, and a low basal *TK2* activity in muscle mitochondria [16]. As a result, *TK2* deficiency can selectively affect skeletal muscle. However, the H126N *Tk2* knock-in mouse demonstrates severely reduced mtDNA content more predominantly in brain than other tissues including skeletal muscle [17]. With these conflicting data, further study needs to be performed to fully elucidate the pathogenesis of this condition.

CLINICAL PEARLS

- *TK2*-related MDS is a chronic, progressive muscle disease mimicking various forms of neuromuscular disorders.
- The diagnosis is supported by elevated serum CK levels, decreased ETC activities, and depletion of mtDNA content in the muscle biopsy sample.
- Molecular genetic testing of *TK2* gene confirms the diagnosis.

REFERENCES

[1] Spinazzola A, Zeviani M. Disorders of nuclear-mitochondrial intergenomic signaling. Gene 2005;354:162–8.

[2] Chanprasert S, Wang J, Weng SW, Enns GM, Boue DR, Wong BL, et al. Molecular and clinical characterization of the myopathic form of mitochondrial DNA depletion syndrome caused by mutations in the thymidine kinase (TK2) gene. Mol Genet Metab 2013;110:153–61.

[3] Gai X, Ghezzi D, Johnson MA, Biagosch CA, Shamseldin HE, Haack TB, et al. Mutations in FBXL4, encoding a mitochondrial protein, cause early-onset mitochondrial encephalomyopathy. Am J Hum Genet 2013;93:482–95.

[4] Al-Hassnan ZN, Al-Dosary M, Alfadhel M, Faqeih EA, Alsagob M, Kenana R, et al. ISCA2 mutation causes infantile neurodegenerative mitochondrial disorder. J Med Genet 2015;52(3):186–94.

[5] Wedatilake Y, Plagnol V, Anderson G, Paine S, Clayton P, Jacques T, et al. Tubular aggregates caused by serine active site containing 1 (SERAC1) mutations in a patient with a mitochondrial encephalopathy. Neuropathol Appl Neurobiol 2015;41(3):399–402.

[6] Chanprasert S, Wong LJC, Wang J, Scaglia F. TK2-Related mitochondrial DNA depletion syndrome, myopathic form. In: Pagon RA, Adam MP, Ardinger HH, Bird TD, Dolan CR, Fong CT, et al., editors. GeneReviews®. 1993. Seattle (WA).

[7] Marti R, Nascimento A, Colomer J, Lara MC, Lopez-Gallardo E, Ruiz-Pesini E, et al. Hearing loss in a patient with the myopathic form of mitochondrial DNA depletion syndrome and a novel mutation in the TK2 gene. Pediatr Res 2010;68:151–4.

[8] Lesko N, Naess K, Wibom R, Solaroli N, Nennesmo I, von Dobeln U, et al. Two novel mutations in thymidine kinase-2 cause early onset fatal encephalomyopathy and severe mtDNA depletion. Neuromuscul Disord 2010;20:198–203.

[9] Oskoui M, Davidzon G, Pascual J, Erazo R, Gurgel-Giannetti J, Krishna S, et al. Clinical spectrum of mitochondrial DNA depletion due to mutations in the thymidine kinase 2 gene. Arch Neurol 2006;63:1122–6.

[10] Behin A, Jardel C, Claeys KG, Fagart J, Louha M, Romero NB, et al. Adult cases of mitochondrial DNA depletion due to TK2 defect: an expanding spectrum. Neurology 2012;78:644–8.

[11] Tyynismaa H, Sun R, Ahola-Erkkila S, Almusa H, Poyhonen R, Korpela M, et al. Thymidine kinase 2 mutations in autosomal recessive progressive external ophthalmoplegia with multiple mitochondrial DNA deletions. Hum Mol Genet 2012;21:66–75.

[12] Garone C, Garcia-Diaz B, Emmanuele V, Lopez LC, Tadesse S, Akman HO, et al. Deoxypyrimidine monophosphate bypass therapy for thymidine kinase 2 deficiency. EMBO Mol Med 2014;6:1016–27.

[13] Zhang S, Li FY, Bass HN, Pursley A, Schmitt ES, Brown BL, et al. Application of oligonucleotide array CGH to the simultaneous detection of a deletion in the nuclear TK2 gene and mtDNA depletion. Mol Genet Metab 2010;99:53–7.
[14] Dimmock D, Tang LY, Schmitt ES, Wong LJ. Quantitative evaluation of the mitochondrial DNA depletion syndrome. Clin Chem 2010;56:1119–27.
[15] Bestwick RK, Moffett GL, Mathews CK. Selective expansion of mitochondrial nucleoside triphosphate pools in antimetabolite-treated HeLa cells. J Biol Chem 1982;257:9300–4.
[16] Saada A, Shaag A, Elpeleg O. mtDNA depletion myopathy: elucidation of the tissue specificity in the mitochondrial thymidine kinase (TK2) deficiency. Mol Genet Metab 2003;79:1–5.
[17] Akman HO, Dorado B, Lopez LC, Garcia-Cazorla A, Vila MR, Tanabe LM, et al. Thymidine kinase 2 (H126N) knockin mice show the essential role of balanced deoxynucleotide pools for mitochondrial DNA maintenance. Hum Mol Genet 2008;17:2433–40.

FURTHER READING

[1] Chanprasert S, Wang J, Weng SW, et al. Molecular and clinical characterization of the myopathic form of mitochondrial DNA depletion syndrome caused by mutations in the thymidine kinase (*TK2*) gene. Mol Genet Metab September–October 2013;110(1–2):153–61.
[2] Parikh S, Goldstein A, Koenig MK, et al. Diagnosis and management of mitochondrial disease: a consensus statement form Mitochondrial Medicine Society. Genet Med December 11, 2014.

Chapter 24

Autosomal Dominant Optic Atrophy

Patrick Yu-Wai-Man[1,2], Patrick F. Chinnery[3,4]
[1]Wellcome Trust Centre for Mitochondrial Research, Institute of Genetic Medicine, Newcastle University, UK; [2]Newcastle Eye Center, Royal Victoria Infirmary, Newcastle upon Tyne, UK; [3]Department of Clinical Neuroscience, University of Cambridge, Cambridge, UK; [4]MRC Mitochondrial Biology Unit, Cambridge Biomedical Campus, Cambridge, UK

CASE PRESENTATION

Clinical Case

A 7-year-old girl was referred to the Ophthalmology Department with a 6-month history of progressive difficulty reading the blackboard at school despite moving closer to the front of the classroom to see better. Her local optometrist was unable to improve her vision, and she had no significant refractive errors. She was born at full-term after an uneventful pregnancy, and she achieved normal developmental milestones in infancy. There were no other health issues, and until recently, she was making good progress in her studies at school.

At the case's baseline visit, her best corrected visual acuity was noted to be 20/40 bilaterally, and confrontation visual field testing suggested a symmetrical central field defect. Anterior segment examination was normal, but subtle bilateral temporal optic disc pallor was observed on dilated fundoscopic examination (Figure 1). Her neurological examination was normal, and brain MR imaging excluded an underlying compressive or infiltrative lesion involving the anterior visual pathways. Visual electrophysiology was consistent with a primary retinal ganglion cell pathology with a normal electroretinogram, but markedly attenuated visual evoked potentials bilaterally. The girl managed to remain in mainstream schooling with the use of appropriate magnification aids, but she was eventually registered partially sighted at the age of 22 years when her visual acuity deteriorated to 20/120 bilaterally. Her pupillary light reflexes remained brisk, and no relative afferent papillary defect were noted. Optical coherence tomography (OCT) showed significant thinning of the retinal nerve fiber layer in keeping with the optic disc appearances (Figure 2). In her late 30s, she noted increased difficulty hearing conversations in crowded environments, and a pure tone audiogram revealed

Mitochondrial Case Studies. http://dx.doi.org/10.1016/B978-0-12-800877-5.00024-3

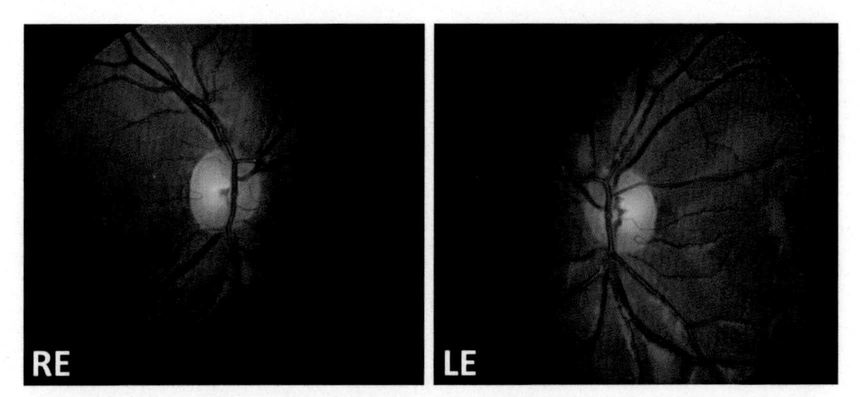

FIGURE 1 Baseline optic disc photographs of the proband taken when she was 7 years old.

high-frequency hearing loss. The case is currently in her early 40s, and she remains fully independent despite being registered legally blind with best corrected visual acuities of 20/800 and dense central scotomas in both eyes (Figure 3).

The proband's father had an almost identical clinical presentation in early childhood. He developed further gradual deterioration in his vision throughout his teenage years, and he was registered legally blind in his early 50s. He attended a follow-up visit with his daughter as part of familial contact tracing. His vision was reduced down to count fingers in both eyes, which was reflected by the generalized pallor of his optic discs (Figure 4). The father also developed progressive hearing difficulties in the fourth decade of life and pure tone audiography confirmed a predominantly high-frequency pattern of hearing loss (Figure 5). The proband's two younger siblings had a normal neuro-ophthalmological examination, and there were no other affected family members. The clinical features in the proband and her father raised the suspicion of autosomal dominant optic atrophy (DOA), and molecular genetic testing eventually confirmed the presence of a pathogenic missense *OPA1* mutation within the catalytic GTPase domain in exon 14 (c.1334G>A, p.Arg445His).

Discussion

Both the proband daughter and her father describe a classical clinical presentation for DOA, which typically presents in childhood, usually before 10 years of age, with a painless and slowly progressive bilateral deterioration in visual acuity. There are usually no other systemic health problems, and the diagnosis is relatively straightforward when there is a family history of early-onset visual loss consistent with an autosomal dominant mode of transmission. Cases with DOA will also exhibit other typical features consistent with a primary optic neuropathy, namely, impaired color vision and a predominantly central or centrocecal scotoma on visual field testing. Temporal pallor of the optic disc is a characteristic feature in DOA, and this segmental involvement due to the preferential involvement of the papillomacular bundle can be formally visualized with OCT imaging (Figure 2).

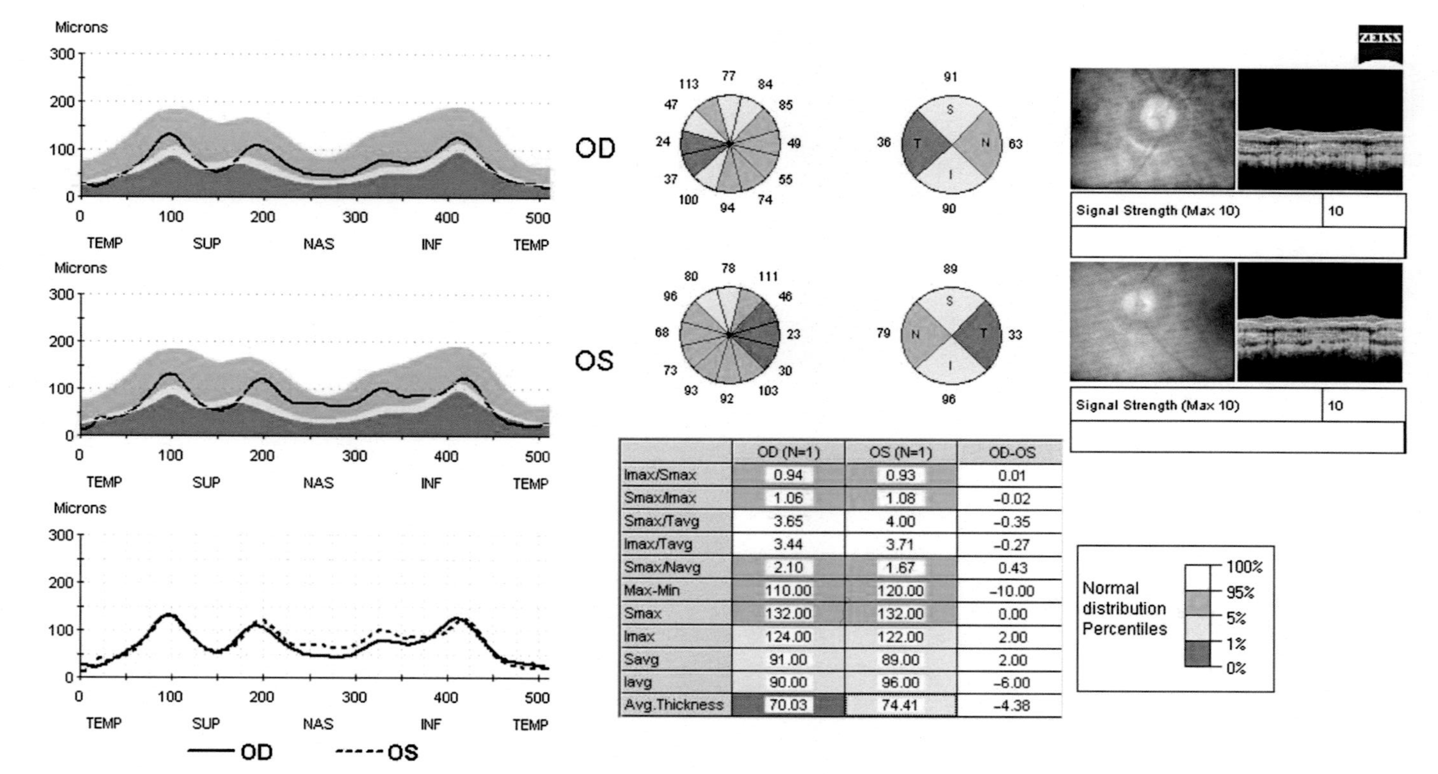

	OD (N=1)	OS (N=1)	OD-OS
Imax/Smax	0.94	0.93	0.01
Smax/Imax	1.06	1.08	-0.02
Smax/Tavg	3.65	4.00	-0.35
Imax/Tavg	3.44	3.71	-0.27
Smax/Navg	2.10	1.67	0.43
Max-Min	110.00	120.00	-10.00
Smax	132.00	132.00	0.00
Imax	124.00	122.00	2.00
Savg	91.00	89.00	2.00
Iavg	90.00	96.00	-6.00
Avg.Thickness	70.03	74.41	-4.38

FIGURE 2 Optical coherence tomography confirmed significant retinal nerve fiber layer thinning, which was more prominent in the temporal quadrant corresponding to the papillomacular bundle.

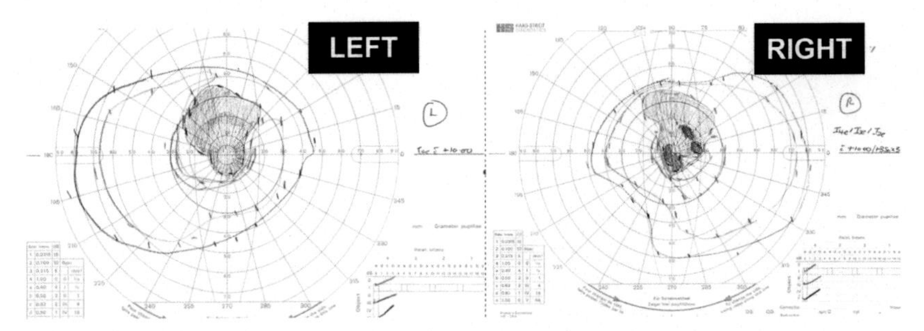

FIGURE 3 Dense central scotomas were documented in both eyes with kinetic Goldmann perimetry.

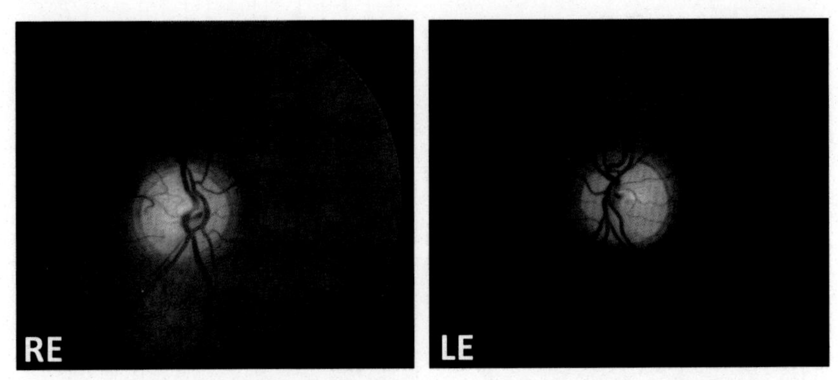

FIGURE 4 The proband's father had marked bilateral optic disc pallor when these dilated fundus photographs were taken at the age of 65 years old.

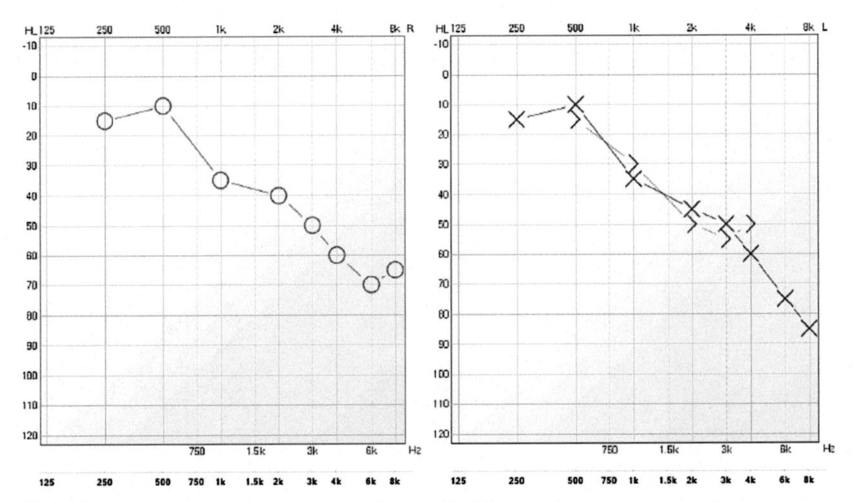

FIGURE 5 Pure tone audiogram was consistent with bilateral high-frequency hearing loss seen in pathogenic *OPA1* mutation carriers.

The majority (50–65%) of families with DOA harbor pathogenic mutations within the *OPA1* gene (3q28-q29), which codes for a 960 amino acid long, dynamin-related GTPase protein that localizes to the inner mitochondrial membrane (OMIM 165500). DOA is therefore a nuclear-encoded mitochondrial optic neuropathy explaining some of the shared pathological features with Leber hereditary optic neuropathy (LHON), especially the marked preferential early loss of retinal ganglion cells within the temporal papillomacular bundle. Interestingly, recent prospective studies have identified additional clinical features besides optic atrophy in ~20% of *OPA1* mutation carriers. Although progressive visual failure remains the defining feature of DOA, with greater access to molecular genetic testing, some *OPA1* mutations have been found to have a strong association with sensorineural deafness, in particular the c.1334G>A (p.Arg445His) *OPA1* mutation. In addition to optic atrophy and sensorineural deafness, some cases will develop more severe neurological features including bilateral ptosis and external ophthalmoplegia, proximal myopathy, peripheral neuropathy, and cerebellar ataxia. These extended *OPA1* disease phenotypes have been referred to as dominant optic atrophy plus (DOA+), and from a clinical perspective, it should be emphasized that these more severe syndromic features usually develop in mid- to late adult life, and they may not be apparent at the time of the initial diagnosis of optic atrophy (Figure 6).

DIFFERENTIAL DIAGNOSIS

Cases presenting with slowly progressive bilateral visual failure require brain MR imaging to exclude a compressive or infiltrative lesion involving the anterior visual pathways (e.g., an optic nerve glioma) or areas of enhancement secondary to an inflammatory etiology (e.g., neurosarcoidosis). Depending on the clinical history, blood test, and additional neurological investigation, investigations should be considered to remove nutritional (vitamin B12 and thiamine levels), inflammatory, and vasculitic causes from consideration. The slowly progressive nature of the visual loss in DOA somewhat mitigates against a number

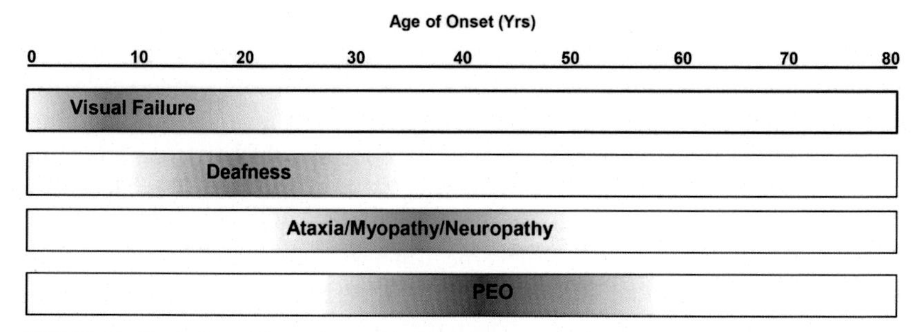

FIGURE 6 Evolution of the major clinical features observed in DOA+ syndromes. Sensorineural deafness is the most frequently observed extraocular feature, present in nearly two-thirds of all cases manifesting DOA+ phenotypes.

of these pathological processes, but the onus is on the clinician to exclude the possibility of a potentially reversible bilateral optic neuropathy.

DIAGNOSTIC APPROACH

DOA is the most common inherited optic neuropathy seen in clinical practice, and the minimum prevalence has been estimated at 1 in 25,000. The evaluation of any case presenting with bilateral visual failure follows the basic principles of a detailed clinical history to provide clues pointing toward the underlying etiology, and importantly, probing questions regarding the family history. Although cases frequently report other affected family members, due to the variable clinical penetrance in DOA, some *OPA1* mutations carriers may only have mild visual loss or even be subclinically affected, thereby masking the dominant mode of transmission and resulting in an apparent sporadic case.

Molecular genetic testing will identify a heterozygous mutation within the *OPA1* gene in 50–65% of cases suspected of having DOA. Over 200 disease-causing *OPA1* variants have been reported so far in this highly polymorphic gene with mutational hotspots in the catalytic GTPase domain (exons 8–15) and the dynamin central domain (exons 16–23). The majority of *OPA1* mutations result in premature termination codons, and the resultant truncated mRNA species are highly unstable, being rapidly degraded by protective surveillance mechanisms operating via nonsense-mediated mRNA decay. Haploinsufficiency is, therefore, a major disease mechanism in DOA, and the pathological consequences of a dramatic reduction in OPA1 protein level is highlighted by those rare families who are heterozygous for micro-deletions spanning the entire *OPA1* coding region. DOA+ phenotypes can be associated with all mutational subtypes, but there is a two- to threefold increased risk of developing multisystem neurological involvement with missense mutations located within the catalytic GTPase domain. The dominant negative explanation for this genotype–phenotype correlation remains speculative.

If *OPA1* mutation analysis is negative, mitochondrial DNA (mtDNA) genetic testing should be considered (providing there is no male transmission), and keeping in mind that LHON has a much more acute presentation and disease evolution compared with DOA. Over the past few years, it has become clear that *WFS1* mutations not only cause classical autosomal recessive Wolfram syndrome, but some mutations can behave in a dominant fashion resulting in a more limited phenotype characterized by optic atrophy in association with diabetes mellitus and/or deafness (OMIM 222300). If the resources are available to do so, *WFS1* testing should, therefore, be screened for *OPA1*-negative cases. A number of genetic loci have been mapped to both autosomal dominant and recessive familial forms of optic atrophy, but the underlying disease-causing genes involved have not yet been identified. Whole exome and genome sequencing will hopeful uncover these remaining unsolved cases—this information is crucial for more accurate genetic counselling both for the case and at-risk family members.

PATHOPHYSIOLOGY

OPA1 resides within the inner mitochondrial membrane, and it mediates multiple cellular functions. It is a critical pro-fusion protein, and pathogenic *OPA1* mutations result in marked mitochondrial network fragmentation. This physical disruption has a knock-on effect on the stability of the mitochondrial respiratory chain complexes, resulting in increased levels of reactive oxygen species and a reduction in overall ATP synthesis. Pro-apoptotic cytochrome *c* molecules are carefully sequestered within the cristae spaces by the zipper-like action of OPA1, and mitochondria are major stores of calcium. Mitochondrial network fragmentation results in the uncontrolled release of these potent pro-apoptotic factors, which ultimately trigger irreversible cell death. Evidence from case cell lines and *Opa1*-mutant mouse models is consistent with an apoptotic mechanism driving the progressive loss of retinal ganglion cells, which eventually results in the development of a clinical apparent optic neuropathy once a critical threshold has been reached. In one *Opa1*-mutant mouse model, loss of dendritic arborization preceded the onset of retinal ganglion cell loss, and the surviving axons had abnormal morphologies with segmental areas of demyelination and myelin aggregation. Another intriguing feature of DOA has recently emerged with the identification of cytochrome *c* oxidase (COX)-negative fibers and multiple mtDNA deletions in skeletal muscle biopsies from cases manifesting both the pure and syndromic forms of the disease. Although the mechanisms still need to be fully elucidated, cases with DOA+ phenotypes harbor significantly higher levels of these somatic mtDNA abnormalities suggestive of a contributory role in the development of the more severe neuromuscular complications. Furthermore, indirect evidence obtained with ultra-deep next-generation sequencing suggests that *OPA1* mutations lead to an increased rate of clonal expansion of somatic mtDNA mutations, triggering the emergence of COX-negative muscle fibers once a critical mutational threshold has been exceeded. The sequence of events linking disturbed mitochondrial dynamics with mtDNA instability and impaired ATP production, and how these mechanisms ultimately precipitate neuronal loss, are of fundamental importance in terms of developing rational treatment strategies.

CASE MANAGEMENT

Supportive Measures

Individuals, in particular young children, can adapt remarkably well to their visual impairment provided that they are appropriately supported. Visual and occupational rehabilitation are essential, and in some countries, cases can get additional help as a result of being registered as severely sight impaired (blind registration) by the ophthalmologist overseeing their care.

Genetic counselling is a key aspect of case management, but it is somewhat complicated by the marked intra- and interfamilial variability observed in DOA. In addition to the wide variability in the progression of the visual loss, it is not possible to predict whether a case will develop other syndromic neuromuscular complications. Despite these limitations, cases with DOA should be informed that the rate of visual deterioration is usually slow, and in the majority of cases, the disease is limited to the optic nerve. Although the overall visual prognosis is better compared with other mitochondrial optic neuropathies such as LHON, more than half of all affected *OPA1* mutation carriers are eventually registered as severely sight impaired by the fifth decade of life. The identification of a pathogenic *OPA1* mutation allows carrier testing for other related family members, and parents can be informed that the recurrence risk for future pregnancies is 50% in keeping with the autosomal dominant mode of inheritance. Knowledge of the specific molecular basis facilitates prenatal and preimplantation diagnosis, which can be used to prevent recurrence.

Specific Therapy

Recent evidence suggests that cases with sensorineural hearing loss benefit from cochlear implantation. A specialist assessment is therefore appropriate for this group of cases to determine their suitability for this procedure. Effective treatments that target the underlying pathological process in DOA are limited. Various combinations of vitamins and compounds with putative neuroprotective effects acting on mitochondrial function have been tried, but none have shown any convincing objective evidence of a marked visual benefit. There is a clear need for properly controlled randomized clinical trials to determine the efficacy of existing and future neuroprotective compounds in rescuing retinal ganglion cell loss in this disorder. Prospective cohort studies are underway to provide the natural history data required to power these clinical trials and to determine the best clinical outcome measures. The eye is an easily accessible organ and visual loss in DOA is invariably progressive. Although gene therapy and stem cells to improve visual function are attractive experimental strategies, these approaches are still at an early stage of preclinical development, and clinicians need to moderate case expectations to reduce the risk of them falling victim to various Internet scams.

CLINICAL PEARLS

- *DOA* typically presents with slowly progressive bilateral symmetrical visual failure in early childhood. Cases will frequently give a history of sitting at the front of the classroom and struggling at school with their studies. Subsequent referral to the hospital eye service is usually prompted by the optometrist not being able to correct the degree of subnormal vision and the absence of any significant refractive error.
- A clear family history is not always evident, and other related family members, especially the parents, should be examined if possible. DOA can show marked intrafamilial disease variability, and up to 20% of *OPA1* mutation carriers are non-penetrant. Subclinical optic atrophy is also common when visual acuity is only mildly reduced to 20/40 or better.

- As for LHON, cases with DOA maintain brisk pupillary light, probably due to the relative preservation of a specific class of melanopsin retinal ganglion cells.
- Visual electrophysiology can be useful in cases of diagnostic uncertainty, and OCT imaging is useful to confirm that the retinal pathology is limited to retinal ganglion cell loss and associated thinning of the nerve fiber layer. Cupping of the optic discs is a recognized feature of DOA, and there are several case reports of individuals being initially misdiagnosed as having normal tension glaucoma and receiving treatment as such. Ophthalmologists, in particular, should rule out the possibility of an inherited optic neuropathy, and pallor of the remaining neuroretinal rim is highly atypical in glaucoma, except in the most advances stages.
- Except for sensorineural deafness, which tends to develop in the first two decades of life, the neuromuscular features associated with DOA+ phenotypes do not usually become apparent until mid- to late adult life. Clinicians should, therefore, maintain a high index of suspicion as appropriate, and timely referral can improve their case's quality of life, especially in the context of preexisting visual impairment.

FURTHER READING

[1] Amati-Bonneau P, Guichet A, Olichon A, et al. *OPA1* R445H mutation in optic atrophy associated with sensorineural deafness. Ann Neurol 2005;58(6):958–63.

[2] Barboni P, Savini G, Cascavilla ML, et al. Early macular retinal ganglion cell loss in dominant optic atrophy: genotype-phenotype correlation. Am J Ophthalmol 2014;158(3):628–36.

[3] La Morgia C, Ross-Cisneros FN, Sadun AA, et al. Melanopsin retinal ganglion cells are resistant to neurodegeneration in mitochondrial optic neuropathies. Brain 2010;133(8):2426–38.

[4] Santarelli R, Rossi R, Scimemi P, et al. *OPA1*-related auditory neuropathy: site of lesion and outcome of cochlear implantation. Brain 2015;138(3):563–76.

[5] Yu Wai Man P, Griffiths PG, Burke A, et al. The prevalence and natural history of dominant optic atrophy due to *OPA1* mutations. Ophthalmology 2010;117(8):1538–46.

[6] Yu Wai Man P, Griffiths PG, Gorman GS, et al. Multi-system neurological disease is common in patients with *OPA1* mutations. Brain 2010;133(3):771–86.

Chapter 25

Childhood-Onset Peripheral Neuropathy with Cognitive Decline

Elizabeth M. McCormick[1], **Russell P. Saneto**[4], **Marni J. Falk**[2,3]

[1]*Divisions of Human Genetics and Child Development and Metabolic Disease, Department of Pediatrics, The Children's Hospital of Philadelphia, Philadelphia, PA, USA;* [2]*Department of Pediatrics, University of Pennsylvania Perelman School of Medicine, Philadelphia, PA, USA;* [3]*Division of Human Genetics, The Children's Hospital of Philadelphia, Philadelphia, PA, USA;* [4]*Department of Neurology/ Division of Pediatric Neurology, Seattle Children's Hospital/University of Washington, Seattle, WA, USA*

CASE PRESENTATION

A 3.5-year-old girl presented for diagnostic evaluation of left foot in-toeing and progressive difficulty walking.

She was the 3.26 kg product of a natural conception who was born at full-term to a primigravida mother. Early developmental milestones were attained on time except for mild motor delay, as she walked at 18 months and ran at 2.5 years. She spoke several single words beginning at 13 months and two-word phrases by age 2 years. Progressive difficulty walking and lower leg weakness were first noted at 3 years, as evidenced by left foot drop, in-toeing, wide-based gait, and increasing difficulty keeping up with her peers during physical activities. She was no longer able to run by 3.5 years due to progressive clumsiness, falling, and weakness. Cognitive concerns including inability to label colors and increasing forgetfulness and confusion were noted at 4 years.

Muscle atrophy was not appreciated on initial examination at age 3 years. However, subsequent examinations revealed atrophy of the bilateral gastrocnemius and distal quadriceps. Axial hypotonia had also developed over this time. At 3 years, deep tendon reflexes were 2/4 in both upper and lower extremities. At 5 years, deep tendon reflexes were 2/4 in the upper extremities and absent in the lower extremities, with no sensation below the knees. She had no facial or bulbar weakness.

Electrophysiologic testing at 5 years revealed moderate to severe sensorimotor axonal neuropathy, greater on the left side. Right sural nerve biopsy revealed decreased numbers of myelinated nerve fibers without discrete onion bulb formation, reported as consistent with a chronic axonal neuropathy.

Mitochondrial Case Studies. http://dx.doi.org/10.1016/B978-0-12-800877-5.00025-5

Her review of systems was otherwise normal except for a history of left equinocavovarus repair at 5 years. Her weight and height have remained at the 50th percentile and head circumference at the 45th percentile. Dilated ophthalmology examination was normal, with no evidence of optic atrophy or retinal dystrophy.

Her family history was non-contributory.

Prior blood-based metabolic testing including blood lactate, copper, thyroid function screen, B vitamin levels, and vitamin E level were all normal. Urine organic acid analysis was also normal. Neuroimaging and electroencephalogram were deferred, as electromyogram findings had significantly narrowed the differential diagnosis.

DIFFERENTIAL DIAGNOSIS

The two features of particular importance that drive the differential diagnosis in this case are sensorimotor axonal neuropathy in a child and early cognitive decline.

Spinal muscular atrophy (SMA) is a common cause of childhood-onset neuropathy that is characterized by progressive muscle weakness resulting from degeneration of the lower motor neurons. In addition, Brown-Vialetto-Van Laere syndrome is a rare condition characterized by sensorimotor axonal neuropathy, facial muscle weakness, cranial nerve and bulbar palsies, sensorineural hearing loss, optic atrophy, and respiratory insufficiency. Some cases with this latter condition have recently been found to harbor pathogenic mutations in the riboflavin transporter genes *SLC52A2* and *SLC52A3*, with a resultant riboflavin transporter deficiency. Remarkably, riboflavin (vitamin B2) supplementation has been shown to slow the progression and even improve symptoms in affected individuals [1]. While the reported child's presentation here of sensorimotor axonal neuropathy is consistent with those with a riboflavin transporter deficiency, she did not have the characteristic facial muscle weakness, cranial nerve and bulbar palsies, sensorineural hearing loss, or optic atrophy. Additionally, intellectual decline, as seen in this case, has not been commonly reported in those with riboflavin transporter deficiency nor in cases with SMA.

The moderate to severe sensorimotor neuropathy involving axonal degeneration in a length-dependent manner in this case was more suggestive of a separate class of conditions known as Charcot-Marie-Tooth (CMT) disease. CMT diseases are inherited motor and sensory neuropathies that comprise a subgroup of nuclear-encoded progressive sensorimotor neuropathies. Many different types of CMT exist, with classification based on the type of neuropathy. CMT type 1 (CMT1) is characterized by demyelinating neuropathy. CMT type 2 (CMT2) is characterized by axonal neuropathy. Additional subtypes are classified by having some combination of these two neuropathy types, including CMT4 that is characterized by both axonal and demyelinating neuropathies. Given the electrophysiologic findings of axonal neuropathy identified in this case, we considered the diagnostic possibilities of CMT2 and CMT4. However, considering that CMT4 is quite rare while CMT2 causes 10–15% of all CMT cases [2], we focused on CMT2. CMT2 can be inherited in either an autosomal recessive or autosomal dominant matter, with many known causative genes (summarized in Ref. [3]) that may variably cause CMT2 with onset across the age range from infancy to adulthood. While most of the causative genes identified to date each account for only a small percentage of cases with CMT2, pathogenic mutations in *MFN2* occur in up to 20% of CMT2 cases. A recent cohort study [4] revealed that most CMT2 cases with pathogenic mutations in *MFN2* present with axonal neuropathy in childhood that progresses over the course of several years. Over time, progressive distal muscle weakness and atrophy occurs most notably involving the lower extremities, although upper extremity involvement can be seen. Additional clinical features of CMT2 disease can include optic atrophy, cognitive decline, and foot deformity. This child's history of neuropathy and cognitive decline was consistent with CMT2. Pathogenic mutations in *MFN2* can be inherited or arise new (de novo), which would be consistent with the probandis negative family history.

MFN2 encodes mitofusin-2, a protein that localizes to the outer mitochondrial membrane and is required for mitochondrial fusion [5–9]. Therefore, CMT2 caused by mutations in *MFN2* is considered a primary mitochondrial disease. Indeed, sensorimotor axonal neuropathy is a finding common in select primary mitochondrial diseases [10,11]. Mitochondrial diseases are highly heterogeneous, with disease-causing mutations already having been identified in more than 250 nuclear or mitochondrial DNA genes [12,13].

While this child's history was consistent with CMT, it was also possible she harbors a mutation in a different gene that causes mitochondrial disease. Of the more than 200 nuclear genes already shown to harbor pathogenic mutations causative of mitochondrial disease [12], perhaps the most common single etiology is *POLG*-related mitochondrial disease. *POLG* encodes the catalytic subunit of the only mitochondrial DNA polymerase, an enzyme necessary for mitochondrial DNA synthesis and high-fidelity replication. *POLG* mutations can cause a wide spectrum of conditions that range in onset from infancy to late adulthood. The more severe end of the clinical spectrum includes conditions in the myocerebrohepatopathy spectrum (MCHS) and Alpers-Hüttenlocher syndrome (AHS). While MCHS is predominantly characterized by myopathy, developmental delay, and liver dysfunction, AHS is characterized by refractory epilepsy and liver dysfunction. Onset of these conditions typically occurs between 2 and 4 years of age. Later-onset presentations of *POLG*-related mitochondrial disease may occur anywhere from teenage years to adulthood, characteristically involving epilepsy, neuropathy, myopathy, and ophthalmoplegia, although isolated ophthalmoplegia can also occur. Additional medical concerns that may be caused by pathogenic *POLG* mutations include failure to thrive, movement disorders, cortical blindness, sensorineural hearing loss, Parkinsonism, mood disorders, and male infertility [14]. Neuropathy is a feature that can be seen at any age and can be the first sign of *POLG* disease, with presenting symptoms that typically include distal weakness, hyporeflexia, and ataxia [14,15]. Cognitive regression can also be seen in *POLG* cases, often occurring at the time of illness or when seizures are severe and in poor control [14]. While the child in this case did not have seizures or liver dysfunction that frequently occur in children with *POLG* pathogenic mutations, her progressive neuropathy and cognitive decline would be consistent with this condition. Furthermore, childhood-onset *POLG* disease is inherited in an autosomal recessive manner, which would be consistent with her negative family history for similarly symptomatic individuals.

While a mutation in a nuclear-encoded gene that encodes a mitochondrial protein could reasonably explain this child's medical history, pathogenic mutations in mitochondrial DNA could also be causative. Several conditions caused by mitochondrial DNA mutations need to be considered as possible genetic etiologies in the differential diagnosis for a child presenting with progressive neurocognitive decline. Mitochondrial encephalomyopathy, lactic acidosis, and stroke-like episodes (MELAS) is a syndrome caused by pathogenic mutations in mitochondrial DNA that is characterized by progressive involvement of multiple systems, including neurologic (seizures or stroke-like episodes, exercise intolerance, headaches, learning difficulty, dementia, sensorineural hearing loss), gastrointestinal (recurrent vomiting), ophthalmologic (optic atrophy, pigmentary retinopathy), cardiac (Wolff-Parkinson-White arrhythmia, conduction block, congestive heart failure), and endocrine or metabolic (diabetes mellitus, lactic acidosis). In addition, affected individuals can develop neuropathy, having onset typically before 10 years of age. In an early study in MELAS cases with the most common m.3243A>G mutation within the *tRNA-Leu* gene, 22% (7 of 32 individuals studied) had neuropathy. Six of these seven neuropathy cases had mixed axonal and demyelinating sensorimotor neuropathy, while one individual had demyelinating polyneuropathy [16]. A second case series by Kaufmann et al. [17] found that 77% of MELAS cases (23 of 30 individuals studied) had abnormal nerve conduction velocity studies, although only half of these individuals had symptoms of neuropathy such as poor balance or limb paresthesia. Electrophysiologic testing in these individuals revealed axonal neuropathy in more than half of those with abnormal testing (12/23), followed by mixed neuropathy in 30% (7/23), and demyelinating neuropathy in four of the individuals (17%). While the child described in this case had neuropathy and presented during childhood as is classic in MELAS, her family history is not characteristic for this condition given there were no reports of similar medical concerns in her mother or maternal relatives. However, it is possible that the mutation arose de novo in the affected child, as is known to occur for the 10 most common mtDNA mutations that collectively have an incidence of one in 200 livebirths [18]. Given the heteroplasmic nature of mitochondrial DNA mutations, it is also possible that a mitochondrial DNA mutation present at a high heteroplasmy load caused the symptoms in the child, whereas lower heteroplasmy levels in her maternal relatives accounted for their absence of clinical symptoms. However, there was no report of older maternal relatives with sensorineural hearing loss or diabetes mellitus, as can occur in adults with low-level heteroplasmy for the common m.3243A>G mutation that causes MELAS [19–21].

Neuropathy can also be seen in neurogenic muscle weakness, ataxia, and retinitis pigmentosa (NARP) syndrome, which is a mitochondrial DNA cytopathy caused by the m.8993T>C or m.8993T>G mutation within the *ATP6* gene. Onset typically occurs in early childhood with presenting features of sensory neuropathy, ataxia, or developmental delay. Additional features may include seizures, sensorineural hearing loss, and cardiac conduction block [22–24]. The presentation of the child in this case could well be consistent with a pathogenic mtDNA mutation leading to a phenotype in the NARP spectrum. Similarly as can occur in MELAS, it is possible she had a de novo mutation or a higher heteroplasmy load than did her unaffected mother. However, her visual function and ophthalmologic examination were normal, with no evidence of a pigmentary retinopathy.

While this child's medical history was concerning for mitochondrial disease, it is also possible a different etiology caused her medical concerns. Neuronal ceroid lipofuscinosis (NCL) is a serious disorder caused by pathogenic mutations in one of several nuclear genes that should always be considered in a child presenting with neurocognitive decline. NCLs are lysosomal storage disorders that present following a typically normal early developmental period with progressive motor and cognitive decline, seizures, and progressive vision loss due to retinal dystrophy. Mutations in *TPP1* are the most common cause for late-infantile NCL, where children present with neurologic decline between 2 and 4 years of age. However, mutations in *PPT1* that can cause both infantile and late-infantile NCL should also be considered. While this little girl's cognitive and motor decline could be consistent with NCL, she did not yet have its characteristic vision loss or epilepsy, although these symptoms can develop later in the course of disease [25]. While neuropathy has only rarely been associated with this condition, there have been reports of sensorimotor neuropathy in cases with NCL [26].

While a genetic etiology was most likely in this case, it was also possible that a vitamin deficiency was the cause of this child's medical problems. Vitamin B12 deficiency results in varying neurological symptoms including extremity numbness and tingling, fatigue, weakness, poor balance, mood changes, and memory impairment as well as megaloblastic anemia, constipation, and weight loss [27]. Vitamin E deficiency can result in neurological symptoms such as peripheral neuropathy and ataxia, as well as retinopathy and immune deficiency [28]. Blood-based screening for these etiologies was normal in the girl presented here.

DIAGNOSTIC APPROACH

A careful neurologic examination is essential to characterize weakness, deep tendon reflexes, and sensory perception. A detailed family history is always important to identify family members who may be similarly affected to aid in refinement of the possible inheritance patterns. As was done in this case, obtaining electromyography (EMG) and nerve conduction velocity studies in individuals presenting with symptoms of neuropathy is important to characterize the type of neuropathy and focus the differential diagnosis.

Obtaining a thorough medical history and thorough review of systems is necessary to assess for multisystem involvement to clarify whether the affected individual has an isolated versus syndromic neuropathy. It is important to systematically screen for potential multisystem involvement, as these features may not always be readily volunteered by the family. Annual ophthalmology, audiology, and cardiology evaluations are important to obtain in any individual with definite or suspected mitochondrial disease, as these high-energy demand organs commonly become symptomatic with time in mitochondrial disease cases. Detailed ophthalmologic examination should include dilated fundus examination as well as electroretinogram and optical coherence tomography to evaluate for potential retinal and optic nerve involvement. Identification of a retinal dystrophy, optic atrophy, ophthalmoplegia, hearing loss, or cardiac abnormality would directly alter the differential diagnosis, refine the molecular testing approach, and potentially alter clinical management.

While thoroughly screening for multisystem involvement is critical, identifying the underlying genetic cause of this child's medical concerns is needed to gain valuable insight into the prognosis and value of pursuing potential therapies. Additionally, establishing a precise molecular etiology provides her family with accurate recurrence risk information [29]. It also opens the door to enable parents, other relatives, and individuals themselves, to prevent future children from being affected with that particular disorder by pursuing genetic testing that can be performed either by preimplantation genetic diagnosis in the setting of in vitro fertilization, or after a pregnancy has occurred in the setting of chorionic villus sampling, amniocentesis, or even newer noninvasive maternal blood test-based methods. However, prenatal diagnostic options only exist when the precise genetic etiology for the disease in a given family is known.

Whole genome mitochondrial DNA sequencing that was performed in the proband's blood by Next-Generation Sequencing analysis was unrevealing of pathogenic point mutations or deletions. However, panel-based sequencing of multiple nuclear genes revealed a known pathogenic heterozygous missense mutation in *MFN2* that had been previously reported as causative of *MFN2*-related CMT. The c.310C>T (p.R104W) mutation identified in this case is located in exon 4 within the GTPase domain. Unfortunately, due to insurance limitations and high out-of-pocket costs, the proband's parents were unable to undergo testing for the familial mutation.

This same *MFN2* mutation was first reported in a family with an affected father and two affected sons, as is consistent with autosomal dominant inheritance [30]. The father first presented with leg stiffness and frequent falls at age 14 years. Over time, he developed axonal motor polyneuropathy as evidenced on EMG, lower extremity distal muscle atrophy, pes cavus, nyctalopia (night blindness), and sensorineural hearing loss. One son presented with an abnormal gait, frequent falls, and generalized muscle atrophy at age 2 years. His other son presented with frequent falls at age 4 years, lower extremity distal weakness, and absent muscle stretch reflexes in the lower limbs. Electrophysiologic testing in both children revealed axonal sensorimotor polyneuropathy. The father's IQ was estimated to be 80, while his sons were estimated to be 70 and 65. Both children were noted to have fine motor difficulty onset between 6 and 7 years of age.

The c.310C>T (p.R104W) mutation in *MFN2* was also reported in two unrelated individuals, where it was known to occur de novo in one child, but the other child's parents were unavailable for segregation testing. Both individuals presented at age 3 years with sensorimotor polyneuropathy and, over time, developed impaired fine motor function and optic atrophy [31].

PATHOPHYSIOLOGY

MFN2 encodes mitofusin-2, a protein that localizes to the outer mitochondrial membrane and is required for mitochondrial fusion [5–9]. Mitochondria are highly dynamic organelles, constantly moving within cells along the cytoskeleton. In addition, mitochondria continually undergo fusion and fission, which are physiologic processes that need to occur in a proper balance to maintain mitochondrial morphology and internal architecture [32–35], as well as to enable formation of functional mitochondrial networks in certain cells [36].

The processes of mitochondrial fission and fusion were first described in yeast and *drosophila*, which illuminated understanding of these critical processes in humans. Fzo and Fzo1p, GTPases in *drosophila* and yeast, respectively, are outer membrane proteins that interact with other proteins, including Mgm1p of the inner membrane, to coordinate fusion across the four combined membranes of two mitochondria [8,37–41]. This process also occurs in mammals [5,42], where these GTPases have two mammalian homologs, MFN1 and MFN2 [5,6], and the homolog of Mgm1p is OPA1. MFN2, an outer membrane protein, and OPA1, an inner membrane protein, coordinate mitochondrial fission and fusion. OPA1 coordinates inner membrane fusion while MFN2 and MFN1 dock and tether neighboring mitochondria to facilitate outer membrane fusion [43]. Disruption of either OPA1 or MFN2 leads to decreased mitochondrial fusion, reduced mitochondrial movement, and small, fragmented and clustered mitochondria [9,37]. These effects are hypothesized to prevent mitochondrial movement to the distal portions of neurons, where there exists a high energy demand [44]. Thus, disorders of mitochondrial fusion often manifest with axonal neuropathies, particularly involving the long nerves in the extremities, the optic nerve, and/or the auditory nerve.

The GTPase domain is of particular importance for MFN2 function [5,8,9]. Pathogenic mutations in *MFN2* were reported in seven families with CMT2A [45], where six of these families had mutations that cluster within, or upstream of the GTPase domain in MFN2.

CLINICAL PEARLS

- To refine the broad differential diagnosis in a child presenting with neuropathy and cognitive decline, it is important to characterize the neuropathy type, assess multisystem involvement, and obtain a detailed family history.
- Molecular testing is key to identify the underlying genetic etiology, which provides important prognostic and genetic counseling information.

REFERENCES

[1] Foley RA, et al. Treatable childhood neuronopathy caused by mutations in riboflavin transporter RFVT2. Brain 2014;137(Pt 1):44–56.

[2] Saporta AS, Sottile SL, Miller LJ, Feely SM, Siskind CE, Shy ME. Charcot-Marie-Tooth disease subtypes and genetic testing strategies. Ann Neurol 2011;69(1):22–33.

[3] Tazir M, et al. Hereditary motor and sensory neuropathies or Charcot-Marie-Tooth diseases: an update. J Neurol Sci 2014;347(1–2):14–22.

[4] Feely SM, et al. *MFN2* mutations cause severe phenotypes in most patients with CMT2A. Neurology 2011;76(20):1690–6.

[5] Santel A, Fuller MT. Control of mitochondrial morphology by a human mitofusin. J Cell Sci 2001;114(Pt 5):867–74.

[6] Rojo M, et al. Membrane topology and mitochondrial targeting of mitofusins, ubiquitous mammalian homologs of the transmembrane GTPase Fzo. J Cell Sci 2002;115(Pt 8):1663–74.

[7] Bach D, et al. Mitofusin-2 determines mitochondrial network architecture and mitochondrial metabolism. A novel regulatory mechanism altered in obesity. J Biol Chem 2003;278(19): 17190–7.

[8] Hales KG, Fuller MT. Developmentally regulated mitochondrial fusion mediated by a conserved, novel, predicted GTPase. Cell 1997;90(1):121–9.

[9] Chen H, et al. Mitofusins Mfn1 and Mfn2 coordinately regulate mitochondrial fusion and are essential for embryonic development. J Cell Biol 2003;160(2):189–200.

[10] Mizusawa H, et al. Peripheral neuropathy of mitochondrial myopathies. Rev Neurol (Paris) 1991;147(6–7):501–7.

[11] Yiannikas C, et al. Peripheral neuropathy associated with mitochondrial myopathy. Aust Paediatr J 1988;24(Suppl. 1):62–3.

[12] Koopman WJ, Willems PH, Smeitink JA. Monogenic mitochondrial disorders. N Engl J Med 2012;366(12):1132–41.

[13] McCormick E, Place E, Falk MJ. Molecular genetic testing for mitochondrial disease: from one generation to the next. Neurotherapeutics 2013;10(2):251–61.

[14] Cohen BH, Naviaux RK. The clinical diagnosis of POLG disease and other mitochondrial DNA depletion disorders. Methods 2010;51(4):364–73.

[15] Stumpf JD, Saneto RP, Copeland WC. Clinical and molecular features of POLG-related mitochondrial disease. Cold Spring Harb Perspect Biol 2013;5(4):a011395.

[16] Karppa M, et al. Peripheral neuropathy in patients with the 3243A>G mutation in mitochondrial DNA. J Neurol 2003;250(2):216–21.

[17] Kaufmann P, et al. Nerve conduction abnormalities in patients with MELAS and the A3243G mutation. Arch Neurol 2006;63(5):746–8.

[18] Elliott HR, et al. Pathogenic mitochondrial DNA mutations are common in the general population. Am J Hum Genet 2008;83(2):254–60.

[19] Remes AM, et al. Adult-onset diabetes mellitus and neurosensory hearing loss in maternal relatives of MELAS patients in a family with the tRNA(Leu(UUR)) mutation. Neurology 1993;43(5):1015–20.
[20] van den Ouweland JM, et al. Maternally inherited diabetes and deafness is a distinct subtype of diabetes and associates with a single point mutation in the mitochondrial tRNA(Leu(UUR)) gene. Diabetes 1994;43(6):746–51.
[21] Oshima T, et al. Bilateral sensorineural hearing loss associated with the point mutation in mitochondrial genome. Laryngoscope 1996;106(1 Pt 1):43–8.
[22] Holt IJ, et al. A new mitochondrial disease associated with mitochondrial DNA heteroplasmy. Am J Hum Genet 1990;46(3):428–33.
[23] Santorelli FM, et al. Heterogeneous clinical presentation of the mtDNA NARP/T8993G mutation. Neurology 1997;49(1):270–3.
[24] Sembrano E, et al. Polysomnographic findings in a patient with the mitochondrial encephalomyopathy NARP. Neurology 1997;49(6):1714–7.
[25] Mole SE, Williams RE. Neuronal ceroid-lipofuscinoses. In: Pagon RA, et al., editor. GeneReviews(R). Seattle (WA); 1993.
[26] Sinha S, et al. Neuronal ceroid lipofuscinosis: a clinicopathological study. Seizure 2004;13(4):235–40.
[27] http://ods.od.nih.gov/factsheets/VitaminB12-HealthProfessional/#h5.
[28] http://ods.od.nih.gov/factsheets/VitaminE-HealthProfessional/.
[29] Falk MJ. Neurodevelopmental manifestations of mitochondrial disease. J Dev Behav Pediatr 2010;31(7):610–21.
[30] Del Bo R, et al. Mutated mitofusin 2 presents with intrafamilial variability and brain mitochondrial dysfunction. Neurology 2008;71(24):1959–66.
[31] Brockmann K, et al. Cerebral involvement in axonal Charcot-Marie-Tooth neuropathy caused by mitofusin2 mutations. J Neurol 2008;255(7):1049–58.
[32] Bereiter-Hahn J, Voth M. Dynamics of mitochondria in living cells: shape changes, dislocations, fusion, and fission of mitochondria. Microsc Res Tech 1994;27(3):198–219.
[33] Nunnari J, et al. Mitochondrial transmission during mating in *Saccharomyces cerevisiae* is determined by mitochondrial fusion and fission and the intramitochondrial segregation of mitochondrial DNA. Mol Biol Cell 1997;8(7):1233–42.
[34] Griparic L, van der Bliek AM. The many shapes of mitochondrial membranes. Traffic 2001;2(4):235–44.
[35] Legros F, et al. Mitochondrial fusion in human cells is efficient, requires the inner membrane potential, and is mediated by mitofusins. Mol Biol Cell 2002;13(12):4343–54.
[36] Amchenkova AA, et al. Coupling membranes as energy-transmitting cables. I. Filamentous mitochondria in fibroblasts and mitochondrial clusters in cardiomyocytes. J Cell Biol 1988;107(2):481–95.
[37] Hermann GJ, et al. Mitochondrial fusion in yeast requires the transmembrane GTPase Fzo1p. J Cell Biol 1998;143(2):359–73.
[38] Rapaport D, et al. Fzo1p is a mitochondrial outer membrane protein essential for the biogenesis of functional mitochondria in *Saccharomyces cerevisiae*. J Biol Chem 1998;273(32): 20150–5.
[39] Wong ED, et al. The dynamin-related GTPase, Mgm1p, is an intermembrane space protein required for maintenance of fusion competent mitochondria. J Cell Biol 2000;151(2):341–52.
[40] Wong ED, et al. The intramitochondrial dynamin-related GTPase, Mgm1p, is a component of a protein complex that mediates mitochondrial fusion. J Cell Biol 2003;160(3):303–11.

[41] Sesaki H, et al. Mgm1p, a dynamin-related GTPase, is essential for fusion of the mitochondrial outer membrane. Mol Biol Cell 2003;14(6):2342–56.
[42] Tang Y, et al. Maintenance of human rearranged mitochondrial DNAs in long-term cultured transmitochondrial cell lines. Mol Biol Cell 2000;11(7):2349–58.
[43] Koshiba T, et al. Structural basis of mitochondrial tethering by mitofusin complexes. Science 2004;305(5685):858–62.
[44] Baloh RH, et al. Altered axonal mitochondrial transport in the pathogenesis of Charcot-Marie-Tooth disease from mitofusin 2 mutations. J Neurosci 2007;27(2):422–30.
[45] Zuchner S, et al. Mutations in the mitochondrial GTPase mitofusin 2 cause Charcot-Marie-Tooth neuropathy type 2A. Nat Genet 2004;36(5):449–51.

Chapter 26

Brain-Specific Mitochondrial Aminoacyl-tRNA Synthetase Disorders: Mitochondrial Arginyl-Transfer RNA Synthetase Deficiency

Russell P. Saneto
Department of Neurology/Division of Pediatric Neurology, Seattle Children's Hospital/University of Washington, Seattle, WA, USA

CASE PRESENTATION

Our case was the fourth child born to non-consanguineous parents of Northern European descent. She was a term delivery, and due to postdelivery problems with peri-delivery hypothermia and metabolic acidosis, she required a 4-day intensive care stay. She was discharged to home in stable condition without concerns. She was first brought to medical attention at 4 months of age, as she was not making developmental progress. She did not have a social smile nor did she visually track. Her pediatrician also noted truncal hypotonia for the first time. She had a normal head circumference (45%) at birth but over time developed a progressive microcephaly; at 16 years she is about 4 standard deviations below the 50% for age. She never developed independent ambulation, but before her first status epilepticus event, she could sit independently and currently cannot even roll over. She is verbal, but never developed expressive language. She continues to require gastrostomy feeding for nutrition. She has cortical visual impairment. She displays limb spasticity with contractures at the wrists and ankles. Her hips are dislocated bilaterally. Her deep tendon reflexes are brisk and cross-adduct at the patellar tendons, bilaterally.

She underwent the Vineland II Adaptive Behavior Scales, Parent/Caregiver Rating Form neuropsychological testing at age 16 years, due to her severe encephalopathy. All of her scores were at less than first percentile for age. Adaptive level is classified as low, and age equivalent scores were less than the eighth-month level on all scales.

Mitochondrial Case Studies. http://dx.doi.org/10.1016/B978-0-12-800877-5.00026-7

At 2 years, she developed seizures, myoclonic jerks involving her extremities, both generalized and focal. In some of her events, eyelid fluttering was associated with her jerks. Multiple electroencephalograms (EEG) have demonstrated both generalized and independent focal discharges over both hemispheres, with background slowing. Neither seizure semiology nor EEG findings have changed over the years. She has developed generalized and status epilepticus partialis continua multiple times. She is intractable to seizure medications, both as monotherapy and combination therapy, including the Ketogenic Diet. Myoclonus of her extremities remains prominent.

An initial nuclear magnetic resonance image (MRI) scan of the head at the age of 2 years showed cerebellar hypoplasia with normal appearing cerebrum. Over time, the MRI scans have shown a progressive cerebellar atrophy involving the vermis and hemispheres, cystic encephalomalacia involving the bilateral caudate heads and putamina, and cerebral atrophy involving the frontal lobes and right posterior lobe and thinning of the corpus callosum (Figure 1). The optic nerves are small and thinned. Although the pons is small and mildly flattened, it has not undergone significant atrophy compared to the progression seen in the cerebellum. The proton magnetic spectroscopy (MRS) scan revealed a doublet at 1.33 ppm (lactate) in multiple voxels over the basal ganglia bilaterally (image not shown).

At age 7, she underwent a muscle biopsy without abnormalities in histochemical staining and electron microscopy structural evaluation. Enzyme activities of complex I–IV of the respiratory chain were normal. Her mitochondrial DNA was sequenced without pathological mutation found. Metabolic testing

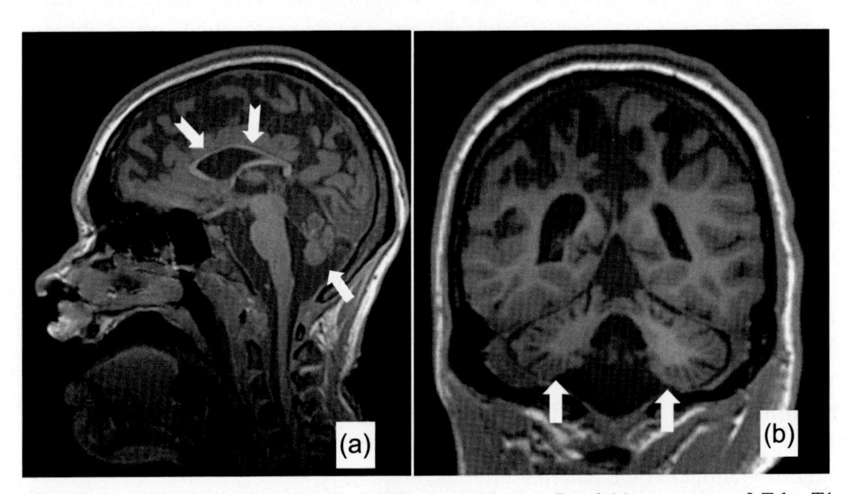

FIGURE 1 Magnetic resonance image (MRI) at age 14 years. Panel (a) represents a 3 Telsa T1 MPRAGE sagittal image (Seimans Trio). The solid white arrow shows the small cerebellum, vermis. The notched white arrows demonstrate the thin corpus callosum. There is mild hypoplasia of the pons and significant atrophy of the cerebrum. Panel (b) represents a 3 Telsa T1 MPRAGE coronal image (Seimans Trio). The solid white arrows show reduction in cerebellar hemispheres in a butterfly pattern. Atrophy of the cerebrum can be seen.

for urine organic acids, plasma amino acids, ammonia, acyl carnitine profile, and isoelectric focusing of transferrin returned without abnormality. She had multiple elevations of arterial lactate of >3 mM (normal < 1.6 mM).

She underwent next generation gene sequencing of an expanded gene panel and compound heterozygote mutations were found in the arginyl-tRNA synthetase, *RARS2* gene. The mutations were in trans, c.472_474delAAA and c.772C>G. The three adenine nucleotide deletion has previously been reported in association with pontocerebellar hypoplasia (PCH) type 6 [1]. The missense mutation, c.772C>G (p. Arg258Cys), has not been previously reported but is predicted to be deleterious by multiple in-silico analysis programs. The location in the protein is a well-conserved region among mammals and has not been reported as a benign polymorphism.

DIFFERENTIAL DIAGNOSIS

The tincture of time is often needed help narrow the diagnosis in mitochondrial diseases. There are so many extrinsic variables that can alter the phenotype in the evolution of a particular mitochondrial disorder. Thus, the differential diagnosis could change depending on the expressed phenotype at the time of evaluation. At 2 years of age with the case demonstrating only a small cerebellum on MRI, developmental delay, and myoclonic seizures without significant biochemical abnormalities, the differential diagnosis is broad. Possible diagnoses would include ataxia telangiectasia, congenital disorders of glycosylation, spinocerebellar ataxia autosomal recessive with axonal neuropathy, progressive encephalopathy with edema, hypsarrhythmia and optic atrophy, and mitochondrial disease. Clinically, the presentation of our case would eliminate some of the above possible diagnoses from consideration. But, diagnosing the etiology of isolated cerebellar atrophy can be difficult. In a study of 63 cases with isolated cerebellar atrophy, 44 were cases of unknown diagnosis, 11 had ataxia telangiectasia, three had ataxia with oculomotor apraxia, three had late-onset GM2 gangliosidosis, and two had *CACNA1A* mutations [2]. In another study of 113 cases with definite mitochondrial disease, 18 cases had cerebellar atrophy/hypoplasia due to respiratory chain defects [3]. But, the clinical and biochemical findings would distinguish our case from this group.

However, as the case aged, the MRI demonstrated progressive cerebellar atrophy, and cerebral involvement became evident with bilateral cystic encephalomalacia and corpus callosum thinning. She developed a progressive microcephaly. She had intractable seizures and severe global developmental delay. Pontine involvement was present, although minor. On MRS, a lactic acid peak at 1.33 ppm in voxels from the basal ganglia suggested an oxidative phosphorylation defect. Other system involvement was minimal. These features now increase the likelihood of a PCH disorder and, in particular, mitochondrial pontine cerebellar hypoplasia (PCH) type 6.

There are at least 10 types of PCH, PCH1–PCH10. These are difficult to differentiate based on an MRI scan, although some differences exist. Some have extra-CNS findings; PCH1B demonstrates progressive muscle wasting; PCH7 has micropenis/ambiguous genitalia; and PCH5 has severe prenatal onset. All have microcephaly. Specific gene mutations have been associated with all but PCH3 and PCH7.

An emerging group of mitochondrial disorders are due to defects in the aminoacyl-tRNA synthetases that are involved in the transferring genetic information from mtDNA into the proteins of respiratory chain. These synthetases initiate mitochondrial protein synthesis by charging a specific tRNA to its cognate amino acid. By a mechanism that is not completely understood, of the 17 aminoacyl-tRNA synthetases that are mitochondrial specific (mt-ARS), six are related to brain-specific disorders (Table 1). Mutations in the arginyl-tRNA synthetase *RARS2* gene were first reported in a severe infantile PCH [4]. The *RARS2*-encoded mitochondrial disorders have been found in at least nine cases, and all have flattening or butterfly-shaped cerebellum, progressive (when investigated) cerebral volume loss, and concordant microcephaly. Significant respiratory chain defects were noted in two of the cases. Lactic acidosis was noted in six cases (including our case). All cases had severe global developmental delay. These signs and symptoms are suggestive of a *RARS2* defect in our case.

DIAGNOSTIC APPROACH

The majority of mitochondrial diseases present with multiorgan dysfunction and select biochemical abnormalities. However, the reader should know that there are specific types of mitochondrial disease that only affect a single organ. Features of global developmental delay, episodic loss of developmental milestones, and myoclonic seizures have the flavor of a mitochondrial disease. Together with the MRI findings of cerebellar hypoplasia/atrophy, the clinician should look for mitochondrial dysfunction.

The focal seizure events would likely trigger an MRI scan of the brain to look for possible seizure etiology. However, in this case at age 2 years, the only finding was cerebellar hypoplasia/atrophy. If a metabolic disorder is suspect, our general approach is to get basic metabolic labs, serum lactate, ammonia, plasma amino acids, acylcarnitine profile, liver functions (i.e., AST/ALT, PT (INR), and PTT), creatine phosphate kinase, and first morning void for urine organic acids. In our case, these labs were normal. Given the possible other etiologies, other labs would also be performed, such as alpha-fetal protein levels, vitamin E levels, and isoelectric focusing of transferrin. At age 2 years, all of these tests would have returned normal. There are times when all metabolic testing returns normal, yet the child has a metabolic disorder. If suspicion is high that a metabolic disorder is present, additional time may present new signs and symptoms to make the diagnosis.

The constellation of signs and symptoms that give clinical clues for mitochondrial disease are often varied and heterogeneous in presentation. At times, maturation of the disease is needed for further clues on the evaluation for diagnosis. Our case's persistent seizures, multiple events of status epilepticus, severe

TABLE 1 Mitochondrial-Specific Aminoacyl-tRNA Synthetases that Are Related to Mostly Brain-Specific Disease

Gene	Cognate Acylamino tRNA Synthetase	Clinical Phenotype
EARS2	Glutamyl-tRNA synthetase	Leukoencephalopathy with thalamus brainstem and cerebellar involvement; OXPHOS defects, high lactate; liver (homozygous mutation); infant onset[a]
FARS2	Phenylalanyl-tRNA synthetase	Cerebral atrophy, basal ganglia signal changes, sparing of cerebellum; OXPHOS defects, high lactate; severe epilepsy; infant onset
DARS2	Aspartyl-tRNA synthetase	Leukoencephalopathy with brainstem and spinal cord involvement; high lactate; childhood–adulthood onset[b]
MARS2	Methionyl-tRNA synthetase	Leukoencephalopathy with corpus callosum thinning, cortical and cerebellar atrophy; childhood–adulthood onset[c]
RARS2	Arginyl-tRNA synthetase	Pontocerebellar hypoplasia/atrophy, progressive cerebral-white matter atrophy/loss; high lactate, severe epilepsy and global developmental delay; infancy–childhood onset[d]
CARS2	Cysteinyl-tRNA synthetase	Leukoencephalopathy, cerebral atrophy with brainstem involvement; severe epilepsy, high lactate; progressive developmental decline; childhood onset

OXPHOS, oxidative phosphorylation.
[a]*Leukoencephalopathy with thalamus and spinal cord involvement with lactate elevation: LTBL.*
[b]*Leukoencephalopathy with thalamus and brainstem involvement with lactate elevation: LBSL.*
[c]*Autosomal recessive spastic ataxia with leukoencephalopathy (ARSAL).*
[d]*Pontocerebellar hypoplasia 6 (PCH6).*

global developmental delay, and progressive microcephaly would suggest to the clinician to repeat the MRI and MRS. Both brain scans were repeated at age 14 years. The findings of progressive cerebellar atrophy, presence of pontine changes, cerebral atrophy, basal ganglia cystic encephalomalacia, and lactate peak would/should lead the clinician toward a mitochondrial etiology.

In the best of environments, the use of next generation gene sequencing panels of nuclear genes involved in mitochondrial structure and function or whole exome sequencing may be the next logical step. In our case, there was no maternal family history suggesting maternal mitochondrial disease, so whole mitochondrial genome sequencing would not be of high yield. But, gene tests are expensive and often insurance does not cover costs for testing. One might opt to perform a muscle biopsy and assay the respiratory chain enzyme activities, and do histochemical and electron microscopy analysis. This was done in our case, and we did send for coenzyme Q10 levels in the muscle biopsy material, which came back with normal values.

Eventually, we were fortunate to get gene testing. We elected to do next generation gene sequencing due to sensitivity of coverage. In our differential were the PCH disorders and mt-ARSs.

PATHOPHYSIOLOGY

PCH has been divided into 10 types that have been associated with nine genes. Four of the known genes associated with PCH are essential for messenger RNA (mRNA) translation. Once considered a hallmark feature of PCH6 was reduced respiratory chain enzyme activity without significant biochemical or neurological features typical of mitochondrial diseases [5]. However, as more cases have been described, some had no abnormalities of the respiratory chain [6]. Our case did not have any abnormalities of the respiratory chain function. Our case did have lactic acidosis, progressive MRI findings, and no other extra-neurological features similar to most of the other described cases. Clinically, our case has suffered from multiple bouts of status epilepticus and progressive microcephaly, also common features of PCH6. Taken together, *RARS2* mutations are the most likely etiology of our case.

RARS2 is a nuclear gene that encodes mitochondrial arginyl-tRNA synthetase, one of the families of 36 aminoacyl-tRNA synthetases. Messenger RNA (mRNA) translation is involved in the manufacture of genetic information into proteins. The role of the tRNA synthetases is to charge a tRNA with its cognate amino acid. This is a two-step process that uses one adenosine triphosphate (ATP) to activate the amino acid and then transfer the activated amino acid to its specific tRNA for translation into a polypeptide.

The 36 tRNA synthetases are expressed in all tissues having a nucleus. Protein synthesis occurs in both cell cytoplasm, as well as in the mitochondria matrix. Seventeen of the aminoacyl-tRNA synthetases are solely for mitochondrial protein synthesis, 16 others are active solely in the cytoplasm, and three (glycine-, lysine-, and glutamine-tRNA synthetases) are active in both. The 20 mt-ARSs active in the mitochondria are essential for the synthesis of all 13 protein subunits in the respiratory chain. Since mitochondrial protein synthesis is necessary in all mitochondrial containing cells, failure of any one of the 20 aminoacyl-tRNA synthetases should induce multisystem abnormalities. However, in the 10 described mitochondrial-specific mt-ARSs causing disease, most induce only specific single organ involvement [6]. In these diseases, five are brain specific (Table 1). Evidence suggests that in some unknown way, residual activity of a certain synthetase is compatible with normal functioning, while in certain cells/organ activity is disease causing. It is not clear why in *RARS2* the cerebellar/pons/cerebrum is sensitive and the target of disease.

Neuropathological findings in three cases with *RARS2* mutation demonstrated small and immature cerebella, a basis pontis showing atrophy or regressive changes, and marked inferior olivary hypoplasia [7]. Findings suggest that specific brain regions had normal development to variable developmental stages, and later regressed. The cases used in this study died prenatally, 42–45 weeks gestational age. The mechanisms of developmental cessation are unclear but, viewing our case, likely are variable from case to case.

CLINICAL PEARLS

- The presentation of a particular mitochondrial disease varies between cases, and the full constellation of signs and symptoms may take time to present.
- Mitochondrial dysfunction may be manifested by only single organ dysfunction.
- The use of genetic testing should be well thought out and driven by clinical and biochemical findings (or lack of findings).
- Using clinical judgement, repeat testing can uncover disease or direction of future testing.
- Choices of genetic testing should be driven by the sensitivity of the testing, looking at clinical findings, and biochemical and neuroimaging abnormalities.

REFERENCES

[1] Glamuzina E, Brown R, Hogarth K, et al. Further delineation of pontocerebellar hypoplasia type 6 due to mutations in the gene encoding mitochondrial arginyl-tRNA synthetase, RARS2. J Inherit Metab Dis 2012;35:459–67.

[2] Al-Maawall A, Blaser S, Yoon G. Diagnostic approach to childhood-onset cerebellar atrophy: a 10-year retrospective study of 300 patients. J Child Neurol 2012;27:1121–32.

[3] Scaglia F, Wong L-JC, Vladutiu GD, Hunter JV. Predominant cerebellar volume loss as a neuroradiologic feature of pediatric respiratory chain defects. AJNR 2005;26:1675–80.

[4] Rankin J, Brown R, Dobyns WB, et al. Pontocerebellar hypoplasia type 6: a British case with PEHO-like features. Am J Med Genet Part A 2010;152A:2079–84.

[5] Edvardson S, Shaag A, Kolesnikova O, et al. Deleterious mutation in the mitochondrial arginyl-transfer RNA synthetase gene is associated with pontocerebellar hypoplasia. Am J Hum Genet 2007;81:857–62.

[6] Konovalova S, Tyynismaa H. Mitochondrial aminoacyl-tRNA synthetases in human disease. Mol Genet Metabol 2013;108:206–11.

[7] Joseph JT, Innes AM, Smith AC, et al. Neuropathologic features of pontocerebellar hypoplasia type 6. J Neuropath Exp Neurol 2014;73:1009–25.

Chapter 27

Mitochondrial Aminoacyl-tRNA Synthetase Disorders Not Generally Affecting Brain

Lisa Riley[1], **John Christodoulou**[1,2,3]
[1]Genetic Metabolic Disorders Research Unit, Children's Hospital at Westmead, Westmead, NSW, Australia; [2]Western Sydney Genetics Program, Children's Hospital at Westmead, Westmead, NSW, Australia; [3]Disciplines of Paediatrics and Child Health and Genetic Medicine, Sydney Medical School, University of Sydney, Sydney, NSW, Australia

CASE PRESENTATIONS

Case 1

The proband case was noted to be anemic at 23 years of age, unresponsive to erythropoietin injections, and investigations ultimately revealed that she had ringed sideroblasts. She had reduced exercise tolerance but has otherwise been healthy, although she had a spinal fusion for scoliosis in mid-adolescence and has endometriosis. Her general physical examination was unremarkable, and she was of normal intellect. EKG and echocardiograph have been normal.

Her hemoglobin was relatively stable at 100 g/L, and now aged 28 years, she has never needed a blood transfusion. She has had persistent lactic acidemia (3–5 mmol/L; normal range 0.7–2.0). Because of her clinical story being highly suggestive of MLASA (mitochondrial myopathy, lactic acidosis, and sideroblastic anemia) and because she was of Lebanese background, the tyrosyl-tRNA synthetase two *YARS2* gene was screened, revealing homozygosity for the c.156C>G (p.Phe52Leu) Lebanese founder mutation.

Though non-consanguineous, her parents came from the same village in northern Lebanon. The proband case had an elder sister who was diagnosed with sideroblastic anemia in her mid-20s, and became transfusion dependent. At the time, she was otherwise described as being healthy, with no other health concerns and of normal intellect. However, she died of hepatic cirrhosis induced by iron overload about 10 years later. Two other siblings are healthy and are not anemic.

Mitochondrial Case Studies. http://dx.doi.org/10.1016/B978-0-12-800877-5.00027-9

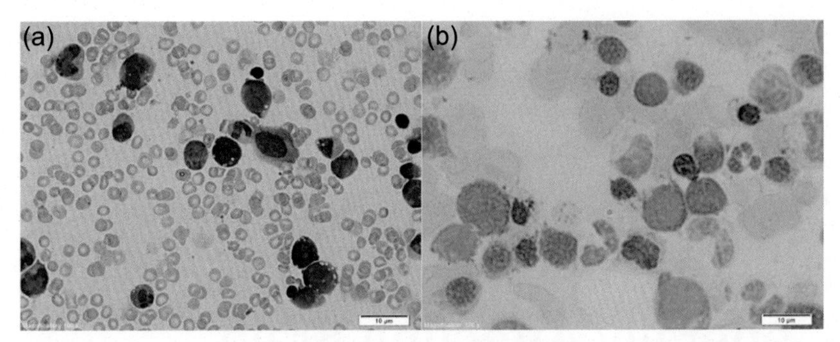

FIGURE 1 **Bone marrow manifestations of YARS2 deficiency.** (a) Bone marrow aspirate (May-Grunwald-Giemsa stain) showing vacuolation of proerythroblasts and myelocytes; (b) bone marrow iron stain showing ringed sideroblasts.

Case 2

The proband case was the second child (older sibling has no health concerns; second pregnancy prior to the proband ended in fetal demise at 12 weeks gestation) of unrelated parents of Lebanese background, and she was born following an uncomplicated pregnancy and delivery at term. Feeding, growth, and early developmental milestones were normal. Following an upper respiratory tract infection, she presented at 8 weeks of age in hypotensive shock associated with profound lactic acidosis (pH 6.56, bicarbonate 4 mmol/L, blood lactate 27 mmol/L). Her acid–base status improved with resuscitative measures, but she had persistent lactic acidemia (3–5 mmol/L). She was found to have a severe anemia (hemoglobin 42 g/L) requiring packed cell transfusions, with a bone marrow aspirate revealing reduced erythropoiesis with prominent red cell vacuolation in the red cell precursors and ringed sideroblasts (Figure 1). In addition she was found to have a severe hypertrophic cardiomyopathy, a pericardial effusion, and hepatomegaly, with deranged liver enzymes and coagulation. She could not be weaned from a ventilator, and with her parents' consent, a decision was made to withdraw active treatment, with demise at 3 months of age.

CSF lactate was normal, and brain proton magnetic resonance spectroscopy (MRS) revealed no lactate peak, while MRI of the brain showed cerebral atrophy. Urine organic acid analysis showed elevations of lactate and ketones. Respiratory chain enzyme (RC) analysis of samples collected in the immediate postmortem period revealed severe deficiencies of complexes I, III, and IV in combination with elevated complex II and citrate synthase in muscle, but not liver. Genetic testing uncovered homozygosity for the c.156C>G mutation in *YARS2*.

DIFFERENTIAL DIAGNOSIS

The presentation of lactic acidosis and sideroblastic anemia is strongly suggestive of a mitochondrial disorder. There are several possible genetic causes. Pearson marrow-pancreas syndrome is the most common cause and results from mtDNA deletions. This disorder typically presents in infancy with failure to thrive, anemia, and lactic acidosis; however exocrine pancreatic insufficiency is also present, whereas this has not been seen in *YARS2* MLASA. Mutations in *PUS1* also cause MLASA, but cases generally have cognitive defects. Recently, a case of MLASA was reported in association with a *MT-ATP6* mutation [1], however this case also had sensorineural hearing loss, agenesis of the corpus callosum, epilepsy, and stroke-like episodes that have not been reported in *YARS2* cases. Lactic acidosis has not been associated with other causes of sideroblastic anemia [2]. If the anemia is severe enough, this could cause lactic acidemia because of severe tissue hypoxemia, but with a transfusion, the lactic acidemia would clear; whereas with YARS2 and related disorders, the lactic acidemia persists.

In the case of other mitochondrial aminoacyl-tRNA synthetases (mtARS) not generally affecting the brain, cases with *AARS2* mutations can present with poor feeding/failure to thrive, delayed motor development and severe generalized muscle weakness, lactic acidosis, and infantile hypertrophic cardiomyopathy [3], similar to more severe *YARS2* cases. Interestingly, there is an alternative *AARS2* presentation involving premature ovarian failure and leukodystrophy and no cardiomyopathy [4]. Cases with *SARS2* mutations also present in infancy with failure to thrive following a premature birth in all cases reported to date [5]. They too have anemia (though not sideroblastic) and lactic acidosis. However, these cases also have hyperuricemia, pulmonary hypertension, and renal failure. Perrault syndrome due to mutations in *HARS2* or *LARS2* is not usually diagnosed until affected females present with premature ovarian failure, with sensorineural hearing loss having been present since childhood in both affected female and male family members [6,7]. Genetic diagnosis is required to distinguish these from other causes of Perrault syndrome.

DIAGNOSTIC APPROACH

YARS2 MLASA has a variable age of onset from birth to adulthood. Most cases present in infancy with lethargy, pallor, and poor feeding/failure to thrive. In older cases, the presentation is usually due to anemia, with some history of reduced exercise capacity or delayed motor milestones. All cases have had elevated serum lactate ranging from mildly to severely elevated levels. Bone marrow aspirate is required to confirm the presence of ringed sideroblasts by iron staining (Figure 1(b)) and vacuolated blood cell precursors (Figure 1(a)). Cases have varying degrees of usually progressive myopathy, exercise intolerance, and in the more severe cases may develop severe respiratory muscle weakness requiring ventilator support. Muscle biopsies display ragged red fibers, and RC enzyme analysis typically shows severe deficiency of complexes I, III, and IV. No RC enzyme deficiency has been detected in patient fibroblasts and liver.

Other presenting features may include hypertrophic cardiomyopathy that has been reported in ~50% of cases, premature ovarian failure, dysphagia, gastrointestinal problems, ptosis, nystagmus and strabismus, limitation of lateral gaze, hepatomegaly, renal tubulopathy, and hypoglycemia. Cognition has been normal in all cases.

In more recently diagnosed cases, muscle biopsy has not been required, with the presence of sideroblastic anemia and lactic acidosis being sufficient to recommend DNA sequencing of *YARS2* or whole exome sequencing to identify the underlying cause. The most commonly reported *YARS2* mutation, c.156C>G, p.Phe52Leu, has been found exclusively in cases of Lebanese origin and is likely to be a founder mutation. There has also been a Lebanese case with a homozygous c.137G>A; p.Gly46Asp mutation [8], a French case with compound heterozygous c.572G>A; p.Gly191Asp and c.1078C>T; p.Arg360* mutations [9], and a Turkish case with a homozygous c.1303A>G; p.Ser435Gly mutation [10].

For other mtARS mutations not generally affecting the brain, lactic acidosis and combined RC deficiency in muscle has been observed in both *AARS2* infantile cardiomyopathy and *SARS2* hyperuricemia, pulmonary hypertension, renal failure in infancy, and alkalosis (HUPRA syndrome) cases. Cases with mtARS mutations are increasingly being diagnosed as a consequence of next-generation sequencing, including *HARS2* and *LARS2* Perrault syndrome cases.

TREATMENT STRATEGY

There are currently no specific therapies targeting the underlying primary genetic or functional defects associated with disorders caused by mutations in the mtARS, and so treatment is essentially symptomatic, with ongoing surveillance for potential late complications of major organ systems.

For individuals with *YARS2* mutations, we did not find mitochondrial vitamin/cofactor cocktails or creatine supplementation to be of any obvious value, either in terms of stabilization of anemia, or improvement of exercise intolerance, nor was a trial of tyrosine supplementation in the original cases beneficial [11]. For those with symptomatic anemia, periodic packed red blood cell transfusions have been of short-term benefit, but the risk of iron overload mandates that some form of chelation must be included as part of ongoing therapy. Pyridoxine appeared to prevent the need for transfusion for sideroblastic anemia for ~10 years in one case [12], and another took a cocktail of supplements including regular intramuscular pyridoxal-5-phosphate and did not require transfusion [11]. Whether this was of therapeutic value remains unknown.

Because gastrointestinal dysfunction may be a complicating issue, careful attention should be given to ensuring optimum nutrition [12]. Respiratory failure and cardiomyopathy are seen in a proportion of affected individuals, and so active vigilance for these health concerns is important.

LONG-TERM OUTCOME

As described above, while distinct stereotypic phenotypes have been associated with mutations in particular members of the mtARS gene family, with expanded molecular genetic screening has come a broadening of the clinical phenotypes [13]. Moreover, prognostication for individuals harboring precisely the same mutation in a given gene can be fraught with difficulties, as has been our experience with *YARS2* [9]. In two cases, the sideroblastic anemia spontaneously resolved, one at 1 year of age following a year of recurrent transfusions and the other at 15 years of age following regular transfusions from infancy. In one of these cases there was also an improvement in all clinical symptoms, while in the other, myopathy did not manifest until several years after the sideroblastic anemia had resolved. Environmental (nutritional status, exposure to infection) and genetic modifiers (mtDNA haplotype), as well as temporal variations in the expression of key genes involved in the regulation of mtDNA transcription and translation, could be possible contributors to this phenotypic variability [9].

Because of these observations, it is not possible to make accurate predictions about long-term prognosis for individuals with mutations in this gene family. Surveillance for known and potentially novel late complications is therefore essential.

PATHOPHYSIOLOGY

mtARS conjugate amino acids to their cognate tRNA and are required for mitochondrial protein translation. It is unclear why mutations in mtARS result in a broad spectrum of phenotypes, despite their similar roles in synthesis of mitochondrial encoded subunits of the respiratory chain complexes. It is interesting to note that mutations in mtARS cognate tRNAs also result in different phenotypes.

mtARS contain a catalytic domain and an anticodon domain involved in tRNA recognition. Mutations in the catalytic domain of mtARS generally result in a mitochondrial protein synthesis defect. The most commonly reported *YARS2* c.156C>G; p.Phe52Leu mutation results in a reduced affinity of YARS2 p.Phe52Leu for $tRNA^{Tyr}$ and a ninefold loss of catalytic activity in an *in vitro* aminoacylation assay [11], while the YARS2 p.Gly191Asp mutation results in a 38-fold loss of catalytic activity [9]. Protein levels of YARS2 p.Phe52Leu were slightly reduced in muscle, while YARS2 p.Gly46Asp resulted in no detectable YARS2 in myoblasts [8]. *In vitro* studies on YARS2 p.Phe52Leu and p.Gly46Asp showed that the mitochondrial protein synthesis defect only manifested in myotubes and could not be detected in myoblasts or fibroblasts. The muscle-specific phenotype appears to be related to the higher requirement for RC complex components in muscle. There was also an increased requirement for YARS2 in differentiated muscle. Consequently, RC complexes containing mitochondrial encoded subunits are affected, with complex I, III, and IV activity and protein levels severely reduced in YARS2 p.Phe52Leu and p.Gly46Asp muscle.

Curiously, mutations in the mitochondrial encoded cognate tRNAs of mtARS do not usually result in the same phenotype as defects in the mtARS. For the cognate tRNA of YARS2, mutations in $tRNA^{Tyr}$ result in a range of symptoms including chronic progressive external ophthalmoplegia, myopathy, exercise intolerance, and focal segmental glomerulosclerosis [14–16], with cases only showing abnormal blood lactate after exercise and no evidence of sideroblastic anemia. This may in part be due to effects of heteroplasmy; however, it is likely that other genetic and environmental factors contribute to the observed phenotypic variability.

Pathophysiology of other mtARS that do not affect the brain show some similar features to *YARS2* MLASA pathology, but also distinct differences. *AARS2* mutations associated with severe infantile cardiomyopathy also have tissue-specific RC complex I & IV deficiency in heart, skeletal muscle and brain but not liver [3]. A mitochondrial protein translation defect could not be detected in an *in vitro* translation assay in patient fibroblasts, myoblasts, or myotubes, suggesting that *AARS2* mutations only manifest in some post-mitotic tissues. More recently, several other *AARS2* mutations have been associated with leukoencephalopathy and premature ovarian failure, with indications that the variations associated with these symptoms did not impact as greatly on mitochondrial protein translation and oxidative phosphorylation [4]. *SARS2* mutations also result in RC complex I, III, and IV deficiency in muscle, suggestive of a mitochondrial protein translation deficiency [5]. SARS2 aminoacylates two mitochondrial tRNAs. Lymphocytes carrying the homozygous *SARS2* c.1169A>G (p.Asp390Gly) mutation displayed an inability to aminoacylate one of these tRNAs. *HARS2* and *LARS2* variations have been associated with sensorineural hearing loss and premature ovarian failure, usually from compound heterozygous mutations resulting in one null allele and one missense mutation with little effect on function, or a homozygous mutation that retained some function [6,7]. We have recently identified *LARS2* mutations with a more severe effect on function in a case with hydrops, lactic acidosis, sideroblastic anemia, and multisystem disease.

These cases suggest that there are gene-specific phenotypes within the mtARS family, but also that for any given gene, there may be very broad phenotypic variation, often with unique clinical features resulting from mutations with more severe effects on protein function. It should be noted that for some cytoplasmic and bacterial ARS, other functions besides their primary role in protein translation have been identified. It may be that some of the gene-specific clinical features resulting from mtARS mutations are due to effects on non-canonical functions of these proteins.

CLINICAL PEARLS

- While the mitochondrial aminoacyl-tRNA synthetase disorders are notable for their distinct recognizable clinical syndromes, broad and overlapping clinical phenotypes may be encountered.
- Biochemical hallmarks include lactic acidemia and reduced, often tissue-specific, functional defects of complexes I, III, and IV.
- A common pathogenic factor is reduced mitochondrial protein synthesis.

REFERENCES

[1] Burrage L, Tang S, Wang J, Donti T, Walkiewicz M, Luchak J, et al. Mitochondrial myopathy, lactic acidosis, and sideroblastic anemia (MLASA) plus associated with a novel *de novo* mutation (m.8969G>A) in the mitochondrial encoded *ATP6* gene. Mol Genet Metab 2014;113:207–12.

[2] Bottomley S, Fleming M. Sideroblastic anemia diagnosis and management. Hematol Oncol Clin North Am 2014;28:653–70.

[3] Gotz A, Tyynismaa H, Euro L, Ellonen P, Hyotylainen T, Ojala T, et al. Exome sequencing identifies mitochondrial alanyl-tRNA synthetase mutations in infantile mitochondrial cardiomyopathy. Am J Hum Genet 2011;88:635–42.

[4] Dallabona C, Diodato D, Kevelam S, Haack T, Wong L, Salomons G, et al. Novel (ovario) leukodystrophy related to *AARS2* mutations. Neurology 2014;82:2063–71.

[5] Belostotsky R, Ben-Shalom E, Rinat C, Becker-Cohen R, Feinstein S, Zeligson S, et al. Mutations in the mitochondrial seryl-tRNA synthetase cause hyperuricemia, pulmonary hypertension, renal failure in infancy and alkalosis, HUPRA syndrome. Am J Hum Genet 2011;88:193–200.

[6] Pierce S, Chisholm K, Lynch E, Lee M, Walsh T, Opitz J, et al. Mutations in mitochondrial histidyl tRNA synthetase HARS2 cause ovarian dysgenesis and sensorineural hearing loss of Perrault syndrome. Proc Natl Acad Sci 2011;108:6543–8.

[7] Pierce S, Gersak K, Michaelson-Cohen R, Walsh T, Lee M, Malach D, et al. Mutations in *LARS2*, encoding mitochondrial leucyl-tRNA synthetase, lead to premature ovarian failure and hearing loss in Perrault syndrome. Am J Hum Genet 2013;92:614–20.

[8] Sasarman F, Nishimura T, Thiffault I, Shoubridge E. A novel mutation in YARS2 causes myopathy with lactic acidosis and sideroblastic anemia. Hum Mutat 2012;33:1201–6.

[9] Riley L, Menezes M, Rudinger-Thirion J, Duff R, de Lonlay P, Rotig A, et al. Phenotypic variability and identification of novel *YARS2* mutations in YARS2 mitochondrial myopathy, lactic acidosis and sideroblastic anaemia. Orphanet J Rare Dis 2013;8:193–203.

[10] Nakajima J, Eminoglu T, Vatansever G, Nakshima M, Tsurusaki Y, Saitsu H, et al. A novel homozygous *YARS2* mutation causes severe myopathy, lactic acidosis, and sideroblastic anemia 2. J Hum Genet 2014;59:229–32.

[11] Riley L, Cooper S, Hickey P, Rudinger-Thirion J, McKenzie M, Compton A, et al. Mutation of the mitochondrial tyrosyl-tRNA synthetase gene, YARS2, causes myopathy, lactic acidosis, and sideroblastic anemia–MLASA syndrome. Am J Hum Genet 2010;87:52–9.

[12] Shahni R, Wedatilake Y, Cleary M, Lindley K, Sibson K, Rahman S. A distinct mitochondrial myopathy, lactic acidosis and sideroblastic anemia (MLASA) phenotype associates with YARS2 mutations. Am J Med Genet 2013;161:2334–8.

[13] Konovalova S, Tyynismaa H. Mitochondrial aminoacyl-tRNA synthetases in human disease. Mol Genet Metab 2013;108:206–11.

[14] Pulkes T, Siddiqui A, Morgan-Hughes J, Hanna M. A novel mutation in the mitochondrial $tRNA^{Tyr}$ gene with associated exercise intolerance. Neurology 2000;8:1210–2.

[15] Raffelsberger T, Rassmanith W, Thaller-Antlanger H, Bitttner R. CPEO associated with a single nucleotide deletion in the mitochondrial $tRNA^{Tyr}$ gene. Neurology 2001;57:2298–301.

[16] Scaglia F, Vogel H, Hawkins E, Vladutiu G, Liu L, Wong L. Novel homoplasmic mutation in the mitochondrial $tRNA^{Tyr}$ gene associated with atypical mitochondrial cytopathy presenting with focal segmental glomerulosclerosis. Am J Med Genet 2003;123A:172–8.

Chapter 28

Defects in Post-Transcriptional Modification of Mitochondrial Transferase RNA: A Patient with Possible Mitochondrial-tRNA Translation Optimization Factor 1, *MTO1* Dysfunction

Russell P. Saneto
Department of Neurology/Division of Pediatric Neurology, Seattle Children's Hospital/University of Washington, Seattle, WA, USA

CASE PRESENTATION

Our case was born to a G2P1>2 mother after an uneventful pregnancy and delivery. Parents are of Northern European descent and unrelated. Until 2.5 months of age, he was developing normally, with a social smile and visually tracking faces. His first seizure was at 2.5 months of age and quickly his development halted. At the time of seizure onset, he was noted to be hypotonic. He has not gained motor or cognitive developmental landmarks since this time. Currently at 8 years of age, he does not sit nor crawl independently. He has cortical visual impairment. He is verbal, but has no expressive language. Nutrition is by gastrostomy tube, due to reflux and severe delayed gastric emptying. His muscle strength is reduced, but he moves all extremities against gravity. He displays a myopathic facies. He has choreoathetoid movements in the upper and lower extremities. Deep tendon reflexes are present but reduced symmetrically.

Seizures began at 2.5 months of age and quickly became intractable to conventional seizure medications. His seizures evolved into infantile spasms at 3–4 months of age. He was eventually started on the ketogenic diet (KD). He continues to have approximately two to three short tonic seizures per day and persistent myoclonus of the upper extremities. During his period of infantile spasms,

Mitochondrial Case Studies. http://dx.doi.org/10.1016/B978-0-12-800877-5.00028-0

the electroencephalogram (EEG) showed a classical hypsarrhythmia pattern. Over time, multiple EEGs have shown evolution into multifocal independent spikes with background slowing. His ongoing tonic seizures have been correlated with generalized paroxysmal fast activity when captured on EEG.

Nuclear magnetic resonance imaging (MRI) studies have showed mild generalized atrophy of the cerebrum. Proton magnetic resonance imaging (MRS) demonstrated reduced N-acetylaspartate in voxels over the frontal lobe white matter. Lactate peak was seen within the ventricular system. Both findings suggest global cortical dysfunction and oxidative phosphorylation defects.

The electrocardiogram (ECG) demonstrated right bundle branch block on multiple occasions. There was no evidence of cardiomyopathy based on early ECG findings and chest X-rays. The most recent ECG findings are suggestive of a right ventricular hypertrophy.

Biochemical testing did not demonstrate abnormalities; plasma amino acids, urine organic acids, ammonia, venous lactate, acylcarnitine profile, very long chain fatty acids, and creatine phosphate kinase (CK) testing returned with normal values. Enzymatic testing of the respiratory chain in muscle biopsy material demonstrated a complex IV deficiency (<26% of controls) with minor defects in complexes I and III. Histochemical analysis did not reveal abnormality and electron microscopy evaluation showed mild variation in fiber size. Total genomic sequencing of the mitochondrial DNA did not reveal pathological mutations. Fibroblast cultures were evaluated for respiratory chain defects, and normal values were found. Pyruvate dehydrogenase complex enzyme activities were normal. Comparative genomic hybridization study was without abnormality. Selective nuclear-encoded genes involved in complex IV function, *SURF1*, *POLG*, and *COX 10*, were sequenced and came back without mutation(s) detected.

We performed next-generation gene sequencing of known and candidate nuclear-encoded genes responsible for mitochondrial disease [1]. There were two heterozygote mutations in trans, c.176G>C and c.922A>G, in the mitochondrial translation optimization factor 1 (*MTO1*). In silico analysis predicted both mutations were deleterious. Parental testing indicated each parent carried one of the two mutations. Both mutations are in highly evolutionary conserved locations within the protein. Taken together, *MTO1* dysfunction is likely the etiology of our case's mitochondrial disorder.

DIFFERENTIAL DIAGNOSIS

There are a variety of disorders that can present with early-onset medically intractable seizures and severe global developmental delay. The combination of seizures and developmental delay suggest a group of disorders, known as epileptic encephalopathy. There are multiple etiologies of epileptic encephalopathy including structural, genetic, and metabolic abnormalities, in particular mitochondrial diseases. Several features of our case would begin to limit the differential diagnosis. Newborn screening did find possible metabolic/genetic etiologies for early-onset epileptic encephalopathy [2]. The EEG findings would select against a few genetic disorders, as burst suppression was not a finding in our case. The differential for etiologies of infantile spasms is broad, but the clinical presentation and lack of biochemical abnormalities would eliminate many of the possible etiologies from consideration [3].

The profound hypotonia, myoclonic seizures, infantile spasms, severe global developmental delay, and significant muscle weakness were suggestive of a possible mitochondrial disorder. The MRS scan demonstrated a large 1.33 ppm lactate peak in the cerebral spinal fluid (CSF) ventricular system, and MRI scan showed global cerebral atrophy and very mild, but present, cerebellar vermal atrophy. There have been mitochondrial syndromes associated with epilepsy, and in particular, infantile spasms. Most cases with infantile spasms have been associated with Leigh syndrome [4], but not all cases with Leigh syndrome have infantile spasms, and many have no seizures at all [5]. There are multiple etiologies of epilepsy in mitochondrial disease. Over a decade ago, we published an abstract on four cases who had respiratory chain mutations and infantile spasms [6]. Other types of epilepsy have been noted in mitochondrial disease with a suggestion of somewhere around 40% of cases having comorbid epilepsy [7].

DIAGNOSTIC APPROACH

In this case, the evaluation was defined by the clinical presentation. Early-onset seizures that evolved into infantile spasms with cessation of development were the overriding clinical features. This presentation suggested possible metabolic or genetic epilepsy. The normal MRI scan eliminated from consideration structural/symptomatic epilepsy. We routinely gather basic metabolic labs to remove from consideration possible other metabolic disorders that give rise to severe epilepsy: lactate, ammonia, CK, plasma amino acids, urine organic acids, very long chain fatty acid profile, and intracellular coenzyme Q10 levels. All of these tests came back without abnormality, suggesting our case did not have a common metabolic reason for seizure and cessation of development. A second MRI revealed mild cerebral atrophy, and an MRS showed lactate in the CSF. The finding of lactate pointed the evaluation to include mitochondrial disease. The failure to gain weight required placement of a gastrostomy tube for nutrition, and since the case would be undergoing general anesthesia, the decision was made to also get a muscle biopsy to look for possible mitochondrial disease.

The muscle biopsy material demonstrated a complex IV defect (<26% of control values) and minor deficiencies in complexes I and III. There was no evidence of maternal lineage mitochondrial disease. The histology of the muscle material was not suggestive of a mitochondrial tRNA abnormality, but total mitochondrial DNA genome was sent on the remote possibility of a mitochondrial protein abnormality. Most of the time, pathological mitochondrial transfer RNA (tRNA) mutations leave a signature of COX (complex IV)-negative fibers and normal to increased succinate dehydrogenase staining. In our case, the finding of homogeneous complex IV or COX staining in the presence of a complex IV enzyme deficiency would be highly unusual for a mitochondrial tRNA mutation. Mitochondrial DNA (mtDNA) sequencing returned without mutations being found, eliminating mtDNA and possible maternal inheritance from consideration. This latter finding may be helpful in genetic counseling, depending on the circumstances of the family.

The clinical picture of epileptic encephalopathy and mitochondrial disease does not have a subgroup of well-defined genetic etiologies. Mitochondrial disease gives rise to many phenocopies, and therefore, the breath of diagnosis needs to have a broad catchment range. At this point in time, next-generation sequencing possesses the most likely approach to identify a possible genetic etiology. The platform used in our case had over 1300 candidate and biologically validated genes associated with mitochondrial disease [1]. Two mutations were found in the MTO1 gene, c.176G>C and c.922A>G. Both were found to segregate in the parents: mother had the c.176G>C mutation, and father possessed the c.922A>G mutation.

The *MTO1* gene has been validated as disease causing using a recombinant yeast model and *MTO1* knockout mice [8,9]. In these models, complex IV was found to be dysfunctional. The mutations found in the *MTO1* gene in our case have not been reported and validated as pathological mutations. They would be classified as variants of unclear significance, insufficient data to establish the likelihood of pathogenicity. For validation, the next steps would be to evaluate computer algorithms and find classic signs of pathogenicity (in silico predictions): gene sequence conservation during evolution, amino acid change altering protein structure (function), and lack of population polymorphism identity. Guidelines for investigating causality of sequence variants in human disease have been suggested [10]. In our case, the computer algorithms predict the mutations to likely be deleterious. In the paper by Vasta et al. [1], the filter used was excluding all variants with a minor allele frequency (MAF) < 0.5%, and the designation of possible pathological for both mutations was made. But, neither mutation has been validated experimentally in bioassay systems. The c.922A>G mutation has been reported as a single nucleotide polymorphism (dbSNP; rs145043138) with a MAF of 0.3%. However, MacArthur et al. [10] suggest a MAF of < 1% to be relevant for variants with relative large effects on disease risk. Taken together, the case reported would likely fall into the possible/probable category.

TREATMENT STRATEGY

Unfortunately, as in most all mitochondrial disease, treatment is based on treating symptoms. Seizures in infantile spasms are best treated by adrenal corticotrophin hormone or vigabatrin, which our case unfortunately failed. As epileptic seizures continued, other seizure medications were tried, but all have had only modest effect at reducing seizure frequency and duration. Owing to his medical intractability, he was placed on the KD at the age of 1 year. Interestingly, a recent article suggests that the KD might be helpful in *MTO1*-deficient mice [8]. Although our case responded to the KD, his seizures continue. Four out of the reported 13 cases with *MTO1* mutations died before 1 year of age, so the influence of the KD cannot be eliminated from consideration.

Most of the *MTO1* cases, $n=8$, were placed on dichloroacetate (DCA) very early in life. The mechanism of DCA action is to inhibit the activity of pyruvate dehydrogenase kinase activity. The enzyme remains active as a result of the persistent unphosphorylated state. Furthermore, in this active state, there is a decrease in the rate of degradation. Several cases have been treated with DCA, and cardiac function was significantly improved [11]. Unfortunately, DCA is not available in the USA, and we were not able to use it in our case.

Other cases with *MTO1* mutations were noted to have feeding difficulties. Our case received a gastrostomy tube for nutrition at age 2 years. We feel this has been helpful to induce proper nutrition and allow easy introduction of the KD and medications.

PATHOPHYSIOLOGY

MTO1 is an evolutionary conserved protein that regulates tRNA modification and mitochondrial translation in a tissue-specific manner. The enzymatic reactions involving activating mitochondrial mRNA translation involve two steps. In the first reaction, MTO1 catalyzes the 5-carboxymethylaminomethylation of the wobble uridine base in the mitochondrial specific tRNAs for glutamine, glutamate, lysine, and leucine. This reaction is coupled to the second reaction, 2-thiouridylase activity encoded by the gene *TRMU*, where a sulfur group is added to uridine. Together, these post-transcriptional modifications increase the accuracy and efficiency of mitochondrial mRNA translation. The final product of inefficient and defective translation creates respiratory chain assembly and stability problems. This is seen in the expression of elevated lactate in serum and/or CSF in all *MTO1*-encoded disease cases.

The published cases of *MTO1*-encoded disease have demonstrated tissue-specific involvement. To date, only 13 cases (including this case) have been reported to have *MTO1* mutations and clinical disease. Both the severity and clinical phenotype have varied. The most common finding is hypertrophic cardiomyopathy, reported in 11 cases. Early death (<1 year) due to cardiac involvement occurred in four cases. Psychomotor delay was found in eight cases, in cases that lived beyond 1 year of life. The involvement of central nervous system was also seen in an additional two cases (including our case) with severe epilepsy. Feeding issues were also seen in five cases. Respiratory involvement was noted in three cases. The reason for tissue specificity is not clear. The finding of tissue specificity is not strictly related to *MTO1* mutations, but it is also seen in other mitochondrial diseases, some of which are related to energy needs during a developmentally sensitive time. However, this simplistic explanation is likely only partially related to the phenotypic expression of mitochondrial disease.

CLINICAL PEARLS

- Disorders of mitochondrial protein synthesis can present with few systemic biochemical abnormalities.
- Lack of biochemical abnormalities does not eliminate a mitochondrial disorder from consideration.
- Analysis of a muscle biopsy may be essential in the diagnostic evaluation of a case.
- The MRS detection of lactate within the brain may help in pursuing a diagnosis.
- Spreading a wide net in genetic testing for mitochondrial diseases may be necessary in cases with uncommon presentations.
- The genetic etiology of our case has allowed parental genetic counseling for possible future children.

REFERENCES

[1] Vasta V, Merritt JL, Saneto RP, Hahn S. Next-generation sequencing for mitochondrial diseases: a wide diagnostic spectrum. Ped Internat 2012;54:585–601.

[2] Papetti L, Parisi P, Leuzzi V, et al. Metabolic epilepsy: an update. Brain Dev 2013;35:827–41.

[3] Paciorkowski AR, Thio LL, Dobyns WB. Genetic and biologic classification of infantile spasms. Ped Neurol 2011;45:355–67.

[4] Sadleir LG, Connolly MB, Applegarth D, Hendson G, Clarke L, Rakshi C, et al. Spasms in children with definite and probable mitochondrial disease. Eur J Neurol 2004;11:103–10.

[5] Ruhoy IS, Saneto RP. The genetics of Leigh syndrome and its implications for clinical practice and risk management. Appl Clin Genet 2014;7:221–34.

[6] Saneto RP, Kotagal P, Cohen BH, Hoppel CL. Treatment and outcome of four patients with mitochondrial cytopathy and infantile spasms. Epilepsia 2001;42:174.

[7] Rahman S. Mitochondrial disease and epilepsy. Dev Med Child Neurol 2012;54:397–406.

[8] Tischner C, Hofer A, Wulff Y, et al. *MTO1* mediates tissue-specificity of OXPHOS defects via tRNA modification and translation optimization, which can be bypassed by dietary intervention. Hum Mol Genet December 30, 2014;24(8):2247–66. http://dx.doi.org/10.1093/hmg/ddu743. [Epub ahead of print].

[9] Baruffini E, Dallobona C, Ivernizzi R, et al. *MTO1* mutations are associated with hypertrophic cardiomyopathy and lactic acidosis and cause respiratory chain deficiency in humans and yeast. Hum Mut 2013;34:1501–9.

[10] MacArthur DG, Manolio TA, Dimmock DP, et al. Guidelines for investigating causality of sequence variants in human disease. Nature 2014;508:469–76.

[11] Ghezzi D, Baruffini E, Haack TB, et al. Mutations of the mitochondrial-tRNA modifier *MTO1* cause hypertrophic cardiomyopathy and lactic acidosis. Am J Hum Genet 2012;90:1079–87.

Chapter 29

Complex I Deficiency

Mark Tarnopolsky, Rashid Alshahoumi
Department of Pediatrics, McMaster University, Hamilton, Ontario, Canada

CASE PRESENTATION

The female case was born by spontaneous vaginal delivery after an uncomplicated pregnancy. Her early developmental milestones were met, and she was generally healthy until 13 months of age when she started showing some signs of regression of these milestones. Specifically, the parents noted that she was having problems with her balance and was slowly losing her ability to walk independently. She also lost some of the words she used to say, and she became less interactive. At 14 months of age, she was not walking, she could no longer hold bottles or pick things up and was no longer talking and started having difficulties swallowing solids. On examination, she did not have any dysmorphic features and had a normal head circumference. She had end-point nystagmus and appeared weak and was hypotonic. Her reflexes were normal and with flexor plantar responses. Blood tests showed normal creatine kinase levels, liver function tests, creatinine, CBC, very long chain free fatty acids, amino acids, and urine organic acids, while total and free carnitine were low and plasma lactate was mildly elevated (3.3 mmol/L, $<$2.2 mmol/L). The MRI of the brain showed diffuse signal abnormalities involving the cerebral white matter felt to be consistent with leukodystrophy (Figure 1). Consequently, she had a skin and muscle biopsy completed during the initial admission with the fibroblasts showing normal activity for arylsulfatase, hexosaminidase, and galactocerebrosidase. The MR spectroscopy showed elevated lactate in the affected white matter, raising the possibility of a mitochondrial cytopathy; consequently, she was started on a mitochondrial cocktail consisting of coenzyme Q10, riboflavin, alpha-lipoic acid, creatine monohydrate, and vitamin E. The parents reported dramatic improvement after these medications were started over a several week period. During this time, she became more interactive, her balance improved, and she started to progress in terms of her developmental milestones. Her overall condition continued to improve slowly, and she started to walk again with a walker and to say more words. When she was evaluated at 5 years of age, her examination showed the development of lower extremity spasticity and dystonic posturing of her right arm, and she was started on Baclofen. The MRI was repeated and

Mitochondrial Case Studies. http://dx.doi.org/10.1016/B978-0-12-800877-5.00029-2

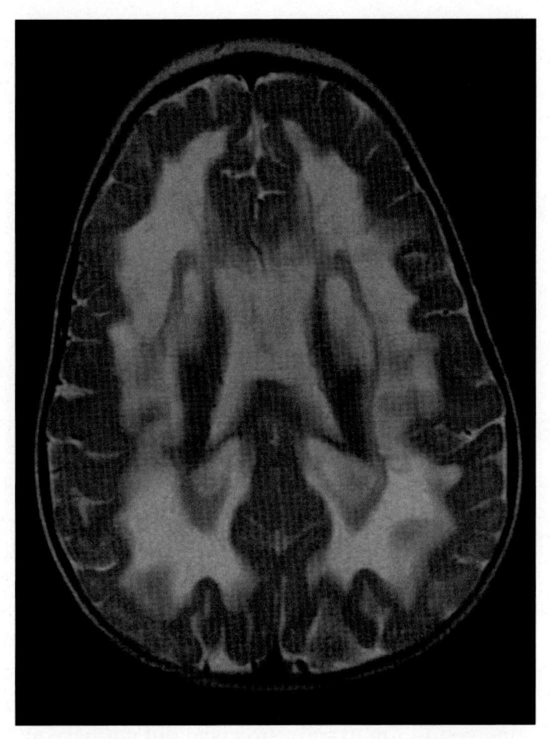

FIGURE 1 T2 1.5 Telsa MRI at presentation at age 15 months showing extensive deep white matter changes.

showed an increase in the white matter involvement together with some cystic lesions in the basal ganglia and possible small infarcts. At age 10 years, her overall condition slightly improved, and she was cognitively at about a 4-year-old level. Although wheelchair bound (possible contribution from a dislocated hip), she was able to ride a tricycle.

DIFFERENTIAL DIAGNOSIS

Neurodevelopmental regression is a rare event in children and can be secondary to a large number of heterogeneous disorders. Neurodevelopmental regression is an ominous phenomena for it often indicates a progressive neurodegenerative disorder. The most common etiologies of neurodegenerative disorders in children include genetic, metabolic, infectious, and inflammatory disorders. Leukoencephalopathies predominantly affect the white matter and represent a common cause of neurodevelopmental regression. The term leukodystrophy refers to heritable myelin disorders affecting the white matter of the central nervous system with or without peripheral nervous system myelin involvement. Table 1 lists some of the more common causes of leukoencephalopathies. The pattern of white matter involvement on MRI is often helpful in narrowing the list of diagnostic possibilities.

TABLE 1 The Major Causes of Leukodystrophy

Demyelinating Leukodystrophy (Inheritance, *gene*)	Hypomyelinating Leukodystrophy (Inheritance, *gene*)	Other Leukoencephalopathies Associated with Metabolic Disorders (Inheritance, *gene*)
• Alexander (AD, de novo/AD, *GFAP*) • Adrenoleukodystrophy (X-linked, *ABCD1*) • Canavan (AR, *ASPA*) • Krabbe (AR, GALC/PSA) • Cerebrotendinous xanthomatosis (AR, *CYP27A1*) • Aicardi-Goutières (AR, *RNASEH2A/B/C, SAMHD1*; AR/AD, *TREX1*) • Metachromatic, metachromatic-like (AR, *ARSA/PSAP*) • Vanishing white matter disease (AR, *EIF2B1–5*) • Adrenoleukodystrophy (AR, *ABCD1*)	• Pelizaeus-Merzbacher disease (X-Linked, *PLP1*) • Pelizaeus-Merzbacher-like disease (AR, *GJC2*) • Oculodentodigital syndrome (AR/AD, *GJA1*) • Hypomyelination with congenital cataracts (AR, *FAM126A*)	**Lysosomal disorders** • Fabry disease (XLR, *GLA*) • Tay-Sach disease, GM2-gangliosidosis (AR, *HEXA*) • GM1-gangliosidosis (AR, *GLB1*) **Peroxisomal disorders** • Zellweger spectrum disorders (AR, multiple *PEX* genes) • D-bifunctional protein deficiency (AR, *HSD17B4*) • Rhizomelic chondrodysplasia punctata type 1, 2, 3 (AR, *PEX7/GNPAT/AGPS*) **Organic acidemias** • Glutaric aciduria type 1 (AR, *GCDH*) • 3-methylglutaconic aciduria type 1 (AR, *AUH*) • L-2-hydroxyglutaric aciduria (AR,*L2HGDH*) **Amino acid disorders** • Phenylketonuria (AR, *PAH*) • Serine synthesis defects (AR, *PSAT1/PHGDH*) **Mitochondrial disorders** • Respiratory chain defects (AR/AD/Maternal, *Multiple genes*) • Leukoencephalopathy with brainstem and spinal cord involvement and lactate elevation: LBSL (AR, *AARS2, DARS2, EARS2*)

AR, autosomal recessive; AD, autosomal dominant; XLR, X-linked recessive.
Note: The specific genes are in *italics*.

The case's T2 images showing white matter hyperintensity with T1 imaging showing corresponding hypointensity point toward this being a primary demyelinating leukodystrophy and not a hypomyelinating condition. There is diffuse, bilateral, confluent, and symmetric white matter involvement without a clear anterior or posterior predominance of disease. Such a pattern can be seen at the end stage of all progressive white matter disease. If seen early in the disease course, the differential includes peroxisomal disorders, molybdenum cofactor deficiency, glutaric aciduria type II, select mitochondrial disorders, and several other more rare primary genetic leukodystrophies. Three of the more common causes of leukodystrophy are the lysosomal storage diseases, Krabbe disease, Tay-Sach disease, and metachromatic leukodystrophy, and most centers would send leukocytes or fibroblasts for specific enzyme analysis as well as urine organic acids, plasma amino acids, and lactate. Peroxisomal disorders such as adrenoleukodystrophy and Zellweger syndrome show elevations of very long chain fatty acids; however, early childhood-onset adrenoleukodystrophy would not be expected in a female (X-linked disease) nor would one suspect Pelizeus–Merzbacher (*PLP1* mutation) in a female for the same reason. In this specific case, the elevated serum and CNS lactate lead to a suspected mitochondrial cytopathy, and an increasing number of specific gene mutations have been discovered that result in neurodevelopmental regression in mitochondrial cytopathies. Leigh disease is one of the more common mitochondrial cytopathies that leads to neurodevelopmental regression. The most common types of Leigh disease are those that affect complex I activity or complex IV activity separately. Other cases of Leigh disease can affect complex II activity or multiple enzymes (maternally inherited Leigh syndrome due to mtDNA mutations). The MRI findings typically present with a symmetric high T2 signal in gray matter regions such as the brain stem, sub-thalamic nuclei, basal ganglia, and cerebellum; however, an increasing number of cases have been reported with leukoencephalopathy changes (~28% of complex I Leigh disease) and no gray matter involvement (as was the current case).

DIAGNOSTIC APPROACH

A child with a neurodegenerative disease can present with a variety of symptoms and signs. The diagnosis can be very challenging as the course can be acute with rapid progression or chronic with slow progression. The first step to make a diagnosis is the detailed history. Knowing about the onset and the course of the disease can lead us to know the category of the etiology, that is, infectious versus genetic versus metabolic. Important points in history should include the following: perinatal history (e.g., prematurity, asphyxia, meningitis,), family history (e.g., consanguinity, neurological problems), developmental history, present history (neurological problems, e.g., seizures, ataxia, hearing, vision, and gastrointestinal). The second step is a detailed neurological examination including dysmorphic features, cranial nerve examination abnormalities (ptosis, ophthalmoplegia), and motor examination (weakness, hypotonia, spasticity, dystonia).

Blood testing should start with plasma amino acids, lactate, and very long chain free fatty acids, and urine should be sent for organic acid and amino acid screening. Further investigations should be done according to the findings from the history, and physical examination including further biochemical studies, brain imaging, and electrophysiology studies (slow nerve conduction in metachromatic leukodystrophy and Krabbe disease) can provide further diagnostic clues. A skin biopsy is helpful to obtain fibroblasts for enzyme testing for lysosomal and peroxisomal storage diseases as well as for mitochondrial respiratory chain analysis. The MRI completed as part of the initial investigation can lead to findings that can be helpful first to determine the general class of CNS disorder and, in the case of a leukodystrophic disorder, whether the process is demyelinating or hypomyelinating.

In the current case, we had suspected a mitochondrial disease given the high lactate and obtained a muscle biopsy and skin biopsy simultaneously. The muscle biopsy should be assessed by both light and electron microscopy for the light microscopy may be normal in pediatric disorders and the electron microscope can first reveal abnormalities (pleomorphic mitochondria, intramitochondrial inclusions, mitochondrial proliferation). The muscle biopsy is also very helpful in that small pieces can be used for enzyme analysis, mtDNA sequencing, and deletion/depletion analysis. In our hands, we can obtain enough muscle from a Bergstrom needle biopsy to perform all of the aforementioned tests, even in neonates.

Although an increasing number of next-generation sequencing panels are available by many laboratories, it is not recommended that these be sent without a careful history, physical examination, and at least one round of blood and urine testing. For example, a very expensive and comprehensive leukodystrophy panel could miss an mtDNA mutation even if some mtDNA mutations were on the panel due to low or undetectable heteroplasmy in blood. An even more common issue is that several variants of uncertain significance will appear with next-generation panel approaches, and it is very difficult to know which of these require further consideration without a very careful evaluation to consider or remove from consideration some of the possible candidate mutations.

In the current case, the case's fibroblasts demonstrated an elevated lactate to pyruvate ratio at 51.2 (simultaneous control = 15.4 ± 1.4) and low complex I + III activity (20.3 nmol/min/mg protein [simultaneous control = 97.8]) with normal complex II + III and IV activity. The muscle biopsy was normal at the light and electron microscopic level, yet complex I + III activity was low (0.18 umol/min/g wet weight [N.R. = 0.53–2.72]), with normal activities for citrate synthase, complex II + III, and complex IV. As a consequence of the very robust and consistent reduction in complex I activity and the clinical picture of regression, we sequenced several of the mitochondrial complex I genes associated with Leigh disease and leukoencephalopathy using Sanger sequencing (many of these are now available as complex I panels using next-generation sequencing and would now be recommended). The NDUFV1 gene showed two sequence variants associated with complex I Leigh disease (c.753delCCCC; p.Ser251fsX44 [allele 1] and c.1156C>T; p.Arg386Cys [allele 2]).

TREATMENT STRATEGY

As is the case with most of the neurodegenerative disorders, supportive measures are the mainstay of care. Many children with Leigh disease experience significant gastrointestinal dysmotility and feeding issues and may require a gastrostomy tube. Such was the case with our case, and her growth curves for height and weight improved after initiating defined formula diet feeds. The other advantage of a feeding tube in children with severe neurological disorders is that it ensures more accurate dosing of all medications (antiseizure medications, anti-spasticity medications, mitochondrial cocktail [see below]). In this case, nutraceutical therapy was initiated with coenzyme Q10, riboflavin, alpha-lipoic acid, creatine monohydrate, and vitamin E, and she had a co-temporal dramatic improvement in function according to parents, family members, and physicians. This combination, without the riboflavin, was shown to reduce the level of lactate and oxidative stress in a randomized controlled trial in adults with MELAS and CPEO; however, it is difficult to assess clinical efficacy in disorders that have a fluctuating course, and it is not clear if there was a cause and effect relationship between the starting of the therapy and the clinical improvement. In spite of the initial clinical improvement, the case did develop lower extremity spasticity that was treated with Baclofen and Botox injections with some improvement. Furthermore, she did develop asymmetric arm dystonia that was somewhat improved with trihexyphenidyl (Artane).

LONG-TERM OUTCOME

The largest study of outcome in children with complex I-associated Leigh disease included analysis of 130 cases from 48 published articles and four new case reports. This study found that most cases presented before 1 year of age, and the most common symptoms were hypotonia, nystagmus, respiratory abnormalities, pyramidal signs, dystonia, psychomotor retardation or regression, failure to thrive, and feeding problems. Twenty-five percent of cases with complex I-associated Leigh disease died before the age of 6 months, ~50% by the age of 2 years, and ~75% by the age of 10 years. The reported range of death for children ($n = 17$) with NDUFV1 mutations was 3.5 months > 13 years of age; consequently, our case is in the upper quintile of this range and doing well by relative comparison. The study of 130 cases with complex I Leigh disease and another study looking at 113 children with mitochondrial disease both showed that the presence of cardiomyopathy was a negative predictor of outcome.

PATHOPHYSIOLOGY

Complex I is the first enzyme complex in the respiratory chain, and it accepts electrons from NADH + H^+ derived from fat, carbohydrate, and amino acids to create an electrochemical gradient across the inner mitochondrial membrane. The build-up of protons in the intermembrane space leads to a proton motive force that drives the re-phosphorylation of ADP > ATP by complex V in the presence of molecular oxygen that is reduced to water at complex IV. Complex I is also involved in shuttling an electron to oxidized coenzyme Q10 (ubiquinone) to reduce it to ubiquinol. Alterations in the ~47 core subunits that form the holo–enzyme complex can reduce the efficiency of proton pumping (and ultimately ATP production) and/or lead to an increase in free radical generation (unpaired electron). There are also a number of known complex I assembly genes (*NDUFAF1,2,4, ACAD9, NUBPL, C20orf7,* and *FOXRED1*), and mutations in these genes can also lead to complex I deficiency Leigh disease by altering the integrity of the holo–enzyme complex. Interestingly, mutations in the core subunits appear to portend a worse clinical outcome than seen in those cases with assembly factor mutations.

CLINICAL PEARLS

- Regression of developmental milestones should alert the physician to a possible neurodegenerative disorder.
- Neurodegenerative disorders can present with many different symptoms and signs and can be secondary to many different causes.
- The approach to a child with a neurodegenerative disorder should begin with a detailed history and examination.
- Further investigations should be decided according to the findings from the history and examination.
- Leukoencephalopathies are a major cause for neurodegeneration.
- MRI is crucial to reach a diagnosis in leukoencephalopathies.
- Leukoencephalopathies can be secondary to a mitochondrial disorder.
- Mitochondrial disorders should be considered in the differential of leukoencephalopathies when the more common leukodystrophies are eliminated from consideration and/or when specific mitochondrial symptoms/signs are found.
- A comprehensive history and examination and targeted testing must proceed next-generation sequencing panels for leukodystrophies.

SUGGESTED READING

[1] Alston CL, Davison JE, Meloni F, van der Westhuizen FH, He L, Hornig-Do HT, et al. Recessive germline SDHA and SDHB mutations causing leukodystrophy and isolated mitochondrial complex II deficiency. J Med Genet 2012;49(9):569–77.

[2] Bray MD, Mullins ME. Metabolic white matter diseases and the utility of MR spectroscopy. Radiol Clin North Am 2014;52(2):403–11.

[3] Costello DJ, Eichler AF, Eichler FS. Leukodystrophies: classification, diagnosis, and treatment. Neurologist 2009;15(6):319–28.

[4] Dallabona C, Diodato D, Kevelam SH, Haack TB, Wong LJ, Salomons GS, et al. Novel (ovario) leukodystrophy related to AARS2 mutations. Neurology 2014;82(23):2063–71.

[5] Ganesan K, Desai S, Udwadia-Hegde A, Ursekar M. Mitochondrial leukodystrophy: an unusual manifestation of Leigh's disease. A report of three cases and review of the literature. Neuroradiol J 2007;20(3):271–7.

[6] Gordon HB, Letsou A, Bonkowsky JL. The leukodystrophies. Semin Neurol 2014;34(3):312–20.

[7] Phelan JA, Lowe LH, Glasier CM. Pediatric neurodegenerative white matter processes: leukodystrophies and beyond. Pediatr Radiol 2008;38(7):729–49.

[8] Kohlschütter A, Eichler F. Childhood leukodystrophies: a clinical perspective. Expert Rev Neurother 2011;11(10):1485–96.

[9] Navarro-Sastre A, Martín-Hernández E, Campos Y, Quintana E, Medina E, de Las Heras RS, et al. Lethal hepatopathy and leukodystrophy caused by a novel mutation in MPV17 gene: description of an alternative MPV17 spliced form. Mol Genet Metabol 2008;94(2):234–9.

[10] Ortega-Recalde O, Fonseca DJ, Patiño LC, Atuesta JJ, Rivera-Nieto C, Restrepo CM, et al. A novel familial case of diffuse leukodystrophy related to NDUFV1 compound heterozygous mutations. Mitochondrion 2013;13(6):749–54.

[11] Saneto RP, Friedman SD, Shaw DW. Neuroimaging of mitochondrial disease. Mitochondrion 2008;8(5–6):396–413.

[12] Scaglia F, Towbin JA, Craigen WJ, Belmont JW, Smith EO, Neish SR, et al. Clinical spectrum, morbidity, and mortality in 113 pediatric patients with mitochondrial disease. Pediatrics 2004;114(4):925–31.

[13] Shevell M, Ashwal S, Donley D, Flint J, Gingold M, Hirtz D, et al. Evaluation of the child with global developmental delay. Neurology 2003;60(3):367–80.

[14] Steenweg ME, Ghezzi D, Haack T, Abbink TE, Martinelli D, van Berkel CG, et al. Leukoencephalopathy with thalamus and brainstem involvement and high lactate "LTBL" caused by EARS2 mutations. Brain 2012;135(Pt 5):1387–94.

[15] Tarnopolsky MA, Pearce E, Smith K, Lach B. Suction-modified Bergström muscle biopsy technique: experience with 13,500 procedures. Muscle Nerve 2011;43(5):717–25.

[16] Timothy J, Geller T. SURF-1 gene mutation associated with leukoencephalopathy in a 2-year-old. J Child Neurol 2009;24(10):1296–301.

[17] Van Berge L, Hamilton EM, Linnankivi T, Uziel G, Steenweg ME, Isohanni P, LBSL Research Group, et al. Leukoencephalopathy with brainstem and spinal cord involvement and lactate elevation: clinical and genetic characterization and target for therapy. Brain 2014;137(Pt 4):1019–29.

[18] Wong LJ, Scaglia F, Graham BH, Craigen WJ. Current molecular diagnostic algorithm for mitochondrial disorders. Mol Genet Metab 2010;100(2):111–7.

Chapter 30

Complex II Deficiency: Leukoencephalopathy Due to Mutated SDHAF1

Simon Edvardson[1], **Ann Saada (Reisch)**[2]
[1]*Pediatric Neurology Unit, Hadassah-Hebrew University Medical Center, Jerusalem, Israel;*
[2]*Department of Genetic and Metabolic Diseases & Monique and Jacques Roboh Department of Genetic Research, Hadassah-Hebrew University Medical Center, Jerusalem, Israel*

CASE PRESENTATION

The first case is a 14-month-old girl (#1) who presented at our neuropediatric clinic with a chief complaint of inability to walk after having achieved independent ambulation a few months earlier.

She was the first child to healthy consanguineous Palestinian parents. Pregnancy, delivery, perinatal course, and family history were unremarkable. The girl had achieved age-appropriate developmental milestones in all domains, had normal growth parameters, and had been generally healthy until a few weeks prior to presentation. The developmental regression had not been associated with any intercurrent illness and had developed over the course of a few weeks. Her initial examination was remarkable for lower limb spasticity, hyperreflexia, and irritability. Cranial nerves, cerebellar function, muscle strength, and cognitive function were intact.

Over the following weeks, spasticity increased and balance was impaired due to what was considered cerebellar dysfunction. Cognitive function regressed over several months with loss of expressive language, and ultimately, no interaction with her surroundings or voluntary action could be elicited. Feeding became increasingly difficult, and a gastrostomy was performed. The patient died of pneumonia at age 5 years.

The second case was the child of the same family, a girl (#2), presented at age 4 months with spasticity and loss of head control and social smiling that had been achieved previously. As with the older sibling, pregnancy and delivery had been uneventful, and the baby had been generally healthy until presentation. Her development did not progress beyond 4 months of age and today, at age 3 years, she has very limited interaction with her surroundings, is fed by gastrostomy,

Mitochondrial Case Studies. http://dx.doi.org/10.1016/B978-0-12-800877-5.00030-9

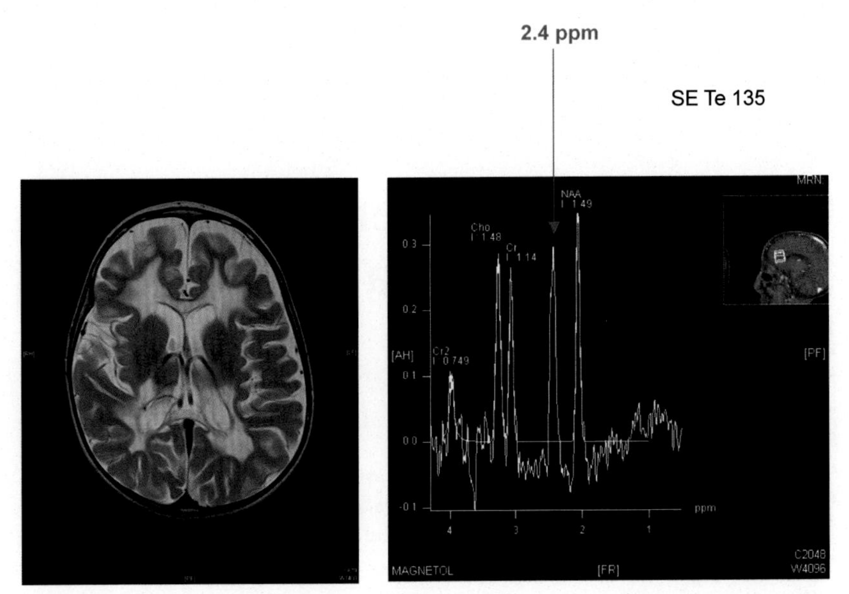

FIGURE 1 Magnetic resonance imaging (MRI) and proton magnetic resonance imaging (MRS) of case #1. Left: T2-weighted sagittal brain MRI of case #1 at age 4 years. Right: MRS of case #1 at age 4 years showing an abnormal peak of succinate.

and has severe contractures. Extensive metabolic investigations were done, and the main finding common to both girls was a leukodystrophic pattern on magnetic resonance imaging (MRI) with increased succinate on magnetic resonance spectroscopy (MRS) (Figure 1). Respiratory chain analysis was not performed in muscle; however, investigation of peripheral lymphocytes revealed decreased activity of the mitochondrial respiratory chain complex II-related activities, succinate dehydrogenase, and succinate cytochrome *c* reductase, normalized to citrate synthase (as mitochondrial tricarboxylic cycle control enzyme) (Table 1). Sequencing of *SDHAF1* gene encoding the complex II-specific assembly factor 1, SDGAF1, revealed a mutation c.170G>A, p.G57E [1].

The identification of the responsible mutation allowed the couple to receive prenatal diagnosis during their next pregnancy resulting in the birth of a healthy daughter.

DIFFERENTIAL DIAGNOSIS

Clinical Diagnosis

In neurology, it is a common didactic approach to approach the differential diagnosis by asking: where is the lesion? Motor regression may be of peripheral (i.e., muscle or nerve) origin or of central origin. The combination of upper motor neuron signs (e.g., hyperreflexia, spasticity) and cognitive impairment would lead to further narrowing of the localization to the level above the brain stem.

TABLE 1 Enzymatic Activities in Lymphocyte Homogenates

Assay[a]	Case #1	Case #2	Controls $n = 15 \pm SD$ (Range)
Succinate dehydrogenase (complex II)	9 [38%]	9 [55%]	19 ± 3 (15–24)
Succinate-cytochrome *c* reductase (complex II + III)	10 [50%]	7 [52%]	15 ± 2 (13–18)
Cytochrome *c* oxidase (complex IV)	107 [78%]	67 [70%]	103 ± 23 (72–156)
Citrate synthase	103	71	80 ± 17 (58–109)

Activities of mitochondrial respiratory chain complexes II–IV were measured by spectrophotometry [6]. Values in brackets [ ] represent the residual activities as a percentage of the control mean, normalized for citrate synthase activity.
[a]*nmol/min per mg protein.*

Age at presentation, family history, ethnic origin, and associated features such as micro- or macrocephaly may contribute important clues to the origin of neurodegenerative disorders in children. In the absence of signs or symptoms pointing to a specific diagnosis or group of disorders, MRI of the brain will be an important tool to guide further investigations. An MRI indicating a white matter disorder will, in conjunction with other variables, lead to a more finite list of diagnostic possibilities.

The term leukodystrophy indicates a deterioration of white matter and implies a genetically determined mechanism causing abnormal synthesis and/or maintenance of myelin sheaths, the cellular substrate of white matter. Leukoencephalopathies are disorders characterized by secondary white matter damage.

The variables to weigh when attempting to further delineate a white matter disorder include the following:

- MRI pattern of white matter changes (e.g., confluent, patchy, cavitating, hypomyelinating)
- Age of onset (e.g., Canavan disease and Krabbe disease present mainly in first year of life)
- Associated features (e.g., nystagmus in Pelizaeus-Merzbacher disease; hair, skin, and nail abnormalities in trichothiodystrophy; peripheral neuropathy in metachromatic dystrophy)
- Multiorgan involvement would certainly increase the suspicion of a mitochondrial disorder.
- Family history and ethnic origin: most of the disorders are inherited in a recessive fashion, though important exceptions are known (e.g., X-linked adrenoleukodystrophy, autosomal dominant Alexander disease). Some of these

disorders have been linked to particular ethnic groups (e.g., Canavan disease among Jewish Ashkenazi). In this context, it is worth noting that mitochondrial disorders may be inherited in a maternal as well as in a Mendelian fashion. An exception is complex II (succinate dehydrogenase, SDH) where all four subunits are nuclear encoded and always autosomal recessive in inheritance [2].

Biochemical Diagnosis

Metabolic Screening

Despite the laudable attempt to formulate a tentative diagnosis by history and physical examination and then proceeding to validate the diagnosis (frequently by molecular genetic testing), this is not always possible, thus leading to screening tests for groups of disorders.

These investigations include, for the most part, metabolic tests such as follows:

- Organic acids in urine
- Lactic acid in body fluids
- Very long chain fatty acids in serum
- Transferrin isoelectric focusing in serum
- Acylcarnitines in blood

Although many disorders can be diagnosed or removed from consideration by the aforementioned parameters, they cannot definitively exclude nor confirm a mitochondrial disease.

This is exemplified by our presented case, case #1, where the only remarkable finding in the routine metabolic evaluation was a slight elevation of urinary methylmalonic acid.

Enzymatic Analysis of the Mitochondrial Respiratory Chain

When a mitochondrial disease is suspected, muscle biopsy may be performed for histochemical, ultrastructural, and respiratory chain enzymatic evaluation.

Notably, two out of five respiratory chain complexes activities may be detected by muscle histochemistry, namely cytochrome *c* oxidase (complex IV) and SDH, however, in a qualitative manner only. Thus, a partial deficiency might remain undetected, and therefore, quantitative spectrophotometric measurements of all five complexes in insolated muscle mitochondrial provide more accurate data.

Occasionally but not always, the enzymatic defect is also present in other tissues. Mitochondrial diseases involving complex II are most probably expressed in multiple tissues because of the autosomal mode of inheritance and are not subjected to the complications of mtDNA heteroplasmy. This was also the case in SDHAF1 defect; the enzymatic defect was expressed in fibroblasts as well as lymphocytes/lymphoblasts [1,3].

Magnetic Resonance Spectroscopy

In the last decade, MRS has also come to be included in the armamentarium of broad screening tests in white matter disorders. This noninvasive neuroimaging modality, allows the *in vivo* biochemical detection of bioenergetic metabolites that are important in disease. Using computer modeling abnormal peaks of metabolites in the brain parenchyma (e.g., lactate in mitochondrial disorders, NAA in Canavan disease) can be identified. In the case of mitochondrial complex II defects, the detection of succinate did certainly direct the biochemical and molecular investigations toward a complex II defect.

Molecular Diagnosis

Diagnostic uncertainty is still prevalent among white matter disorders despite the tools mentioned above, leading to the inclusion of DNA sequencing as an indispensable test to establish a final diagnosis.

Where clinical and laboratory findings are highly indicative (as in the presented case), direct Sanger sequencing of the candidate gene could save time and resources.

Increasingly replacing the candidate gene approach, whole exome sequencing, mitochondrial DNA sequencing, and eventually whole genome sequencing, are quickly becoming tools used for the molecular diagnosis of rare monogenic disorders. The cost for exome sequencing has plummeted, and the software needed for analysis has become increasingly sophisticated, enabling identification, not only of mutations in known genes, but also discovery of new disease-causing genes.

Nevertheless, both new genes and new mutations often require additional functional evaluation for the proof of concept. This is especially the case with missense mutation in single cases where in silico analysis is not sufficient to confirm pathogenicity. Functional analysis is mostly done in research laboratories requiring considerable time and financial investments that are presently not covered by the healthcare system. The amount of work involved is exemplified by the excellent and thorough investigation performed by Ghezzi et al., who were the first to identify SDHAF1 mutations as causative of complex II deficiency [4].

The clinical manifestation of other complex II deficiencies caused by mutations in SDH subunits A–D present with a wide range of symptoms including early-onset encephalomyopathy, cardiomyopathy, and optic atrophy and tumor susceptibility in adults [2,3]. Secondary complex II deficiency may also be a result of mutations in a number of genes affecting iron-sulfur metabolism [3,5].

To date, although a number of complex II deficiencies lack mutations in known SDH assembly factors or SDH structural subunits, most cases of complex II deficiency can be solved by combining the enzymatic and MRS findings with the DNA sequencing data, which together provide sufficient means of diagnosis without any further need for evaluation [1,3,6].

DIAGNOSTIC APPROACH

The regression of the two sisters pointed to an autosomal recessive disorder. A homozygous mutation was suspected in view of parental consanguinity. Numerous metabolic and genetic disorders were considered, but a major reduction in diagnostic possibilities could be achieved by the MRI showing a leukoencephalopathic pattern. The inclusion of MRS in the initial study led us to suspect complex II deficiency. This could be verified by respiratory chain analysis of lymphocytes, and thus, we were able to circumvent the need for a more invasive muscle biopsy. These findings directed the molecular evaluation specifically toward the SDHAF1 gene, subsequently leading us to the identification of the homozygous disease-causing mutation [1].

LONG-TERM OUTCOME AND TREATMENT STRATEGY

In the reported cases of SDHAF1, the prognosis is grave with leukoencephalopathy presenting between birth and 10 months. The psychomotor regression is rapidly progressive after onset. Although most patients succumbed between the ages of 1.5 and 11 years, some cases remained in a stable condition beyond their first decade [1,4]. Remarkably, one patient harboring an SDHAF1 nonsense mutation and low SDH residual activity in fibroblasts presented with milder symptoms and intact cognitive function. He also attended and performed well in school [1]. Thus, presently, it seems that neither residual SDH activity nor the nature of the mutation is able to predict long-term outcome.

No specific treatment strategy other than palliative was documented, although one case was reported to be stabilized on vitamin B2 treatment [4].

PATHOPHYSIOLOGY OF DISEASE

SDH is a unique mitochondrial enzyme that catalyzes the conversion of succinate to fumarate in the tricarboxylic acid cycle (TCA, Krebs cycle) and is also an integral component of respiratory chain complex II carrying electrons from FADH2 to the ubiquinone pool. This multimeric enzyme is composed of four subunits: the largest subunit A (SDHA), a hydrophilic flavoprotein, containing the active site; subunit B (SDHB) is an iron-sulfur protein; and together with the hydrophobic subunits C (SDHC) and D (SDHD), they comprise complex II of the respiratory chain. The SDHAF1 protein is one of three factors essential for the assembly and the activity of SDH and respiratory chain complex II.

Pathological mutations in the genes encoding SDH subunits or its known assembly factors have been associated with a wide spectrum of clinical presentations. Generally, SDHA and SDHAF1 mutations are associated with classical mitochondrial diseases: infantile encephalomyopathy, Leigh syndrome, late-onset neurodegenerative disease, and dilated cardiomyopathy; whereas germline mutations in genes encoding the B–C subunits and in SDH5 (SDH assembly factor 2, SDHAF2) with loss of heterozygosity in tumors are associated with hereditary paragangliomas, pheochromocytomas, and gastrointestinal malignancies [1–4,6–8].

The precise pathomechanism is unclear. Alterations of SDH activity would alter the TCA cycle, accumulation of succinate which is an inhibitor of many under debate as metabolic derangement occurs both in the TCA (with accumulation of succinate, an inhibitor of many α-ketoglutarate-dependent enzymes. Furthermore, ATP synthesis would be disrupted inducing a frank energy deficit. Very recent experiments in yeast and *Drosophila melanogaster* revealed that SDHAF1 and SDHAF3 are involved in the maturation of the iron-sulfur clusters in subunit SDHB. Dysfunction in the maturation of iron-sulfur clusters induced increased oxidative stress and those tissues requiring the most energy were particularly vulnerable, as showed by the benefit of antioxidants. Moreover, the authors also demonstrated the potential benefit of antioxidants [9]. In accord, it makes sense that the brain and muscle are primarily affected by SDHAF1 mutations. Open to discussion are the observations that so far the heart seems to remain unaffected and no malignancies have been reported. Possibly other known and yet to be identified SDH/complex II tissue-specific assembly factors with overlapping activities might play a role in the pathomechanism of SDHAF1 defect, in particular, and complex II deficiencies, in general.

CLINICAL PEARLS

- Infantile leukoencephalopathy with accumulated succinate detected by MRS is highly indicative of mitochondrial complex II deficiency caused by mutated SDHAF1.
- Decreased complex II (succinate dehydrogenase, SDH) activity in muscle is a hallmark of this disorder. This assay can also be performed in lymphocytes or skin fibroblasts circumventing the need for the more invasive muscle biopsy.
- Sequencing of the SDHAF1 gene provides final diagnosis and allows genetic counseling.

REFERENCES

[1] Ohlenbusch A, Edvardson S, Skorpen J, Bjornstad A, Saada A, Elpeleg O, et al. Leukoencephalopathy with accumulated succinate is indicative of SDHAF1 related complex II deficiency. Orphanet J Rare Dis 2012;7:69.

[2] Rustin P, Munnich A, Rötig A. Succinate dehydrogenase and human diseases: new insights into a well-known enzyme. Eur J Hum Genet 2002;10:289–91.

[3] Levitas A, Muhammad E, Harel G, Saada A, Caspi VC, Manor E, et al. Familial neonatal isolated cardiomyopathy caused by a mutation in the flavoprotein subunit of succinate dehydrogenase. Eur J Hum Genet 2010;18:1160–5.

[4] Ghezzi D, Goffrini P, Uziel G, Horvath R, Klopstock T, Lochmüller H, et al. SDHAF1, encoding a LYR complex-II specific assembly factor, is mutated in SDH-defective infantile leukoencephalopathy. Nat Genet 2009;41:654–6.

[5] Spiegel R, Saada A, Halvardson J, Soiferman D, Shaag A, Edvardson S, et al. Elpeleg O.Deleterious mutation in FDX1L gene is associated with a novel mitochondrial muscle myopathy. Eur J Hum Genet 2014;22:902–6.
[6] Brockmann K, Bjornstad A, Dechent P, Korenke CG, Smeitink J, Trijbels JM, et al. Succinate in dystrophic white matter: a proton magnetic resonance spectroscopy finding characteristic for complex II deficiency. Ann Neurol 2002;52:38–46.
[7] Reisch AS, Elpeleg O. Biochemical assays for mitochondrial activity: assays of TCA cycle enzymes and PDHc. Methods Cell Biol 2007;80:199–222.
[8] Hao HX, Khalimonchuk O, Schraders M, Dephoure N, Bayley JP, Kunst H, et al. SDH5, a gene required for flavination of succinate dehydrogenase, is mutated in paraganglioma. Science 2009;325:1139–42.
[9] Na U, Yu W, Cox J, Bricker DK, Brockmann K, Rutter J, et al. The LYR factors SDHAF1 and SDHAF3 mediate maturation of the iron-sulfur subunit of succinate dehydrogenase. Cell Metab 2014;20:253–66.

Chapter 31

BCS1L Mutations as a Cause of Björnstad Syndrome–GRACILE Syndrome Complex III Deficiency

Bruce H. Cohen
Northeast Ohio Medical University, Rootstown, OH, USA; The NeuroDevelopmental Science Center and Divison of Neurology, Department of Pediatrics, Children's Hospital and Medical Center of Akron, Akron, OH, USA

CASE PRESENTATIONS

Case 1

Case 1 has been partially reported previously. She was born at term without apparent issues, although she was small (<second percentile) at birth. By 2 months of age, she was diagnosed with failure to thrive. Her early course was remarkable for poor growth and failure to achieve normal motor, language, and social milestones. Hearing testing in the first year of life confirmed deafness. Her hair was sparse and fractured easily. A metabolic illness was suspected, and testing revealed a significant generalized amino aciduria along with persistent elevation in her blood lactate (3–5 mM, normal range $<$2.2 mM). Hepatic and iron studies were normal. She was referred for a muscle biopsy and study of both electron transport chain enzymology and oxidative phosphorylation function (OXPHOS) using the polarographic method. Her examination revealed sparse hair with microscopic pili corti, a small but proportional body habitus, and no dysmorphism other than the pili corti. Her liver was normal. She had a happy demeanor but did not display meaningful expressive language and had very limited receptive language abilities. Her cranial nerve examination was normal aside from her hearing loss; specifically, there was no optic atrophy, retinitis pigmentosa, ophthalmoplegia, facial diplegia, or severe bulbar dysfunction. She was hypotonic, had low muscle bulk without frank myopathy, was ataxic, and was not ambulatory. The significant biochemical findings (laboratory of Charles Hoppel, MD, CIDEM) included the following:

Electron Transport Chain Enzymology (Freshly isolated mitochondria)

NADH-cytochrome *c* reducatase (Complex I + III)

Patient value: 51 nmol/min/mg mito protein (3.7% control value); Control mean $\pm$ SD $=$ 1377 $\pm$ 554.

Mitochondrial Case Studies. http://dx.doi.org/10.1016/B978-0-12-800877-5.00031-0

Decylubiquinol-cyt *c* reductase (Complex III)

Patient value: 671 nmol/min/mg mito protein (15% control value); Control mean ± SD = 4512 ± 1527.

OXPHOS Polarography (Freshly Isolated Mitochondria)

Duroquinol (Complex III substrate)

State iii: 388 (588 ± 74)
State iv: 64 (182 ± 43)
High ADP: 264 (564 ± 75)
Uncoupled: 306 (901 ± 149)
Activity expressed in atoms oxygen/minute/mg protein; patient value (mean ± SD)

Genetic Studies (Laboratory of Christine E. Seidman, MD)

BCS1L: p.R184C and p.G35R *(in trans)*

This child was followed for a number of years and had a stable clinical course, without evidence of speech, severe restriction in receptive language function, and without the ability to ambulate independently. She always seemed to have a happy disposition.

Case 2

The second child, unrelated to the first case and living over 2000 miles away from my clinic, presented a year later at 3 years of age. Her course was remarkable similar, although her growth delay was not as severe. This child was small at birth and grew along the second percentile. Her mother felt she had poor hearing and low motor tone in the first few days of life. By 15 months of age, she could pull herself into a standing position but soon lost that ability. She never grew more than a few strands of hair. Because of the hearing loss and poor growth, she had a mitochondrial screening evaluation that revealed a blood lactic acid of 4.2 mM and elevated liver enzymes (two times the upper limit of normal). A muscle biopsy did not show any microscopic findings, and a liver biopsy showed mild periportal fibrosis and apoptosis. She was referred for further evaluation.

Because of her clinical presentation being remarkably similar, I chose not to perform a muscle biopsy as planned. Her examination showed no evidence of dysmorphism other than sparse hair growth. Her general examination was normal aside from her small but proportional body habitus (below the second percentile for both height and weight). It was difficult to find a stand of hair long enough to submit for microscopy (which did show pili corti). She had a very happy affect but did not use words or vocalizations and did not show evidence of understanding nonverbal commands or clues. She had no cranial nerve findings other than her hearing loss, and the remainder of her neurological examination was similar to the girl in case 1. Evaluation of liver function (ultrasound, liver enzymes, and liver function including a brief-duration fasting glucose, ammonia, bilirubin, albumin, cholesterol, and coagulation studies) were normal as well as iron studies (CBC, iron, and ferritin). Sanger sequencing of her *BCS1L* gene showed her to have compound heterozygote mutations *in trans* p.R183C and p.R114W. She has been remarkably stable since first meeting her in 2007.

DIFFERENTIAL DIAGNOSIS

Because of the restricted phenotype, specifically the combination of congenital hearing loss and pili corti, along with growth retardation, severe language dysfunction, biochemical findings of lactic acidosis, and amino aciduria, the differential diagnosis is quite narrow.

Pili corti can be seen classically in Menkes kinky hair disease, ectodermal dysplasia, congenital disorder of glycosylation 1a, and dozens of other disorders.

There are several mitochondrial phenotypic syndromes that share some features similar to this case. Infantile tubulopathy-hepatopathy-encephalopathy disease caused by complex III dysfunction is likely part of the spectrum of illness, although it was described biochemically as opposed to the syndromic features of Björnstad syndrome and GRACILE syndrome. Both the mtDNA depletion disorders and specific forms of Leigh syndrome can present in a similar fashion.

DIAGNOSTIC APPROACH

The hallmark features of Björnstad syndrome provide a strong clue to early recognition of *BCS1L*-related disease. In most instances, hearing loss will be identified at birth, as current protocols require newborn hearing screening before discharge from the newborn nursery. The pili corti can be confirmed by microscopic inspection, although most pathology labs do not offer hair examination as a test that can be easily obtained. Any child with failure to gain weight, poor growth, or consideration of failure to thrive should be evaluated for renal tubular acidosis, and the use of urinary amino acid testing in this part of the evaluation will identify the generalized amino aciduria. Likewise, many clinicians use this test, especially in the evaluation of infants and young children, during a mitochondrial evaluation. Because the spectrum of this disorder can be highly variable, it is reasonable to test for liver dysfunction and iron accumulation in any child suspected of having the Björnstad syndrome–GRACILE syndrome spectrum of illness. This suggested evaluation would include but not be limited to an ultrasound of the liver and gall bladder, liver enzymes, liver function tests (fasting glucose, ammonia, bilirubin, albumin, cholesterol, and coagulation studies) being normal as well as iron studies (CBC, iron, and ferritin). The specific diagnostic study is sequencing of the *BCS1L* gene.

TREATMENT STRATEGY

As with most mitochondrial disorders, the most important care we can offer is supporting the dysfunctional body system. As with any child with total hearing loss, language therapy is necessary, although the benefit to an individual child with such restricted intellectual and receptive language functioning is variable. Likewise, the use of physical and occupational therapy techniques is reasonable. All children should have long-term nutritional evaluations, with the goal of giving the child a diet that meets the global nutritional (especially protein) and caloric needs. Most cases have a thin body habitus and adding more calories will result in inappropriate weight gain and not proper growth. If a renal tubular acidosis is present, buffering with bicarbonate therapy is reasonable. Although beyond the scope of these cases, because this is a mitochondrial illness affecting electron transport chain function, the use of mitochondrial supplements (vitamins and co-factors) is reasonable but cannot be supported by the literature. In the two cases (one child was followed for 4 years and the other for now 7 years), there has been no observable benefit, although their courses were stable.

LONG-TERM OUTCOME

Children with the classic Björnstad syndrome have an apparently normal prognosis for life, other than the hearing loss and obvious lack of scalp hair. Infants with severe GRACILE disease, as suggested by the acronym, will usually die in the first few years of life, with their illness marked by repeated interventions required for nutritional support with marked metabolic fragility. A recent study found a syndrome of hypomania and intermittent psychosis in several cases. However, the manifestations of typical Björnstad syndrome could be overlooked in some of the severe cases, as the course in severe GRACILE syndrome or the neonatal complex III tubulopathy-hepatopathy-encephalopathy syndrome is severe and often fatal. As with most mitochondrial disorders caused by mutations in the nuclear DNA, where the spectrum of the illness seems to be base, at least in part, by the specific pairing of pathogenic mutations, the outcome is too variable to predict unless there are data from enough cases with the common mutations. However, some of the specific mutations have been deemed to have less pathogenicity as others.

PATHOPHYSIOLOGY

Björnstad syndrome was described in 1965 as an autosomal recessive disorder presenting with pili torti and severe sensorineural hearing loss. Children with this disorder have normal intellectual functioning, although over the years cases of children meeting the common features of the condition were described with intellectual problems. GRACILE syndrome was described in 1998, and infants with the classic form of this condition were not described to have pili torit or deafness. *BCS1L* encodes for the protein BCS1L, which is protein that localizes to the inner mitochondrial membrane and is necessary to insert the Rieske Fe/S protein into the subunits of complex III during the assembly of complex III and prior to the formation of the I-III-IV supercomplex. Mutations in *BCS1L* were described to cause a neonatal form of complex III deficiency (neonatal renal tubular dysfunction, encephalopathy, and liver failure) and GRACILE syndrome in 2002 and Björnstad syndrome along with the Björnstad syndrome–GRACILE spectrum in 2007. Many of the mutations seem to be specific for GRACILE syndrome, others specific for Björnstad syndrome, with some of the Björnstad-specific mutations having a milder phenotype. In studies of tissue from cases with these disorders, the supercomplex structure (complex I-III-IV) formed, but the intermediate proteins lacked the Rieske Fe/S protein, and this resulted in decreased complex III activity. All mutations in *BCS1L* resulted in increased production of reactive oxygen species, but the amount of increase was a function of the specific mutation, which suggests the nature of wide clinical spectrum of illness. One final conclusion was that those with the restricted Björnstad syndrome had increased production of reactive oxygen species at the level of complex I, leading to oxidative stress in the inner ear and hair follicle.

CLINICAL PEARLS

- Björnstad syndrome, causing pili corti and congenital sensorineural hearing loss, GRACILE syndrome, and infantile complex III phenotype tubulopathy-hepatopathy-encephalopathy syndrome may be caused by mutations in *BCS1L*.
- Because pili corti and congenital deafness are easily identified, consideration for *BCS1L*-related disease should be considered in both classic Björnstad syndrome and Björnstad syndrome with additional features including encephalopathy, developmental disabilities, failure to thrive, and features of multiple organ dysfunction.
- Although the muscle biopsy findings in case 1 demonstrated unequivocal evidence of complex III dysfunction, it should be stressed that this case presented before the link between the phenotype and genotype was established and before commercial testing for this gene was available. Therefore, it is suggested that if the illness is along the spectrum of *BCS1L*-related disease, gene sequencing is less invasive and less costly than a muscle biopsy.

FURTHER READING

[1] Al-Owain M, Colak D, Albakheet A, Al-Younes B, Al-Humaidi Z, Al-Sayed M, et al. Clinical and biochemical features associated with BCS1L mutation. J Inherit Metab Dis September 2013;36(5):813–20. http://dx.doi.org/10.1007/s10545-012-9536-4. Epub September 19, 2012. PubMed PMID: 22991165.

[2] Fellman V. The GRACILE syndrome, a neonatal lethal metabolic disorder with iron overload. Blood Cells Mol Dis 2002;29:444–50.

[3] Hinson JT, Fantin VR, Schönberger J, Breivik N, Siem G, McDonough B, et al. Missense mutations in the BCS1L gene as a cause of the Björnstad syndrome. N Engl J Med February 22, 2007;356(8):809–19. PubMed PMID: 17314340.
[4] Lubianca Neto JF, Lu L, Eavey RD, Flores MA, Caldera RM, Sangwatanaroj S, et al. The Bjornstad syndrome (sensorineural hearing loss and pili torti) disease gene maps to chromosome 2q34-36. Am J Hum Genet May 1998;62(5):1107–12. PubMed PMID: 9545407; PubMed Central PMCID: PMC1377094.
[5] Visapää I, Fellman V, Vesa J, Dasvarma A, Hutton JL, Kumar V, et al. GRACILE syndrome, a lethal metabolic disorder with iron overload, is caused by a point mutation in BCS1L. Am J Hum Genet October 2002;71(4):863–76. Epub September 5, 2002. PubMed PMID: 12215968; PubMed Central PMCID: PMC378542.

Chapter 32

Complex IV

Mark Tarnopolsky, Rashid Alshahoumi
Department of Pediatrics, McMaster University, Hamilton, Ontario, Canada

CASE PRESENTATION

The case was born at term after an unremarkable pregnancy. There were some feeding difficulties after birth, and he was noted to be quite sleepy. His initial newborn screening was suggestive of glutaric aciduria type 1; however, urine organic acids done on day 6 of life showed no elevation of glutaric acid, with a trivial elevation of 3-hydroxyglutaric acid. He was seen in the metabolic clinic on that day as a follow-up to the positive NBS and was difficult to arouse and was hypotonic. Stat blood gases were done and showed a pH of 7.18, PO_2 of 51 mm Hg, CO_2 of 48 mm Hg, bicarbonate of 17 mmol/L, a base deficit of 11, and a lactate concentration of 12.8 mmol/L (N<2.2 mmol/L). Consequently, he was admitted to the pediatric intensive care unit (PICU). In the PICU, he was given intravenous fluids and started on nasogastric feeds. His lactate remained very high for 2 weeks (range = 5.4–15.6 mmol/L), and he had a percutaneous muscle biopsy taken from the vastus lateralis, and a skin biopsy was also taken for fibroblast analysis (see below).

Given that a mitochondrial disease was suspected, he was started on a mitochondrial cocktail consisting of creatine monohydrate, alpha-lipoic acid, L-carnitine, and coenzyme q10. An echocardiogram was normal, and the septic evaluation returned normal. His examination showed no dysmorphic features, he remained somnolent and at times difficult to arise, cranial nerve examination was normal, he was hypotonic but could raise limbs against gravity and had normal muscle stretch reflexes. His overall condition started to stabilize over the 2-week period in the PICU, and he was transferred to the general pediatric ward. At about 3 weeks of age, his lactate levels started to drop, but because of difficulty swallowing, he had a percutaneous endoscopic gastrostomy (PEG) tube inserted at 6 weeks, and he was discharged home at 8 weeks of age. At home, he continued to improve, and his developmental milestones were improving. His lactate levels normalized at about 4 months of age. The neurological examinations of the mother and father were both normal; however, the mother gave a history of a poor suck as a baby (possible milder phenotype).

Mitochondrial Case Studies. http://dx.doi.org/10.1016/B978-0-12-800877-5.00032-2

When he was seen in clinic at 2 years of age, his examination only showed mild hypotonia but otherwise was normal. He had mild developmental delay in his gross motor, speech, and language skills. The PEG tube was removed at 2.5 years of age, and his mitochondrial medications were weaned off at that time. At 3 years of age, he was still having some difficulties with pronunciation of some letters, but otherwise his developmental milestones were normal. His neurological examination continued to show mild hypotonia but was otherwise normal.

DIFFERENTIAL DIAGNOSIS

The child was admitted to the PICU with a diagnosis of neonatal encephalopathy not yet determined. The term neonatal encephalopathy implies a general disturbance of neurologic function in the neonatal period in a child born after 35 weeks of gestation. It can result from a vast number of heterogeneous disorders including brain injury, infections, toxicity, and genetic or metabolic causes. Hypoxic ischemic injury remains a major cause for neonatal encephalopathy. Metabolic causes are rare but are very important in the differential of a neonate with neonatal encephalopathy. In this case, a stat lactate led to the rapid identification of lactic acidosis. Lactic acidosis can be classified into type A or type B according to presence or absence of tissue hypoxia. The most common conditions are mainly type A. Type B1 and B2 would be very rare in neonates, and the presence of fulminant neonatal encephalopathy, ataxia, hypotonia, failure to thrive, hypoglycemia, or aminoaciduria may suggest an inborn error metabolism. The differential diagnosis for lactic acidosis is shown in Table 1. There are very few inborn errors of metabolism where lactic acidosis is severe in infancy and normalizes completely with the rare exception of infantile reversible cytochrome *c* oxidase deficiency and the extremely rare case of thiamine transporter-2 deficiency.

DIAGNOSTIC APPROACH

Given that the child did not have evidence for type A or type B1/B2 lactic acidosis, we had considered the possibility of a mitochondrial cytopathy or other inborn error of metabolism. Several of the other inborn errors of metabolism were ruled out by the normal glucose (GSD I) and organic acid screen. The muscle biopsy was sent for enzymology and electron microscopy, but unfortunately, a second sample was not obtained due to parental concerns about the procedure during the acute event. The muscle biopsy showed normal mitochondrial morphology and a nonspecific increase in nonmembrane bound glycogen. This lowered the likelihood of a mitochondrial DNA depletion syndrome for the mitochondria are often enlarged and have a paucity of cristae in that syndrome. The electron transport chain enzymology showed normal complex I+III, II+III, and citrate synthase activity; however, the cytochrome *c* oxidase (COX) was severely low (0.08 [NR=0.8–6.03] μmol/min/g wet weight). The fibroblast enzymology showed normal lactate/pyruvate ratio, PDH (native and DCA activated), pyruvate carboxylase, complex II+III, and COX activity.

TABLE 1 Causes of Lactic Acidosis in a Child

Type A Lactic Acidosis	Type B Lactic Acidosis		
	Normal O_2 Delivery/Absent Hypoxia		
Impaired O_2 Delivery/Presence of Hypoxia	B1 Underlying Disease/Condition	B2 Drugs/Toxins	B3 Inborn Error of Metabolism
• Cardiac failure • Cardiac arrest • Hypovolemia • Hypotension • Sepsis • Sever hypoxia ($PaO_2 < 30$ mm Hg) • Severe anemia (Hb < 50 g/L) • CO poisoning	• Hepatic failure • Renal failure • Diabetes mellitus • Malignancy (lymphomas, leukemia) • HIV • Acute severe asthma	• Valproic acid • Thiamine deficiency • Biotin deficiency • Metformin • Salicylates • Acetaminophen • Ethanol/methanol • Ethylene glycol • B2 agonists (epinephrine) • Cyanide • Iron • Propofol • Total parenteral nutrition	• Leigh disease (complex I, III, IV, V) • Mitochondrial encephalopathy, lactic acidosis, and stroke-like episodes (MELAS) • Myoclonic epilepsy with ragged red fibers (MERRF) • Kearn-Sayre syndrome • Pearson syndrome • Other mitochondrial diseases: SCO2, reversible COX deficiency, lethal infantile mitochondrial disease (LIMD) • Methylmalonic aciduria • Pyruvate dehydrogenase deficiency • Pyruvate carboxylase deficiency • Glucose-6-phosphatase deficiency (glycogen storage disease 1)

Genetic testing for SURF-1 and SCO2 mutations was normal. Muscle-derived mitochondrial DNA sequencing (Sanger method) showed a mutation at m.14674T>C homoplasmic mutation ($tRNA^{GLU}$), also confirmed in the mother, and no other mtDNA sequence variants of significance. Consequently, the child was diagnosed with infantile reversible cytochrome *c* oxidase deficiency.

TREATMENT STRATEGY

The initial treatment of a child with severe lactic acidosis involves identification of the cause, medical support, and monitoring. Most of the causes of lactic acidosis seen in a pediatric ICU will be due to impaired O_2 delivery/presence of hypoxia (i.e., type A lactic acidosis) in the neonatal period, and those associated with an underlying known disease become more prevalent in older childhood (i.e., type B1, Table 1). Many of these disorders have specific therapies (e.g., insulin, inotropes, fluid administration, blood transfusion, etc.) and require prompt identification to prevent morbidity and mortality. The history and a drug screen will identify most cases of lactic acidosis due to drugs/toxins (i.e., type B2, Table 1). For drug toxicity or toxin exposure, the treatment involves supportive measures and occasionally specific interventions such as L-carnitine for valproate toxicity, ethanol for acute methanol poisoning, N-acetylcysteine for acetaminophen toxicity, and often hemodialysis as a nonspecific measure for many drugs/toxins. Inborn errors of metabolism are considered in the differential diagnosis if these aforementioned disorders are ruled out. Treatment is again supportive with fluids as needed, cardiac and respiratory monitoring and interventions if required, and occasionally, treatment of the lactic acidosis. The approach to treating lactic acidosis per se involves either decreasing the rate of appearance of lactate by treating the generator of tissue hypoxia and/or activating pyruvate dehydrogenase (PDH) with dichloroacetate to reduce the flux of pyruvate > lactate, or buffering the proton using sodium bicarbonate or other buffer. The use of any of these acute interventions to lower lactic acid and/or to buffer the proton must be based on the severity of the situation and with careful monitoring of blood gases and lactate and clinical status in an acute care setting, preferably an intensive care unit with continuous cardiac monitoring. Dichloroacetate may be a good choice in known cases of PDH deficiency but can be considered when the lactate is high (>10 mmol/L) in cases with unknown etiology. Long-term dichloroacetate use should be avoided due to the potential for peripheral nerve toxicity. Long-term sodium bicarbonate therapy can exacerbate tissue acidosis and should be avoided. Nutritional therapy with coenzyme Q10 + alpha-lipoic acid + creatine monohydrate + vitamin E was started in this case and given by nasogastric feeds. It is difficult to judge the clinical efficacy of this therapy in an acute/subacute setting; however, the improvement in some biochemical features of mitochondrial disease with this cocktail and the paucity of side effects would suggest that it is reasonable to start this therapy.

LONG-TERM OUTCOME

Unlike the uniformly fatal outcome in many of the cases of severe lactic acidosis in infancy due to mutations in *SCO2, SURF-1,* and mtDNA mutations associated with lethal infantile mitochondrial disease (LIMD, i.e., m.15,923A>G) and Leigh disease (LD, i.e., m.9176T>G), the outcome in children with the m.14674T>C is optimistic. Some have used the term benign to describe this disorder; however, it is important to note that most cases had severe and potentially life-threatening lactic acidosis in the neonatal period and required ICU support and often nutritional support until the COX enzyme activity normalized and the clinical outcome took on a benign course. Such was the case in our child, who showed gradual improvement in sucking/eating and general development to the point where he was discharged from the developmental pediatric team and appears clinically indistinguishable from other boys his age.

PATHOPHYSIOLOGY

The m.14674T>C mutation affects the last (discriminator) base at the 3′ end of the mitochondrial tRNA for glutamic acid (Glu). This is thought to alter the ability of the Glu to bind to the tRNA and limit amino acid incorporation into the polypeptide chain and also results in low steady-state levels of the $tRNA^{Glu}$. The specificity for the COX subunits has been attributed to the fact that they have a relatively high Glu content and/or the relative location of the Glu residues in the protein. The reason for the apparent spontaneous improvement in COX activity and clinical function is unclear but does not appear to be due to a significant increase in the steady-state level of $tRNA^{Glu}$ molecules. There is some evidence that mtDNA copy number may increase with age, and this leads to a threshold of $tRNA^{Glu}$ molecules sufficient to maintain translation of the COX subunits. Another theory relates to the fact that some of the COX subunits undergo developmental isoform switching including COXVIa and COXVIIa.

CLINICAL PEARLS

- Neonatal encephalopathy can be secondary to a vast number of causes.
- Inborn error of metabolism is rare but very important in the differential of a neonate with encephalopathy.
- Lactic acidosis in the context of fulminant neonatal encephalopathy may suggest an inborn error of metabolism and, in particular, a mitochondrial disorder.
- Fibroblast cultures are important in the evaluation for a child with suspected mitochondrial disease and are extremely helpful to identify disorders of pyruvate metabolism. Low lactate/pyruvate ratio suggests pyruvate carboxylase (PC) deficiency, and a high level suggests primary mitochondrial disease or pyruvate dehydrogenase deficiency (PDH).

- Fibroblast measurements of PC and PDH enzyme activity are accurate in diagnosing these disorders; however, mitochondrial enzyme measurements can be normal in fibroblasts but severely abnormal in muscle or other tissue (as was the case presented herein).
- Given the reversible nature of the m.14674T>C mutation, it is important not to give a grim/fatal prognosis to a neonate with encephalopathy and severe lactic acidosis until the final diagnosis is established.
- The phenotypic heterogeneity can be extreme in mitochondrial disorders with some cases with homoplasmic mutations being asymptomatic or pauci-symptomatic (e.g., the mother in this case) and others presenting with fulminant disease.

FURTHER READING

[1] Avula S, Parikh S, Demarest S, Kurz J, Gropman A. Treatment of mitochondrial disorders. Curr Treat Options Neurol 2014;16(6):292.
[2] Chen TH, Tu YF, Goto YI, Jong YJ. Benign reversible course in infants manifesting clinicopathological features of fatal mitochondrial myopathy due to m.14674 T>C mt-tRNAGlu mutation. QJM 2013;106(10):953–4.
[3] Ezgu F, Senaca S, Gunduz M, Tumer L, Hasanoglu A, Tiras U, et al. Severe renal tubulopathy in a newborn due to BCS1L gene mutation: effects of different treatment modalities on the clinical course. Gene 2013;528(2):364–6.
[4] Federico A, Dotti MT, Fabrizi GM, Palmeri S, Massimo L, Robinson BH, et al. Congenital lactic acidosis due to a defect of pyruvate dehydrogenase complex (E1). Clinical, biochemical, nerve biopsy study and effect of therapy. Eur Neurol 1990;30(3):123–7.
[5] Horvath R, Kemp JP, Tuppen HA, Hudson G, Oldfors A, Marie SK, et al. Molecular basis of infantile reversible cytochrome c oxidase deficiency myopathy. Brain 2008;132(Pt 11): 3165–74.
[6] Kraut JA, Madias NE. Lactic acidosis. N Engl J Med 2014;371:2309–19.
[7] Kaufmann P, Engelstad K, Wei Y, Jhung S, Sano MC, Shungu DC, et al. Dichloroacetate causes toxic neuropathy in MELAS: a randomized, controlled clinical trial. Neurology 2006;66(3):324–30.
[8] Low E, Crushell EB, Harty SB, Ryan SP, Treacy EP. Reversible multiorgan system involvement in a neonate with complex IV deficiency. Pediatr Neurol 2008;39(5):368–70.
[9] Meert KL, McCaulley L, Sarnaik AP. Mechanism of lactic acidosis in children with acute severe asthma. Pediatr Crit Care Med 2012;13(1):28–31.
[10] Mimaki M, Hatakeyama H, Komaki H, Yokoyama M, Arai H, Kirino Y, et al. Reversible infantile respiratory chain deficiency: a clinical and molecular study. Ann Neurol 2010;68(6): 845–54.
[11] Pérez-Dueñas B, Serrano M, Rebollo M, Muchart J, Gargallo E, Dupuits C, et al. Reversible lactic acidosis in a newborn with thiamine transporter-2 deficiency. Pediatrics 2013;131(5): e1670–1675.
[12] Cohen RD, Frank Woods H. Lactic acidosis Revisited. Diabetes 1983;32(2):181–91.
[13] Rodriguez MC, MacDonald JR, Mahoney DJ, Parise G, Beal MF, Tarnopolsky MA. Beneficial effects of creatine, CoQ10, and lipoic acid in mitochondrial disorders. Muscle Nerve 2007;35(2):235–42.

[14] Salo MK, Rapola J, Somer H, Pihko H, Koivikko M, Tritschler HJ, et al. Reversible mitochondrial myopathy with cytochrome c oxidase deficiency. Arch Dis Child 1992;67(8):1033–5.
[15] Uusimaa J, Jungbluth H, Fratter C, Crisponi G, Feng L, Zeviani M, et al. Reversible infantile respiratory chain deficiency is a unique, genetically heterogenous mitochondrial disease. J Med Genet 2011;48(10):660–8.
[16] Vernon C, Letourneau JL. Lactic acidosis: recognition, kinetics, and associated prognosis. Crit Care Clin 2010;26(2):255–83.

Chapter 33

Complex V Disorders

Mark Tarnopolsky, Rashid Alshahoumi
Department of Pediatrics, McMaster University, Hamilton, Ontario, Canada

CASE PRESENTATION

A 2.5-year-old, previously healthy, female child presented to the emergency department with cough, fever, lethargy, and a 6-day history of decreased movement in her legs and a lack of facial expression. One week prior to admission, she had symptoms of an upper respiratory tract infection with fever and was seen by the family doctor and prescribed amoxicillin. She was admitted to the pediatric intensive care unit for investigation and because of chest X-ray changes and the persistent fever, she was switched over to intravenous cephalosporin antibiotics for presumed community-acquired pneumonia. Because of the bradykinesia, she had a CT scan of the head completed and a neurology consultation. On examination, the child was somnolent but followed some simple commands. Cranial nerve examination was positive only for a nearly complete lack of facial expression. Motor examination showed normal muscle bulk, generalized bradykinesia, cogwheeling of the arms, and rigidity of the legs with a flexor plantar response. Respiratory examination revealed only coarse upper airway sounds. The CT scan showed bilateral hypodensities in the basal ganglia. Blood cultures were sterile, and sputum grew nonspecific species that were felt to be contamination. A blood sample taken at the time of admission was negative for carbon monoxide. The cerebral spinal fluid (CSF) showed normal glucose, protein, and cell count, and it did not grow any organisms, and common viral testing was normal; however, the CSF lactate was elevated at 4.4 mmol/L (N < 2.2 mmol/L). A brain MRI showed a high signal on T2-weighted images in the head of the caudate, globus pallidus, and putamen.

The family history revealed that the mother and father were non-consanguineous and of Filipino descent, and she had a few episodes of unsteadiness during illness and experienced gestational diabetes with the current (first and only) pregnancy. One of her brothers died at age 35 years with questionable cardiomyopathy; another sister was healthy at age 50 years; and another brother in his early 50s had type 2 diabetes. The maternal grandmother died in her mid-50s with some undefined cardiac event. The case was tried on L-DOPA/carbi-DOPA with slight improvement in facial expression and speech. She was discharged after 2 weeks but returned to the emergency department 2 months later with severe bradykinesia and hypoventilation

Mitochondrial Case Studies. http://dx.doi.org/10.1016/B978-0-12-800877-5.00033-4

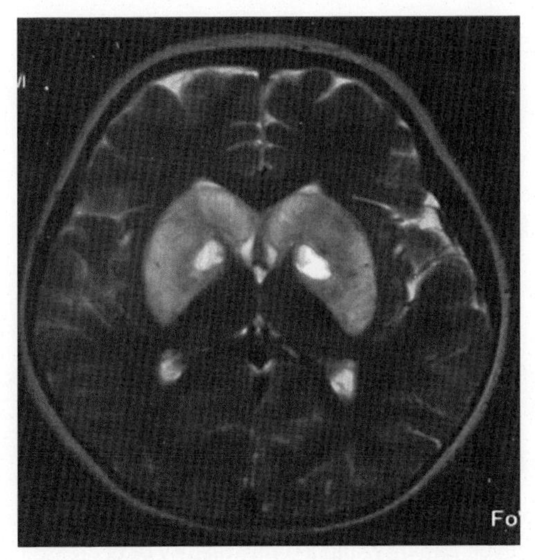

FIGURE 1 T2-weighted 1.5 Telsa MRI showing a high signal in the putamen, globus pallidus, and head of caudate.

(pCO_2 = 90 mm Hg). She was intubated for respiratory failure, and the MRI (Figure 1) showed progression of the previously noted T2 changes but with changes in the globus pallidus suggestive of advanced BSN. She failed several attempts at weaning from the ventilator and passed away ~6 weeks later from pneumonia.

DIFFERENTIAL DIAGNOSIS

Parkinsonism is rare among the pediatric population. Hypokinetic-rigid syndrome is a frequently used term for pediatric cases because of the distinctive features seen in this age group. The most common causes of acute Parkinsonism/hypokinetic-rigid syndrome are infectious or immunomediated disorders and drug toxicity (Table 1). There are, however, many other causes, and these Parkinsonism cases in pediatrics are usually incomplete, more slowly progressive, and often appear in the context of multiple other neurological manifestations such as encephalopathy, seizures, developmental delay, and other neurological findings (Table 1).

DIAGNOSTIC APPROACH

In the current case, the first priority was to eliminate from consideration infectious/treatable causes and carbon monoxide poisoning (that could place others at risk). With the symmetric basal ganglia changes, the possibility of a mitochondrial cytopathy was high on the differential diagnostic list, and testing was arranged. The CSF sample was taken at the time of admission to remove infectious causes from consideration, and a lactate was sent on ice, which was abnormally high. Unfortunately, a pyruvate was not sent at that time, for it

TABLE 1 Cause of Parkinsonism in Pediatrics

- Infections
- Immunomediated disorders such as NMDA-R encephalitis
- Drugs (neuroleptics, anti-nausea agents [metoclopramide], some calcium blockers)
- Hydrocephalus
- Hypoxic-ischemic encephalopathy
- Basal ganglia tumors
- Brain stem malformations
- Hypoparathyroidism and pseudohypoparathyroidism
- Genetic mutations such as *ATP1A3 (DYT12), PRKRA (DYT16)*
- Inborn errors of metabolism:
    - Disorders of monoamine metabolism, for example, PTPS-D (6-Pyruvoyltetrahydropterin synthase deficiency) and DHPR-D (dihydropteridine reductase deficiency).
    - Mitochondrial disorders, for example, pyruvate carboxylase deficiency, pyruvate dehydrogenase deficiency, and respiratory chain deficiencies
- Other metabolic disorders that can have some Parkinsonism signs: ceroid lipofuscinosis, Gaucher disease, pantothenate kinase-associated neurodegeneration PKAN, Lesch-Nyhan disease

can be helpful to evaluate primary mitochondrial disease and pyruvate dehydrogenase deficiency (high L/P ratio) from pyruvate carboxylase deficiency (L/P ratio usually <20). Plasma lactate was sent and was initially 1.6 mmol/L and later only mildly elevated at 2.4 mmol/L (N<2.2 mmol/L). Plasma amino acids (alanine) and urine organic acids were both normal (the latter also to eliminate from consideration the remote possibility of an organic aciduria). A muscle biopsy was completed using a 5-mm modified Bergström needle, and a fragment of skin was taken at that time for fibroblast culture. Muscle histology and ultrastructure were normal, and muscle-derived DNA was screened using PCR-RFLP analysis for mtDNA mutations at positions; m.3243A>G, m.8993T>C, m.8993T>G, m.9176T>C, and m.9185T>C. It should be noted that this case presented in 2002, and the current strategy in such a case would be to sequence the entire mtDNA using next-generation sequencing instead of just targeted mutation analysis for common (m.3243A>G) and likely (m.8993 and m.9176T>C) mutations. The muscle-derived mtDNA heteroplasmy was 89% for the maternally inherited Leigh Disease mutation at m.8993T>C. The mother had a mutant load of 16% in blood. The child's fibroblasts showed a complex V activity (oligomycin sensitive ATPase activity) of 17.7 nmol/min/mg protein = 3.9 (ratio/citrate synthase [CS]) (control = 19.6 = 6.1 ± 1.7 [ratio/CS] [n = 3]).

TREATMENT STRATEGY

Due to the severe and rapidly progressive nature of maternally inherited Leigh syndrome (MILS)/bilateral striatal necrosis (BSN), the treatments are generally supportive. The orofacial and pharyngeal rigidity in the current case required the placement of a gastrostomy tube for energy maintenance and airway protection. Nutritional therapy with coenzyme Q10 + alpha-lipoic acid + creatine monohydrate + vitamin E was started. This therapy may improve some biochemical features of the disorder and provide some clinical benefit in less severe cases; however, in rapidly progressive and severe disorders, any potential clinical benefit is overshadowed by the inevitable decline in function. Dopamine and dopamine agonists may be considered in the case of Parkinsonian features; however, these are not likely to be of benefit in advanced disease with MILS or BSN.

LONG-TERM OUTCOME

Unfortunately, the progression of children presenting with rapidly progressive psychomotor regression due to most mitochondrial diseases is usually grim, and many children have long-term neurological deficit or die with secondary infections or respiratory failure. A word of caution is that there has been a case report of child with the m8993T > C mutation who presented with classical MILS, albeit at age 4 years, and showed surprising spontaneous improvements with no neurological signs of symptoms at age 18 years. Somewhat surprisingly, the clinical outcome in 113 children presenting with primary mitochondrial cytopathies in childhood (mean age = 40 mos) was much better for children without cardiomyopathy (95% survival) versus those with cardiomyopathy (18%) by age 16 years. It should be noted that in the latter study, the smallest proportion of cases were those with complex V deficiency (8%). The clinical phenotype and age of onset of disease associated with the 8993 mutations are strongly dependent upon the percentage mutant heteroplasmy in fibroblasts or muscle. Mutational burdens of >85% heteroplasmy are usually associated with an infantile onset of MILS or BSN. Mutational burdens between ~50 and 85% are associated with neurogenic ataxia and retinitis pigmentosa (NARP). Cases with lower levels of heteroplasmy can be asymptomatic or develop mild, later-onset NARP or peripheral neuropathy.

PATHOPHYSIOLOGY

The m.8993T > C mutation affects the complex V structural gene, *ATPase6*. This is one of two complex V (ATP synthase) structural subunits encoded for by the mtDNA. This specific mutation changes a highly conserved leucine to a proline residue at position 156 of the human mtDNA genome. Yeast models of both the T > C and T > G mutations at position 8993 show that the ATP synthase activity was reduced by ~40 and ~20%, respectively. Of interest, both yeast models showed a significant reduction in cytochrome *c* oxidase activity, suggesting control of COX expression (likely via subunit Cox1p) by ATP synthase activity.

CLINICAL PEARLS

- Cases with severe mitochondrial disease may have normal blood lactate values, and other screening methods such as plasma amino acids and organic acids can also be normal.
- Testing for complex V function must be done on live cells; consequently, doing a skin biopsy for fibroblast culture at the time of other procedures (muscle biopsy, L/P) is recommended if a mitochondrial disease is suspected.
- Mitochondrial enzyme activity can be normal unless expressed relative to the total number of mitochondria (i.e., relative to citrate synthase).
- The severity of the complex V deficiency can be quite mild compared to the severe clinical symptoms seen in the 8993 mutations.
- A sudden decline in neurological function during an apparently mild viral or bacterial infection in a previously healthy child can be the first manifestation of a neurometabolic disorder.
- Rapidly progressive Parkinsonism/hypokinetic-rigid syndrome is usually due to infectious or para-infectious disorders or drug toxicity.

FURTHER READING

[1] Debray FG, Lambert M, Lortie A, Vanasse M, Mitchell GA. Long-term outcome of Leigh syndrome caused by the NARP-T8993C mtDNA mutation. Am J Med Genet A 2007;143A(17):2046–51.

[2] Kucharczyk R, Rak M, di Rago JP. Biochemical consequences in yeast of the human mitochondrial DNA 8993T>C mutation in the ATPase6 gene found in NARP/MILS patients. Biochim Biophys Acta 2009;1793(5):817–24.

[3] Rak M, Tetaud E, Duvezin-Caubet S, Ezkurdia N, Bietenhader M, Rytka J, et al. A yeast model of the neurogenic ataxia retinitis pigmentosa (NARP) T8993G mutation in the mitochondrial ATP synthase-6 gene. J Biol Chem November 23, 2007;282(47):34039–47.

[4] Rodriguez MC, MacDonald JR, Mahoney DJ, Parise G, Beal MF, Tarnopolsky MA. Beneficial effects of creatine, CoQ10, and lipoic acid in mitochondrial disorders. Muscle Nerve 2007;35(2):235–42.

[5] Rojo A, Campos Y, Sánchez JM, Bonaventura I, Aguilar M, García A, et al. NARP-MILS syndrome caused by 8993 T>G mitochondrial DNA mutation: a clinical, genetic and neuropathological study. Acta Neuropathol 2006;111(6):610–6.

[6] Scaglia F, Towbin JA, Craigen WJ, Belmont JW, Smith EO, Neish SR, et al. Clinical spectrum, morbidity, and mortality in 113 pediatric patients with mitochondrial disease. Pediatrics 2004;114(4):925–31.

[7] Tarnopolsky MA, Pearce E, Smith K, Lach B. Suction-modified Bergström muscle biopsy technique: experience with 13,500 procedures. Muscle Nerve 2011;43(5):717–25.

[8] Tatuch Y, Christodoulou J, Feigenbaum A, Clarke JT, Wherret J, Smith C, et al. Heteroplasmic mtDNA mutation (T>G) at 8993 can cause Leigh disease when the percentage of abnormal mtDNA is high. Am J Hum Genet 1992;50(4):852–8.

[9] Tatuch Y, Pagon RA, Vlcek B, Roberts R, Korson M, Robinson BH. The 8993 mtDNA mutation: heteroplasmy and clinical presentation in three families. Eur J Hum Genet 1994;2(1):35–43.

[10] Uziel G, Moroni I, Lamantea E, Fratta GM, Ciceri E, Carrara F, et al. Mitochondrial disease associated with the T8993G mutation of the mitochondrial ATPase 6 gene: a clinical, biochemical, and molecular study in six families. J Neurol Neurosurg Psychiatry 1997;63(1):16–22.

Chapter 34

Primary Cerebellar CoQ10 Deficiency

Catarina M. Quinzii, Michio Hirano, Emanuele Barca
Columbia University Medical Center, Department of Neurology, New York, NY, USA

CASE PRESENTATION

This case presented with cerebellar ataxia and upper motor neuron dysfunction.

The family history was unremarkable; his non-consanguineous parents and 12-year-old sister are healthy. Birth and early development were normal, and in retrospect, his parents noted that at age 4 years, he did not run as well as his peers. At about age 7, he manifested difficulty walking. He has difficulty playing sports but is able to swim and shoot basketball. His handwriting is impaired, and his speech is also mildly abnormal. At age 8, he was evaluated neurologically and noted to have ataxia with cerebellar atrophy on brain MRI.

At age 16, another neurological evaluation was performed. The case was alert with a normal mental status examination. He had mild slurring of his speech but normal language function. Square wave jerks were present on primary gaze. Extraocular movements were full, but gaze pursuit was jerky, and his saccadic movements showed mild undershoot.

He had no alteration of muscle bulk, tone, or strength. A coarse arm sustention tremor was present, more evident on the right than left. No sensory alterations including proprioception were evident. Cerebellar functions were impaired; on finger-nose movements, he had slight overshoot and intention tremor, more evident in the right arm, as well as side-to-side excursion on heel-to-shin test. The gait was normal, but he was unable to tandem. Muscle stretch reflexes were increased at the biceps, triceps, knees, and ankles with nonsustained clonus. Hoffmann sign was present bilaterally, but Babinski sign was absent.

DIFFERENTIAL DIAGNOSIS

The first step in the differential diagnosis of cerebellar ataxias is to distinguish acquired forms from neurodegenerative-inherited conditions, because the former are often treatable, and genetic counseling is important for the case and relatives.

Mitochondrial Case Studies. http://dx.doi.org/10.1016/B978-0-12-800877-5.00034-6

In general, acute onset of cerebellar ataxia does not suggest a neurodegenerative disease, with some rare exceptions (e.g., pyruvate dehydrogenase complex deficiency). Neurodegenerative diseases should be considered in cases with progressive cerebellar ataxia.

Our case presented with early-onset, slowly progressive cerebellar involvement, which suggested a genetic disorder. Common and potentially treatable conditions that can cause ataxias and need to be excluded include autoimmune diseases such as celiac disease, anti-glutamic acid decarboxylase antibody syndrome, and vitamin E deficiency. Markers for some neurodegenerative cerebellar disease are easy to test such as ceruloplasmin, which is reduced in Wilson disease, alpha-fetoprotein, which can be elevated in ataxia-telangiectasia and ataxia-oculomotor apraxia (AOA1), and phytanic acid, a biomarker of Refsum disease. Other rare metabolic disorders such as lysosomal storage diseases were excluded in our case. Brain MRI is mandatory since cerebellar syndrome can be secondary to neoplastic disorders of posterior cranial fossa. Our case had diffuse cerebellar atrophy, which was consistent with a neurodegenerative disorder. Although dominant inheritance was not evident in our case, the most common causes of spinocerebellar ataxias (SCAs) were investigated and excluded. However, SCAs are extremely rare in childhood; the average age of onset of SCA 1, 2, and 3 is 30 years, and the average age of onset of SCA 6 is about 50. On the contrary, recessive ataxias are a heterogeneous group of disorders often with early onset and slowly progressive course, variably associated with peripheral neuropathy, movement disorders, oculomotor abnormalities, and pyramidal tract dysfunction. The most common autosomal recessive form is Friederich ataxia, but the early and prominent cerebellar atrophy and the retained tendon reflexes in our case argues against the diagnosis.

DIAGNOSTIC APPROACH

The normal diagnostic tests included vitamins E and B12 levels, long chain fatty acids, lysosomal enzyme activity studies, phytanic acid, immunoglobulins, alpha-fetoprotein, and ceruloplasmin levels. In addition, genetic tests for SCA-1, 2, 3, 6, and 7, DRPLA, and aprataxin were normal. Muscle and nerve biopsy was performed, and histological studies revealed mild nonspecific neurogenic and myogenic changes. CoQ10 levels were reduced in muscle (12.6 μg/gm-tissue, normal 23 ± 6.9) and fibroblasts (29.7 ngr/mg prot, normal 58.5 ± 4.1).

Genetic analysis of genes required for CoQ10 biosynthesis revealed two heterozygous *ADCK3* (*Coq8/CABC1*) mutations, c.1541A/G (p.Tyr514Cys) in exon 13 and 1750_1752 del ACC (Thr584del) in exon 15 [1].

A subgroup of cases with juvenile-onset cerebellar ataxia have primary CoQ10 deficiency caused by mutations in *ADCK3* (autosomal recessive cerebellar ataxia 2 (ARCA2)) [2]. Analysis of muscle biopsy is useful in *ADCK3* mutant cases, as muscle histology often reveals lipid accumulation, mitochondrial proliferation, and COX-deficient fibers, with typical ragged red fibers being usually absent.

Reduced biochemical activities of respiratory chain complexes, in particular complexes I+III (nicotinamide adenine dinucleotide–cytochrome *c* oxidoreductase) and complexes II+III (succinate-cytochrome *c* oxidoreductase) in muscle suggest CoQ10 deficiency, although activities of these enzymes may be normal particularly when the deficiency is mild. Reduction of these enzyme activities and deficiency of CoQ10 in skin fibroblasts can be an important confirmation of ubiquinone deficiency; however, normal levels do not exclude deficiency in muscle. Direct measurement of CoQ10 in skeletal muscle by high-performance liquid chromatography is the most reliable test for the diagnosis. In contrast, plasma concentrations of ubiquinone are significantly influenced by dietary uptake, therefore not reliable. Measurements of CoQ10 in peripheral blood mononuclear cells have detected deficiency in a small number of cases; however, correlations with muscle CoQ10 measurements in a larger cohort of cases will be necessary to assess clinical use of mononuclear cell ubiquinone levels [3].

Secondary forms of CoQ10 deficiency in cerebellar ataxia have been described in association with mutations in *APTX* [3], cause of AOA1, characterized by adult onset or childhood onset of severe cerebellar ataxia, oculomotor apraxia, seizures and axonal neuropathy, and mutations in *ANO10* [4], whose phenotype includes corticospinal tract signs and second motoneuron involvement.

TREATMENT STRATEGY

The case initiated multivitamin nutritional supplements including CoQ10 up to 900 mg per day. The case reports subjective stabilization of ataxia on CoQ10 supplementation.

In cases harboring *ADCK3* mutations, dosages and response to CoQ10 therapy have been variable. There is no consensus regarding drug formulations (e.g., ubiquinone versus ubiquinol) and daily dosage, although several studies suggest that doses of ubiquinone up to 2400 mg/day in adults and up to 30 mg/kg/day in children are therapeutically optimal. There is a highly variable clinical response of the cerebellar form of CoQ10 deficiency in cases after 6 months of treatment with high doses (10–30 mg/kg/day) [3]. Five cases showed long-lasting improvement of dystonia and myoclonus, balance, and academic performance with CoQ10 supplementation. Interestingly, a study published in 2010 in 14 cases with recessive ataxia showed that CoQ10 therapy led to an overall significant improvement in the international cooperative ataxia rating scale scores (ICARS), primarily enhancing posture and gait, but only in the seven cases with muscle CoQ10 deficiency [5]. Notable clinical improvements in strength, seizures, and some improvement of cerebellar function were noticed in the first cases described with AOA1 and CoQ10 deficiency [6]. In addition, postural stability, gait, speech articulation, and normalization of hormonal blood values improved with CoQ10 in two brothers with ataxia and hypogonadism [7]. Mild improvement was also observed in the cases carrying *ANO10* mutations [4].

LONG-TERM OUTCOME

The course of ARCA2 is usually mild, without the inexorable severe progression observed in most of the other recessive ataxias. Usually cases have a chronic evolution starting at around age 20 years, until death. A few cases of ARCA2 have manifested an infantile degenerative encephalopathy demonstrating an extreme presentation of this disease. In some cases, acute complications such as stroke-like episodes have contributed to the clinical variability of the disease. In addition, cases may manifest rapid deterioration of cognitive functions with or without the precipitous development of untreatable seizures and status epilepticus [2].

PATHOPHYSIOLOGY/NEUROBIOLOGY OF DISEASE

ADCK3 is the human homolog of the yeast CoQ8, a kinase thought to regulate stability of the Q-biosynthesis complex [1,8]. In humans, CoQ10 shuttles electrons in the mitochondrial respiratory chain, and it serves several additional cellular functions as antioxidant, membrane stabilizer, and modulator of apoptosis. Consequently, deficiency of CoQ10 potentially disrupts multiple vital cellular functions [9]. In vitro studies showed that the mitochondrial phenotype of cultured fibroblasts with *ADCK3* mutations depends on the level of CoQ10; fibroblasts with normal CoQ10 behaved like controls, whereas fibroblasts with mild CoQ10 deficiency had slightly increased ROS production and elevated levels of protein and lipid oxidation when stressed with glucose-free medium. Although decreased cell growth may, in part, be due to increased cell death triggered by *ADCK3* modulation of the P53-dependent apoptotic pathway, cell death was not increased in these fibroblasts. To account for the disease pathogenesis in the absence of oxidative stress and cell death in *ADCK3* mutant fibroblasts, it is important to consider the possibility of tissue-specific effects as well as potential resistance of cultured skin fibroblasts to the effects of CoQ10 deficiency. In addition, factors other than CoQ10 deficiency may be contributing to *ADCK3* mutant cells phenotype [10].

CLINICAL PEARLS

- It is important to identify CoQ10 deficiency as this condition often responds to supplementation.
- Individuals and their family members may benefit from genetic counseling if a confirmed genetic finding is discovered.
- Diagnosis of CoQ10 deficiency can be made by direct measurement of CoQ10, preferably in muscle; however, since muscle biopsy is not routinely performed in ataxic cases, skin fibroblast, but not plasma, can be an alternative tissue for diagnosis.
- Reduced biochemical activities of respiratory chain complexes I+III and II+III are suggestive of CoQ10 deficiency.

REFERENCES

[1] Lagier-Tourenne C, Tazir M, López LC, et al. ADCK3, an ancestral kinase, is mutated in a form of recessive ataxia associated with coenzyme Q10 deficiency. Am J Hum Genet 2008;82(3):661–72.

[2] Mignot C, Apartis E, Durr A, et al. Phenotypic variability in ARCA2 and identification of a core ataxic phenotype with slow progression. Orphanet J Rare Dis 2013;8:173.

[3] Emmanuele V, López LC, Berardo A, et al. Heterogeneity of coenzyme Q_{10} deficiency: patient study and literature review. Arch Neurol August 2012;69(8):978–83.

[4] Balreira A, Boczonadi V, Barca, et al. *ANO10* mutations cause ataxia and coenzyme Q_{10} deficiency. J Neurol November 2014;261(11):2192–8.

[5] Pineda M, Montero R, Aracil A, et al. Coenzyme Q(10)-responsive ataxia: 2-year treatment follow-up. Mov Disord 2010;25(9):1262–8.

[6] Musumeci O, Naini A, Slonim AE, et al. Familial cerebellar ataxia with muscle coenzyme Q10 deficiency. Neurology April 10, 2001;56(7):849–55.

[7] Gironi M, Lamperti C, Nemni R, et al. Late-onset cerebellar ataxia with hypogonadism and muscle coenzyme Q10 deficiency. Neurology March 9, 2004;62(5):818–20.

[8] Mollet J, Delahodde A, Serre V, et al. *CABC1* gene mutations cause ubiquinone deficiency with cerebellar ataxia and seizures. Am J Hum Genet 2008;82(3):623–30.

[9] Crane FL. The evolution of coenzyme Q. Biofactors 2008;32(1–4):5–11.

[10] Quinzii CM, López LC, Gilkerson RW, et al. Reactive oxygen species, oxidative stress, and cell death correlate with level of CoQ_{10} deficiency. FASEB J October 2010;24(10):3733–43.

Chapter 35

Multisystemic Infantile CoQ10 Deficiency with Renal Involvement

Catarina M. Quinzii, Mariana Loos
Columbia University Medical Center, Department of Neurology, New York, NY, USA

CASE PRESENTATION

A boy was born at term by normal vaginal delivery after an uneventful pregnancy. Birth weight, length, and head circumference were at the 50th percentile, and the Apgar scores were 9 and 10. He was the fourth child from healthy non-consanguineous parents. His family history was unremarkable except for a sister diagnosed with migraine.

At age 5 months of life, he was admitted for evaluation of severe global developmental delay and failure to thrive, which had been first suspected during the second month of life. The physical examination revealed a non-dysmorphic and lethargic child, who could not fixate nor track objects, and he did not respond to sounds. Head circumference, weight, and height were below the 10th percentile. No visceromegaly was present. Cranial nerve examination showed normal ocular movements. Funduscopy revealed bilateral optic atrophy. He had swallowing difficulties, requiring a nasogastric tube. Motor examination showed severe axial hypotonia with increased tone in the limbs. Hyperreflexia with bilateral clonus and Babinski signs were present. He never achieved head control, social smile, nor the ability to babble.

The initial laboratory evaluation showed metabolic acidosis with increased plasma lactate (53 mg/dL, normal 5–30) and alanine (73 μL/dL, normal < 44). Urinary organic acid profile revealed increased lactate and Krebs cycle metabolites. Complete blood count, liver enzymes, creatine kinase, ammonia levels, and acylcarnitine profile were normal. Urine analysis showed increased protein, suggestive of renal glomerular dysfunction. The initial brain CT scan showed bilateral frontal atrophy.

At 8 months of age, he was hospitalized for generalized status epilepticus. CSF analysis revealed increased lactic acid (39.2 mg/dL, normal 10.8–18.4)

Mitochondrial Case Studies. http://dx.doi.org/10.1016/B978-0-12-800877-5.00035-8

with normal glucose and protein levels. The interictal electroencephalogram (EEG) showed a poorly organized background without epileptiform discharges. Brain magnetic resonance imaging disclosed bilateral and symmetric T2 hyperintense lesions involving the cerebellar white matter and brain stem consistent with Leigh syndrome (Figure 1). On the echocardiogram, mild hypertrophy of the left ventricle was detected. Oral therapy with L-carnitine (50 mg/kg/d), coenzyme Q10 (10 mg/kg/d), riboflavin (100 mg/d), and thiamine (50 mg/d) was started without a clinical response.

The child's neurological status deteriorated; the child developed persistent focal motor seizures that partially responded to antiepileptic drugs, and he developed dystonic movements.

At age 19 months, a brain CT scan revealed extensive white matter involvement with bilateral lesions in the basal ganglia, cerebellum, and brain stem (Figure 1). A diagnostic muscle biopsy revealed mildly increased subsarcolemmal mitochondrial aggregates but no ragged red or cytochrome *c* oxidase-negative fibers. Respiratory chain enzyme activities in muscle showed increased citrate synthase activity. Isolated activities of complexes I, II, and IV were normal, whereas the combined activity of complexes I and III and II and III were below the limit of detection. These findings suggested CoQ10 deficiency, which was confirmed in muscle (2.3 μg/g tissue, normal 32.1 ± 6.8). His medical condition worsened and, during a febrile illness, the boy died.

PDSS2 was sequenced postmortem. *PDSS2* encodes decaprenyl diphosphate synthase subunit 2, the first enzyme dedicated to CoQ10 biosynthesis, and identified two novel heterozygous missense mutations, a heterozygous C>A transition at nucleotide 590, changing amino acid 197 from alanine to glutamic acid and a heterozygous T>C transition at nucleotide 932, changing amino acid 311 from phenylalanine to serine. Mutations were in trans, and both were predicted as damaging with a score of 1.00 (Polyphen 2).

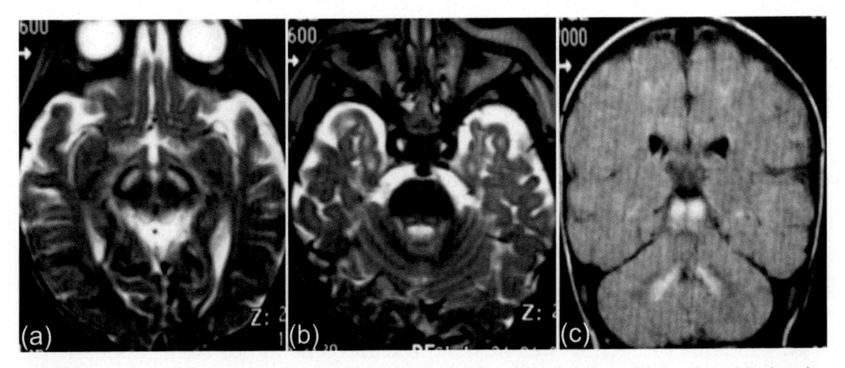

FIGURE 1 Axial T2-weighted (a and b) and coronal FLAIR (c) images (8 months old) showing bilateral and symmetrical lesions involving substantia, tegmental tracts, and cerebellar white matter.

DIFFERENTIAL DIAGNOSIS

Renal tubular dysfunction is a frequent childhood presentation of mitochondrial diseases caused by mitochondrial translation defects, mitochondrial DNA (mtDNA) depletion, defects of respiratory chain assembly factors, as well as MELAS and mtDNA multiple deletion syndromes [1]. In cases of mtDNA depletion or mitochondrial translation defect, measurement of mitochondrial respiratory chain enzymes activity would show combined enzymatic defects, associated with increased mitochondrial mass in mtDNA depletion. In cases with primary mtDNA defect, family history often suggests maternal inheritance.

Notably, primary CoQ10 deficiency is one of the few mitochondrial diseases often manifesting with glomerular nephropathy, isolated or associated with other symptoms/signs. Pearson syndrome can also manifest with glomerular nephropathy, although not typically.

Leigh syndrome is a bilateral subacute necrotizing encephalopathy, affecting basal ganglia, sub-thalamic nuclei, and the brain stem. MR spectroscopy typically shows lactate elevation, indicative of lactic acidosis. Other causes of bilateral injury of the deep gray matter are acquired metabolic disorders, as hypoxia and hypoglycemia, inherited myelin disorders, degenerative disorders, and infectious or iatrogenic disorders. Wilson disease is another inherited metabolic disorder associated with basal ganglia injury. However, Leigh syndrome is the most common pediatric presentation of mitochondrial disease. It is associated with clinical, biochemical, and molecular heterogeneity. The most common biochemical defect is complex I deficiency, which can be associated with mutations in the mitochondrial ND genes, or with mutations in the nuclear encoded subunits and assembly factors of complex I.

The most common maternally inherited Leigh syndrome (MILS) results from mutations in the ATPase 6 gene, and it is associated with complex V defect.

In these cases, biochemical assessment of the respiratory chain reveals reduction of complex I, whereas measurement of ATP synthesis in muscle, if performed, would detect complex V deficiency. However, Leigh syndrome can be associated also with defects in complex II, IV (COX), and pyruvate dehydrogenase complex (PDHC).

Deficiency of SURF1, a complex IV assembly factor, is the most frequent cause of COX-deficient Leigh syndrome. The majority of cases with *SURF1* mutations present in infancy with poor weight gain, hypotonia, poor feeding, vomiting, developmental delay or regression, movement disorders, oculomotor involvement, and central respiratory failure.

Another autosomal recessive COX-deficient Leigh syndrome is the French-Canadian Leigh syndrome (LRPPRC).

Defects of the β subunit of succinyl-CoA synthetase (SUCLA2) have been associated with mtDNA depletion and the triad of hypotonia, dystonia/Leigh-like syndrome, and deafness.

In our case, family history suggested autosomal recessive inheritance, but the biochemical findings excluded all these diseases.

DIAGNOSTIC APPROACH

The majority of mitochondrial disorders are multisystem diseases. Kidney involvement, in particular glomerulonephrosis, in isolation or as part of a multisystemic disorder, is a common feature of CoQ10 deficiency [2]. Initial clinical evaluation of cases with suspected CoQ10 deficiency should include blood lactate measurement, although normal values do not exclude ubiquinone deficiency. In cases with the infantile multisystemic form of CoQ10 deficiency, muscle biopsies have typically revealed abnormal mitochondrial proliferation (ragged red fibers or excessive succinate dehydrogenase histochemical activity) and lipid accumulation as well as reduced biochemical activities of respiratory chain enzyme complexes I+III and II+III, and decreased CoQ10 levels. Fibroblasts CoQ10 levels are frequently but not invariably low [2].

Most cases with the infantile-onset multisystemic variant have genetically confirmed primary CoQ10 deficiency, associated with mutations in *COQ2*, *PDSS2*, *COQ9*, or *PDSS1*. Cases with isolated nephrotic syndrome have had *ADCK4*, *COQ2*, or *COQ6* mutations. Cases with *COQ2* mutations have presented with either infantile multisystemic syndrome or isolated nephropathy. In contrast, mutations in the *COQ6* gene have been associated with kidney involvement (nephrotic syndrome and nephrolithiasis) and sensorineural hearing loss [3].

TREATMENT STRATEGY

Cases with CoQ10 deficiency showed variable responses to CoQ10 treatment. We recommend oral supplementation doses up to 2400 mg daily in adults and up to 30 mg/kg daily in children, divided into three doses per day.

In some cases with the infantile multisystemic form, CoQ10 supplementation has halted progression of the encephalopathy and improved the myopathy [2]. One case with a homozygous *COQ2* mutation on CoQ10 therapy showed neurologic but not renal improvement and underwent kidney transplant; however, his sister with isolated nephropathy received CoQ10 and has had progressive recovery of renal function, reduced proteinuria, and no neurologic manifestations. In contrast, a case with a homozygous *COQ9* mutation during therapy had a reduction of blood lactate but due to neurologic and cardiac worsening died at age 2 years. Similarly, despite treatment, a case with *PDSS2* mutations developed intractable seizures and died at age 8 months, and another case with infantile-onset Leigh syndrome, hepatopathy, and hypertrophic cardiomyopathy had an initial clinical improvement but died at age 3 years. Two cases with *COQ6* mutations and one with *ADCK4* mutations showed decreased proteinuria after CoQ10 treatment.

LONG-TERM OUTCOME

The multisystemic infantile-onset form of CoQ10 deficiency, especially in presence of encephalopathy, is associated with a worse prognosis than other presentations of CoQ10 deficiency [2]. In contrast, isolated nephrotic syndrome has been the more responsive phenotype of primary CoQ10 deficiency to CoQ supplementation [2,4]. However, because CoQ10 deficiency is a relatively new disease with a small number of reported cases, it is not possible to make firm genotype–phenotype correlation or accurately predict long-term prognosis.

PATHOPHYSIOLOGY/NEUROBIOLOGY OF DISEASE

CoQ10 is a lipid-soluble component of virtually all cell membranes, synthesized in the mitochondrial matrix, and composed of a benzoquinone ring with an isoprenyl side-chain. In the mitochondrial respiratory chain, CoQ10 carries electrons from complexes I and II to complex III. Besides shuttling electrons in the mitochondrial respiratory chain, CoQ10 serves several additional cellular functions, and consequently, deficiency of CoQ10 potentially disrupts multiple vital cellular functions [5]. In fact, studies of tissues and cells from cases with CoQ10 deficiency have revealed multiple pathological consequences. Skeletal muscle biopsies from cases with all clinical forms of CoQ10 deficiency have shown variable defects of respiratory chain enzyme activities (complexes I+III and II+III). In addition, muscles from cases with myopathic CoQ10 deficiency have revealed signs of apoptosis [6]. Studies of pathogenic mechanisms using cultured fibroblasts from cases have revealed correlations between degree of CoQ10 deficiency and oxidative stress. Whereas severe or mild deficiencies are not associated with oxidative stress, moderate deficiency (30–50% of normal) is associated with oxidative stress, impaired growth, and cell death [7]. Evidence of autophagy associated with oxidative stress has been shown in mammalian cells [7]. Intriguingly, ultrastructural evidence of autophagy has also been found in kidneys of mice with a homozygous mutation in *PDSS2* (*PDSS2*$^{KD/kd}$) [8]. Rapid improvements of renal function in *PDSS2* mice after CoQ10 or probucol supplementation suggest that autophagy might be triggered by oxidative stress [9], and the roles of oxidative stress and impaired ATP synthesis in the pathogenesis of CoQ deficiency have been confirmed in *PDSS2* mutant mice [10].

CLINICAL PEARLS

- It is important to identify CoQ10 deficiency because this condition, particularly kidney dysfunction, often responds to supplementation.
- Molecular diagnosis is important because individuals and their family members may benefit from genetic counseling.
- Kidney disease, especially steroid-resistant nephrotic syndrome in isolation or as part of an early childhood-onset multisystemic disease, is suggestive of primary CoQ10 deficiency.
- The combination of kidney disease, Leigh syndrome, and CoQ10 deficiency is suggestive of mutations in *PDSS2*.
- Diagnosis of CoQ10 deficiency can be made by direct measurement of CoQ10 in muscle or fibroblasts, but not in blood, and it can be reinforced by the presence of reduced biochemical activities of complexes I+III and II+III.

REFERENCES

[1] O'Toole JF. Renal manifestations of genetic mitochondrial disease. Int J Nephrol Renovasc Dis January 31, 2014;7:57–67.

[2] Emmanuele V, López LC, Berardo A, Naini A, Tadesse S, Wen B, et al. Heterogeneity of coenzyme Q10 deficiency: patient study and literature review. Arch Neurol August 2012;69(8):978–83.

[3] Doimo M, Desbats MA, Cerqua C, et al. Genetics of coenzyme q10 deficiency. Mol Syndromol July 2014;5(3–4):156–62.

[4] Montini G, Malaventura C, Salviati L. Early coenzyme Q10 supplementation in primary coenzyme Q10 deficiency. N Engl J Med June 26, 2008;358(26):2849–50.

[5] Turunen M, Olsson J, Dallner G. Metabolism and function of coenzyme Q. Biochim Biophys Acta January 28, 2004;1660(1–2):171–99.

[6] Di Giovanni S, Mirabella M, Spinazzola A, et al. Coenzyme Q10 reverses pathological phenotype and reduces apoptosis in familial CoQ10 deficiency. Neurology August 14, 2001;57(3):515–8.

[7] López LC, Luna-Sánchez M, García-Corzo L, et al. Pathomechanisms in coenzyme q10-deficient human fibroblasts. Mol Syndromol July 2014;5(3–4):163–9.

[8] Peng M, Falk MJ, Haase, et al. Primary coenzyme Q deficiency in Pdss2 mutant mice causes isolated renal disease. PLoS Genet April 25, 2008;4(4):e1000061.

[9] Falk MJ, Polyak E, Zhang Z, et al. Probucol ameliorates renal and metabolic sequelae of primary CoQ deficiency in Pdss2 mutant mice. EMBO Mol Med July 2011;3(7):410–27.

[10] Quinzii CM, Garone C, Emmanuele, et al. Tissue-specific oxidative stress and loss of mitochondria in CoQ-deficient Pdss2 mutant mice. FASEB J February 2013;27(2):612–21.

Index

'*Note*: Page numbers followed by "f" indicate figures and "t" indicate tables.'

N

O

U

V

W

X

Y

Z

Printed in the United States
By Bookmasters